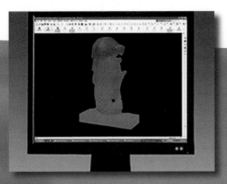

THIRD EDITION RAPID
PROTOTYPING

Principles and
Applications

C K CHUA   K F LEONG   C S LIM

Nanyang Technological University, Singapore

World Scientific

NEW JERSEY · LONDON · SINGAPORE · BEIJING · SHANGHAI · HONG KONG · TAIPEI · CHENNAI

Published by

World Scientific Publishing Co. Pte. Ltd.

5 Toh Tuck Link, Singapore 596224

USA office: 27 Warren Street, Suite 401-402, Hackensack, NJ 07601

UK office: 57 Shelton Street, Covent Garden, London WC2H 9HE

British Library Cataloguing-in-Publication Data
A catalogue record for this book is available from the British Library.

RAPID PROTOTYPING: PRINCIPLES AND APPLICATIONS (3rd Edition)

1006034880

ISBN-13 978-981-277-897-0
ISBN-10 981-277-897-7
ISBN-13 978-981-277-898-7 (pbk)
ISBN-10 981-277-898-5 (pbk)

Typeset by Stallion Press
Email: enquiries@stallionpress.com

Printed in Singapore by Mainland Press Pte Ltd.

My wife, Wendy and children, Cherie, Clement and Cavell, whose forbearance, support and motivation have made it possible for us to finish writing this book.

Chee Kai

Soi Lin, for her patience and support and Qian who brings us cheers and joy.

Kah Fai

My wife, Eugenia and children, Lisa and Ian, for their support and motivation.

Chu Sing

CONTENTS

Foreword   **xi**

Preface   **xiii**

Acknowledgments   **xvii**

About the Authors   **xxi**

List of Abbreviations   **xxiii**

Chapter 1. Introduction   **1**

 1.1. Prototype Fundamentals   1
 1.2. Historical Development   7
 1.3. Fundamentals of Rapid Prototyping   10
 1.4. Advantages of Rapid Prototyping   13
 1.5. Commonly Used Terms   17
 1.6. Classification of Rapid Prototyping Systems   18
   References   21
   Problems   22

Chapter 2. Rapid Prototyping Process Chain   **25**

 2.1. Fundamental Automated Processes   25
 2.2. Process Chain   26
   References   33
   Problems   33

Chapter 3. Liquid-Based Rapid Prototyping Systems   **35**

 3.1. 3D Systems' Stereolithography Apparatus (SLA)   35
 3.2. Objet Geometries Ltd.'s Polyjet   52
 3.3. D-MEC's Solid Creation System (SCS)   63

3.4.   Envisiontec's Perfactory   68
3.5.   Autostrade's E-Darts   76
3.6.   CMET's Solid Object Ultraviolet–Laser Printer   80
3.7.   EnvisionTec's Bioplotter   84
3.8.   Rapid Freeze Prototyping (RFP)   89
3.9.   Microfabrication   98
3.10. Microfabrica®'s EFAB® Technology   102
3.11. D-MEC's ACCULAS   106
3.12. Other Notable Liquid-Based RP Systems   110
       References   131
       Problems   135

Chapter 4.   Solid-Based Rapid Prototyping Systems   137

4.1.   Stratasys' Fused Deposition Modeling (FDM)   137
4.2.   Solidscape's Benchtop System   148
4.3.   Cubic Technologies' Laminated Object
       Manufacturing (LOM™)   153
4.4.   3D Systems' Multi-Jet Modeling System (MJM)   165
4.5.   Solidimension's Plastic Sheet Lamination (PSL)/
       3D System's Invision LD Sheet Lamination   171
4.6.   Kira's Paper Lamination Technology (PLT)   176
4.7.   CAM-LEM's CL 100   182
4.8.   Other Notable Solid-Based RP Systems   186
       References   194
       Problems   196

Chapter 5.   Powder-Based Rapid Prototyping Systems   199

5.1.   3D System's Selective Laser Sintering (SLS)   199
5.2.   Z Corporation's Three-Dimensional Printing (3DP)   213
5.3.   EOS's EOSINT Systems   222
5.4.   Optomec's Laser Engineered Net Shaping (LENS)   236
5.5.   Arcam's Electron Beam Melting (EBM)   243
5.6.   Concept Laser GmbH's LaserCUSING   248
5.7.   MCP-HEK Tooling GmbH's Realizer II
       (Selective Laser Melting)   257

5.8.	Phenix Systems's PM Series (LS)	262
5.9.	Sintermask Technologies AB's Selective Mask Sintering (SMS)	267
5.10.	3D-Micromac AG's MicroSINTERING	273
5.11.	Therics Inc's Theriform Technology	276
5.12.	The Ex One Company's ProMetal	278
5.13.	Voxeljet Technology GmbH'S VX System	283
5.14.	Other Notable Powdered-Based RP Systems	286
	References	293
	Problems	298
Chapter 6.	**Rapid Prototyping Data Formats**	**301**
6.1.	STL Format	301
6.2.	STL File Problems	302
6.3.	Consequences of Building Valid and Invalid Tessellated Models	306
6.4.	STL File Repair	308
6.5.	Other Translators	336
6.6.	Newly Proposed Formats	338
6.7.	Standard for Representing Layered Manufacturing Objects	350
	References	352
	Problems	354
Chapter 7.	**Applications and Examples**	**357**
7.1.	Application–Material Relationship	357
7.2.	Finishing Processes	359
7.3.	Applications in Design	360
7.4.	Applications in Engineering, Analysis and Planning	361
7.5.	Applications in Manufacturing and Tooling	365
7.6.	Aerospace Industry	379
7.7.	Automotive Industry	383
7.8.	Jewelry Industry	387
7.9.	Coin Industry	388

7.10.   Tableware Industry   390
7.11.   Geographic Information System Applications   393
7.12.   Arts and Architecture   394
        References   396
        Problems   400

Chapter 8.   Medical and Bioengineering Applications   403

8.1.   Planning and Simulation of Complex Surgery   403
8.2.   Customized Implants and Prosthesis   413
8.3.   Design and Production of Medical Devices   423
8.4.   Forensic Science and Anthropology   427
8.5.   Visualization of Biomolecules   429
        References   432
        Problems   435

Chapter 9.   Evaluation and Benchmarking   437

9.1.   Using Bureau Services   437
9.2.   Setting Up a Service Bureau   438
9.3.   Technical Evaluation Through Benchmarking   453
9.4.   Industrial Growth   470
9.5.   Recent and Future Development Trends   470
        References   476
        Problems   481

Appendix   List of RP Companies   483

CD-ROM Attachment   491

CD-ROM User Guide   493

Index   501

FOREWORD

I was pleased to hear that co-authors Chua, Leong and Lim were planning a third edition of their book on additive fabrication. A growing number of colleges and universities are adding equipment, coursework and projects to support coverage of the subject. Making the experience a good one for students can be a challenge without quality textbooks and hands-on activities. *Rapid Prototyping: Principles and Applications* not only aims to satisfy this need, it delivers.

The book is filled with information that even industry veterans will find useful. The chapters include detailed descriptions of the available additive processes, giving readers an overview of what's available commercially. The photographs, illustrations and tables make these chapters visually appealing and straightforward. References and problems at the end of each chapter help readers expand their understanding of the topics presented.

The chapter on data formats provides detail that new and advanced users alike will appreciate. It dives into related formats, such as IGES and SLC, but its in-depth coverage of the STL file format, including its limitations, problems and solutions, makes this chapter shine.

The book's sections on applications in manufacturing explain why and how organizations are putting the technology to work. Examples range from building flight-ready metal castings at Bell Helicopter to producing a human skull to aid in brain surgery at Keio University Hospital in Japan.

The depth and breadth presented in this book make it clear that the authors have a strong understanding of additive fabrication technologies and applications. The information is written and presented in an easy-to-follow format. And the book's accompanying CD is icing on the cake.

Thank you, Chua, Leong and Lim, for producing an outstanding book and for giving me the opportunity to share my thoughts on it.

Terry Wohlers
President
Wohlers Associates, Inc.

PREFACE

The focus on productivity has been one of the main concerns of industries worldwide since the early 1990s. To increase productivity, industries have attempted to apply more computerized automation in manufacturing. Amongst the latest technologies to have significant stride over the past two decades are the *Rapid Prototyping Technologies*, otherwise also known as *Solid Freeform Fabrication, Desktop Manufacturing* or *Layer Manufacturing Technologies.*

The revolutionary change in factory production techniques and management requires a direct involvement of computer-controlled systems in the entire production process. Every operation in this factory, from product design, to manufacturing, to assembly and product inspection, is monitored and controlled by computers. CAD–CAM or Computer-Aided Design and Manufacturing has emerged since the 1960s to support product design. Up to the mid-1980s, it has never been easy to derive a physical prototype model, despite the existence of Computer Numerical Controlled (CNC) Machine Tools. Rapid Prototyping Technologies provide the bridge from product conceptualisation to product realization in a reasonably fast manner, without the fuss of NC programming, jigs and fixtures.

With this exciting promise, the industry and academia have internationally established research centers for Rapid Prototyping (RP), with the objectives of working in this leading edge technology, as well as of educating and training more engineers in the field of RP. An appropriate textbook is therefore required as the basis for the development of a curriculum in RP. The purpose of this book is to provide an introduction to the fundamental principles and application areas in RP. The book traces the development of RP in the arena of Advanced Manufacturing

Technologies and explains the principles underlying each of the RP techniques. Also covered are the detail descriptions of the RP processes and their specifications. In this third edition, new RP techniques are introduced and existing ones updated, bringing the total number of RP techniques described to more than 30. The book would not be complete without emphasizing the importance of RP applications in manufacturing and other industries. In addition to industrial examples provided for each of the vendors, an entire chapter is devoted to application areas. As RP has expanded its scope of applications one whole chapter that focuses on the biomedical area has been added.

One key inclusion in this book is the use of multimedia to enhance understanding of the technique. In the accompanying compact disc (CD), animation is used to demonstrate the working principles of major RP techniques such as Stereolithography, Polyjet, Laminated Object Manufacturing, Fused Deposition Modeling, Selective Laser Sintering and Three-Dimensional Printing.

In addition, the book focuses on some of the very important issues facing RP today and these include, but are not limited to:

(1) The problems with the *de facto* STL format.
(2) The range of applications for tooling and manufacturing, including biomedical engineering.
(3) The benchmarking methodology in selecting an appropriate RP technique.

The material in this book has been used for more than 40 times for professional courses conducted for both academia and industry audiences since 1991. Certain materials were borne out of research conducted in the School of Mechanical and Aerospace Engineering at the Nanyang Technological University, Singapore. To be used more effectively for graduate or final year (senior year) undergraduate students in Mechanical, Aerospace, Production or Manufacturing Engineering, problems have been included in this textbook. For university professors and other tertiary-level lecturers, the subject RP can be combined easily with other topics such as: CAD, CAM, Machine Tool Technologies and Industrial Design.

Chua C. K.
Associate Professor

Leong K. F.
Associate Professor

Lim C. S.
Associate Professor

School of Mechanical and Aerospace Engineering
Nanyang Technological University
50 Nanyang Avenue
Singapore 639798

ACKNOWLEDGMENTS

First, we would like to thank God for granting us his strength throughout the writing of this book. Second, we are especially grateful to our respective spouses, Wendy (Chua), Soi Lin (Leong) and Eugenia (Lim) and our respective children, Cherie, Clement, Cavell (Chua); Qianyu (Leong); Lisa and Ian (Lim) for their patience, support and encouragement throughout the year it took to complete this edition.

We wish to thank the valuable support from the administration of Nanyang Technological University (NTU), especially the School of Mechanical and Aerospace Engineering (MAE). In addition, we would like to thank our current and former students An Jia, Chan Lick Khor, Chong Fook Tien, Chow Lai May, Esther Chua Hui Shan, Derrick Ee, Foo Hui Ping, Witty Goh Wen Ti, Ho Ser Hui, Ketut Sulistyawati, Kwok Yew Heng, Angeline Lau Mei Ling, Jordan Liang Dingyuan, Liew Weng Sun, Lim Choon Eng, Liu Wei, Phung Wen Jia, Mahendra Suryadi, Mohamed Syahid Hassan, Anfee Tan Chor Kwang, Tan Leong Peng, Toh Choon Han, Wee Kuei Koon, Yap Kimm Ho, Yim Siew Yen, Yun Bao Ling and Zhang Jindong, our colleague Mr. Lee Kiam Gam for their valuable contributions to make the multimedia CD which demonstrates some major RP techniques possible. We would also like to express sincere appreciation to our special assistants Deborah Cheah Meng Lin, Alex Tan Kok Wai, Howard Tang Ho Hwa and Vu Trong Thien for their selfless help and immense effort in the coordination and timely publication of this book.

Much of the research work which have been published in various journals and now incorporated into some chapters in the book can be attributed to our former students, Dr. May Win Naing, Dr. Yeong Wai Yee, Dr. Florencia Edith Wiria, Tong Mei, Micheal Ko, Ang Ker Chin, Tan Kwang Hui, Liew Chin Liong, Gui Wee Siong, Simon Cheong, William Ng, Chong Lee Lee, Lim Bee Hwa, Chua Ghim Siong, Ko Kian Seng, Chow Kin Yean, Verani, Evelyn Liu, Wong Yeow Kong, Althea Chua, Terry Chong, Tan Yew Kwang, Chiang Wei Meng, Ang Ker Ser, Lin Sin

Chew, Ng Chee Chin and Melvin Ng Boon Keong and colleagues Professor Robert Gay, Mr. Lee Han Boon, Dr. Jacob Gan, Dr. Du Zhaohui, Dr. M. Chandrasekeran, Dr. Cheah Chi Mun and Dr. Georg Thimm.

We would also like to extend our special appreciation to Mr. Terry Wohlers for his foreword and permission for the use of his executive summary of the Wohler's report, Dr. Ming Leu of the University of Missouri-Rolla, USA, Dr. Amba D Bhatt of the National Institute of Standards and Technology, Gaithersburg, MD, USA, Dr. Philip Coane and Dr. Goettert of the Institute for Micromanufacturing, Ruston, LA, USA and Prof. Fritz Prinz of Stanford University, USA for their assistance.

The acknowledgments would not be complete without the contributions of the following companies for supplying and helping us with the information about their products they develop, manufacture or represent:

(1) 3D Systems Inc., USA
(2) 3D-Micromac AG, Germany
(3) Acram AB., Sweden
(4) Aeromet Corp., USA
(5) Alpha Products and Systems Pte. Ltd., Singapore
(6) Autostrade Co., Ltd., Japan
(7) CAM-LEM Inc., USA
(8) Carl Zeiss Pte Ltd., Singapore
(9) Champion Machine Tools Pte Ltd., Singapore
(10) CMET Inc., Japan
(11) Concept Laser GmbH, Germany
(12) Cubic Technologies Inc., USA
(13) Cubital Ltd., Israel
(14) Cybron Technology (S) Pte Ltd., Singapore
(15) D-MEC Corporation, Japan
(16) Ennex Corp., USA
(17) EnvisionTec., Germany
(18) EOS GmbH, Germany
(19) Fraunhofer-Institute for Applied Materials Research, Germany
(20) Fraunhofer-Institute for Manufacturing Engineering and Automation, Germany
(21) Hong Chek Co. Pte Ltd., Singapore

(22) Innomation Systems and Technologies Pte Ltd., Singapore
(23) Kira Corporation Ltd., Japan
(24) MCP-HEK Tooling GmbH, Germany
(25) Meiko's RPS Co., Ltd., Japan
(26) MicroFabrica Inc., USA
(27) MIT, USA
(28) Objet Geometries Ltd., Israel
(29) Optomec Inc., USA
(30) Phenix Systems, France
(31) Sintermask Technologies AB, Sweden
(32) Solidimension Ltd., Isreal
(33) Solidscape Inc., USA
(34) Soligen Inc., USA
(35) Stratasys Inc., USA
(36) Teijin Seiki Co., Ltd., Japan
(37) The Ex One Company, USA
(38) Therics Inc., USA
(39) Voxeljet Technology GmbH, Germany
(40) Z Corp., USA
(41) Zugo Technology Pte Ltd., Singapore

We would also like to express our special gratitude to Cherie Chua who has so painstakingly poured through our manuscript and help us polish the language.

Last but not least, we also wish to express our thanks and apologies to the many others not mentioned above for their suggestions, corrections and contributions to the success of the previous editions of the book. We would appreciate your comments and suggestions on this third edition book.

Chua C. K.
Associate Professor

Leong K. F.
Associate Professor

Lim C. S.
Associate Professor

ABOUT THE AUTHORS

Chua Chee Kai is an Associate Professor of the Systems and Engineering Management Division at the School of Mechanical and Aerospace Engineering, Nanyang Technological University, Singapore. Dr. Chua has extensive teaching and consulting experience in Rapid Prototyping (RP) including being advisor to RP bureau start-ups and conducted RP courses more than 40 times for many companies. His research interests include the development of new techniques for RP, tissue engineering scaffold design and fabrication, as well as their engineering and manufacturing applications in industry. He was the keynote speaker for conferences on "Trends of RP" and has sat on more than 10 programme committees for international RP conferences. Dr. Chua sits on the Editorial Advisory Board of Rapid Prototyping Journal, Journal of Materials Processing Technology and International Journal of Advanced Manufacturing Technology. He is also the co-editor-in-chief of the Virtual and Physical Prototyping Journal. He is the author of over 210 publications, including books, book chapters, international journals and conference proceedings and has one patent in his name. From his research work, he has won many academic prizes and his publications have been cited more than 1400 times based on Science Citation Index. Dr. Chua can be contacted by email at mckchua@ntu.edu.sg.

Leong Kah Fai is an Associate Professor at the School of Mechanical and Aerospace Engineering, Nanyang Technological University, Singapore. He is the Programme Director for the Master of Science Course in Smart Product Design and was the Founding Director of the Design Research Center at the University. He graduated from the National University of Singapore and Stanford University for his undergraduate and graduate degrees, respectively. He is actively involved in local design, manufacturing and standards work. He has chaired several committees in the Singapore Institute of Standards and Industrial Research and Productivity

Standards Board, receiving Merit and Distinguished Awards for his serv-ices in 1994 and 1997, respectively. He has delivered keynote papers at several conferences in application of RP in biomedical science and tissue engineering. He has authored over 110 publications, including books, book chapters, international journals and conferences and has one patent to his name. He has won several academic prizes and his publications have been cited more than 1100 times based on the Science Citation Index. Mr. Leong can be contacted by email at mkfleong@ntu.edu.sg.

Lim Chu-Sing, Daniel is an Associate Professor at the Manufacturing Engineering Division, School of Mechanical and Aerospace Engineering, Nanyang Technological University (NTU), Singapore. He holds an adjunct appointment with the A*STAR Brenner Center for Molecular Medicine. Dr. Lim received the National Science and Technology Post-doctoral Fellowship in 1995. In 1997, he was secretary to the Task Force to establish Biomedical Engineering at NTU. In 1999, he was appointed Deputy Director of the University's Biomedical Engineering Research Center. In 2001, he received the Tan Chin Tuan Fellowship as Visiting Scientist in Biophotonics to Duke University's Department of Biomedical Engineering. In 2005, Dr. Lim became the center's Director and held a concurrent appointment as Director of the University's Biomedical and Pharmaceutical Engineering Cluster. From 2005 to 2007, Dr. Lim was seconded to the newly established School of Chemical and Biomedical Engineering as Vice-Dean (Admin). Dr. Lim currently chairs the Bioengineering Technical Committee and serves as Council Member for the Institute of Engineers, Singapore. He has authored more than 150 jour-nal and conference publications, including books and book chapters. For his work, Dr. Lim received several awards including the JCCI Singapore Foundation Award for Education, Prime Minister's Office TEC Innovator Award, Tan Kah Kee Young Inventor's Award, National Academy of Science Young Scientist Award (Certificate of Finalist) and the Rotary Club Young Executive of the Year (1st Runner-up). Dr. Lim can be con-tacted by email at mchslim@ntu.edu.sg.

LIST OF ABBREVIATIONS

2D	Two-dimensional
3D	Three-dimensional
3DP	Three-dimensional printing
ABS	Acrylonitrile butadiene styrene
ACSII	American standard code for information interchange
AIM	ACES injection molding
AOM	Acoustic optical modulator
BDM	Beam delivery system
CAD	Computer-aided design
CAE	Computer-aided engineering
CAM	Computer-aided manufacturing
CAM-LEM	Computer-aided manufacturing of laminated engineering materials
CBC	Chemically bonded ceramics
CD	Compact disc
CIM	Computer-integrated manufacturing
CLI	Common layer interface
CMM	Coordinate measuring machine
CNC	Computer numerical control
CSG	Constructive solid geometry
CT	Computerized tomography
DLP	Digital light processing
DMD	Direct metal deposition; digital mirror device
DMLS	Direct metal laser sintering
DSP	Digital signal processor
DSPC	Direct shell production casting
EB	Electron beam
EBM	Electron beam melting
EDM	Electric discharge machining

EMS	Engineering modeling system
FDM	Fused deposition modeling
FEA	Finite element analysis
FEM	Finite element method
GIS	Geographic information system
GPS	Global positioning system
HPGL	Hewlett–Packard graphics language
HQ	High quality
HR	High resolution
HS	High speed
IGES	Initial graphics exchange specification
LAN	Local area network
LCD	Liquid crystal display
LEAF	Layer exchange ASCII format
LED	Light emitting diode
LENS	Laser engineered net shaping
LMT	Laser manufacturing technologies
LOM	Laminated object manufacturing
LS	Laser sintering
M^{3}D	Maskless mesoscale material deposition
M-RPM	Multi-functional RPM
MEM	Melted extrusion modeling
MEMS	Micro-electro-mechanical systems
MIG	Metal inert gas
MJM	Multi-jet modeling system
MJS	Multiphase jet solidification
MRI	Magnetic resonance imaging
NASA	National Aeronautical and Space Administration
NC	Numerical control
PC	Personal computer; polycarbonate
PCB	Printed circuit board
PDA	Personal digital assistant
PLT	Paper lamination technology
PPSF	Polyphenylsulfone
PSL	Plastic sheet lamination
RDM	Resin delivery module

RFP	Rapid freeze prototyping
RM&T	Rapid manufacturing and tooling
RP	Rapid prototyping
RPI	Rapid prototyping interface
RPM	Rapid prototyping and manufacturing
RPS	Rapid prototyping systems
RPT	Rapid prototyping technologies
RSP	Rapid solidification process
SAHP	Selective adhesive and hot press
SCS	Solid creation system
SDM	Shaped deposition manufacturing
SFF	Solid freeform fabrication
SFM	Solid freeform manufacturing
SGC	Solid ground curing
SHR	Single head replacement
SLA	Stereolithography apparatus
SLC	Stereolithography contour
SLM	Selective laser melting
SLS	Selective laser sintering
SMS	Selective mask sintering
SSM	Slicing solid manufacturing
SOUP	Solid object ultraviolet–laser plotting
STL	Stereolithography file
TTL	Toyota Technical Centre
UV	Ultraviolet

Chapter 1
INTRODUCTION

The competition in the world market for manufactured products has intensified tremendously in recent years. It has become important, if not vital, for new products to reach the market as early as possible, before the competitors.[1] To bring products to the market swiftly, many of the processes involved in the design, test, manufacture and market of the product have been squeezed, both in terms of time and material resources. The efficient use of such valuable resources calls for new tools and approaches in dealing with them and many of these tools and approaches have evolved. They are mainly technology-driven, usually involving the computer. This is mainly a result of the rapid development and advancement in such technologies over the last few decades.

In product development,[2] time pressure has been a major factor in determining the direction and success of developing new methods and advanced technologies. These also have a direct impact on the age old practice of prototyping in the product development process. This book will introduce and examine in a clear and detailed way one such development, namely, that of Rapid Prototyping (RP).

1.1. PROTOTYPE FUNDAMENTALS

1.1.1. Definition of a Prototype

A prototype is an important and vital part of the product development process. In any design practice, the word "prototype" is often not far from the things that the designers will be involved in. In most dictionaries, it is defined as a noun, e.g., the Oxford Advanced Learner's Dictionary of Current English[3] defines it as shown in Fig. 1.1.

However, in design, it often means more than just an artifact. It has often been used as a verb, e.g., prototype an engine design for engineering evaluation, or as an adjective, e.g., build a prototype printed circuit board (PCB).

1

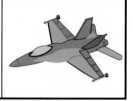

A prototype is the first or original example of something that has been or will be copied or developed; it is a model or preliminary version; e.g.: A prototype supersonic aircraft.

Fig. 1.1. A general definition of a prototype.

To be general enough to be able to cover all aspects of the meaning of the word prototype for use in design, it is very loosely defined here as:

An approximation of a product (or system) or its components in some form for a definite purpose in its implementation.

This very general definition departs from the usual accepted concept of the prototype being physical. It covers all kinds of prototypes used in the product development process, including objects like mathematical models, pencil sketches, foam models and of course the functional physical approximation or exact replica of the product. Prototyping is the process of realizing these prototypes. Here, the process can range from just an execution of a computer program to the actual building of a functional prototype.

1.1.2.   Types of Prototypes

The general definition of the prototype contains three aspects of interests:

(1) the implementation of the prototype; from the entire product (or system) itself to its subassemblies and components,
(2) the form of the prototype; from a virtual prototype to a physical prototype and
(3) the degree of the approximation of the prototype; from very rough representation to exact replication of the product.

The implementation aspect of the prototype covers the range of prototyping the complete product (or system) to prototyping part of, or a subassembly or a component of the product. The complete prototype, as its name suggests, models most, if not all, the characteristics of the product. It is usually implemented full-scale as well as being fully functional. One example of such a prototype is one that is given to a group of carefully selected people with special interest, often called a focus group, to examine and identify outstanding problems before the product is committed to its final design. On the other hand, there are prototypes that are needed to study or investigate special problems associated with one component, subassemblies or simply a particular concept of the product that require close attention. An example of such a prototype is a test platform that is used to find the comfortable rest angles of an office chair that will reduce the risk of spinal injuries after prolonged sitting on such a chair. Most of the time, subassemblies and components are tested in conjunction with some kind of test rigs or experimental platform.

The second aspect of the form of the prototype takes into account how the prototype is being implemented. On one end, virtual prototypes that refer to prototypes that are nontangible; usually represented in some form other than physical, e.g., a mathematical model of a control system.[4] Usually such prototypes are studied and analyzed. The conclusions drawn are purely based upon the assumed principles or science that underwrote the prototype at that point of time. An example is the visualization of airflow over an aircraft wing to ascertain lift and drag on the wing during supersonic flight. Such prototype is often used when either the physical prototype is too large and therefore takes too long to build, or the building of such a prototype is exorbitantly expensive. The main drawback of this kind of prototypes is that they are based on current understanding and thus they will not be able to predict any unexpected phenomenon. It is very poor or totally unsuitable for solving unanticipated problems. The physical model, on the other hand, is the tangible manifestation of the product, usually built for testing and experimentation. Examples of such prototypes include mock-up of a cellular telephone that looks and feels very much like the real product but without its intended functions. Such a prototype may be used purely for aesthetic and human factors evaluation.

The third aspect covers the degree of approximation or representativeness of the prototype. On one hand, the model can be a very rough representation of the intended product, like a foam model, used primarily to study the general form and enveloping dimensions of the product in its initial stage of development. Some rough prototypes may not even look like the final product, but are used to test and study certain problems of the product development. An example of this is the building of catches with different material to find the right "clicking" sound for a cassette player door. On the other hand, the prototype can be an exact full scale replication of the product that models every aspect of the product, e.g., the pre-production prototype that is used to satisfy customer-needs evaluation as well as to address manufacturing issues and concerns. Such "exact" prototypes are especially important towards the end-stage of the product development process. This aspect is sometimes referred to as "representativeness".

Figure 1.2 shows the various kinds of prototypes placed over the three aspects of describing the prototype. Each of the three axes represents one aspect of the description of the prototype. Note that this illustration is not meant to provide an exact scale to describe a prototype, but serves to demonstrate that prototypes can be depicted along these three aspects.

RP typically falls within the range of physical prototypes, which are usually fairly accurate and can be implemented on a component level or at a system level. This is shown as the shaded volume in Fig. 1.2. The versatility and range of different prototypes, from complete systems to individual components, that can be produced by RP at varying degrees of approximation makes it an important tool for prototyping in the product development process. Adding the major advantage of speed in delivery, it has become an important component in the prototyping arsenal that cannot be ignored.

1.1.3.   Roles of the Prototypes

Prototypes play several roles in the product development process. They include the following:

(1) experimentation and learning,
(2) testing and proofing,
(3) communication and interaction,
(4) synthesis and integration and
(5) scheduling and markers.

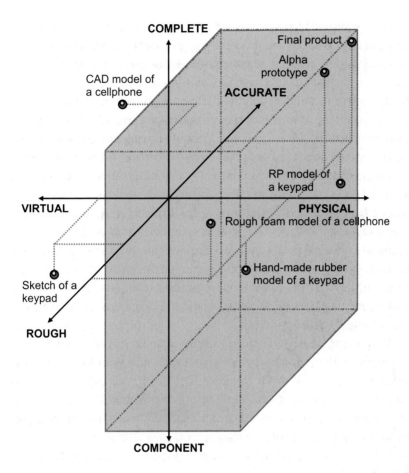

Fig. 1.2. Types of prototypes described along the three aspects of approximation, form and implementation.

To the product development team, prototypes can be used to help the thinking, planning, experimenting and learning processes whilst designing the product. Questions and doubts regarding certain issues of the design can be addressed by building and studying the prototype. For example, in designing the appropriate arm support of an office chair, several physical prototypes of such supports can be built to learn about the "feel" of the support when performing typical tasks on the office chair.

Prototypes can also be used for testing and proofing of ideas and concepts relating to the development of the product. For example, in the early design of folding reading glasses for the elderly, concepts and ideas of the folding mechanism can be tested by building rough physical prototypes to test and prove these ideas to check if they work as intended.

The prototype also serves the purpose of communicating information and demonstrating ideas, not just within the product development team, but also to the management and clients (whether in-house or external). Nothing is clearer for the explanation or communication of an idea than a physical prototype where the intended audience can have the full experience of the visual and tactile feel of the product. A 3D representation is more often than not superior to that of a 2D sketch of the product. For example, a physical prototype of a cellular phone can be presented to carefully selected customers. Customers can handle and experiment with the phone and give feedback to the development team on the features of and interactions with the phone, thus providing valuable information for the team to improve its design.

A prototype can also be used to synthesize the entire product concept by bringing the various components and subassemblies together to ensure that they will work together. This will greatly help in the integration of the product and surface any problems that are related to assembling the product together. An example is a complete or comprehensive functional prototype of a personal digital assistant (PDA). When putting the prototype together, all aspects of the design, including manufacturing and assembly issues, will have to be addressed, thus enabling the different functional members of the product development team to understand the various problems associated with putting the product together.

Prototyping also serves to help in the scheduling of the product development process and is usually used as markers for the start or end of the various phases of the development effort. Each prototype usually marks a completion of a particular development phase and with proper planning, the development schedule can be enforced. Typically in many companies, the continuation of a development project often hinges on the successful execution of the prototypes to provide impetus for the management to forge ahead with it.

It should be noted that in many companies, prototypes do not necessarily serve all these roles concurrently, but they are certainly a necessity in any product development project.

The prototypes created with RP technologies will serve most if not all of these roles. Being accurate physical prototypes that can be built with speed, many of these roles can be accomplished quickly, effectively and together with other productivity tools, e.g., CAD, repeatedly with precision.

1.2. HISTORICAL DEVELOPMENT

The development of RP is closely tied in with the development of applications of computers in the industry. The declining cost of computers, especially of personal computers, has altered the way a factory works. The increase in the use of computers has spurred the advancement in many computer-related areas including computer-aided design (CAD), computer-aided manufacturing (CAM) and computer numerical control (CNC) machine tools. In particular, the emergence of RP systems could not have been possible without the existence of CAD. However, through careful examinations of the numerous RP systems in existence today, it can be easily deduced that other than CAD, many other technologies and advancements in other fields such as manufacturing systems and materials have also been crucial in the development of RP systems. Table 1.1 traces the historical development of relevant technologies related to RP from the estimated date of inception.

Table 1.1. Historical development of RP and related technologies.

Year of inception	Technology
1770	Mechanization[4]
1946	First computer
1952	First numerical control (NC) machine tool
1960	First commercial laser[5]
1961	First commercial robot
1963	First interactive graphics system (early version of computer-aided design)[6]
1988	First commercial rapid prototyping system

1.2.1. Three Phases of Development Leading to RP

Prototyping or model making in the traditional sense is an age-old practice. The intention of having a *physical* prototype is to realize the conceptualization of a design. Thus, a prototype is usually required before the start of the full production of the product. The fabrication of prototypes is experimented in many forms — material removal, castings, molds, joining with adhesives, etc. and with many material types — aluminum, zinc, urethanes, wood, etc.

Prototyping processes have gone through three phases of development, the last two of which have emerged only in the last 20 years.[7] Like the modeling process in computer graphics,[8] the prototyping of physical models is growing through its third phase. Parallel phases between the computer modeling process and prototyping process can be drawn as seen in Table 1.2. The three phases are described as follows.

1.2.1.1. *First Phase: Manual Prototyping*

Prototyping had begun as early as humans started to develop tools to help them live. However, prototyping as applied to products in what is considered to be the first phase of prototype development began several centuries ago. In this early phase, prototypes typically are not very sophisticated and fabrication of prototypes takes on average about four weeks, depending on the level of complexity and representativeness.[9] The techniques used in making these prototypes tend to be craft-based and are usually extremely labor intensive.

1.2.1.2. *Second Phase: Soft or Virtual Prototyping*

As applications of CAD/CAE/CAM become more widespread, the early 1980s saw the evolution of the second phase of prototyping — *Soft or Virtual Prototyping*. Virtual prototyping takes on a new meaning as more computer tools become available — computer models can now be stressed, tested, analyzed and modified as if they were physical prototypes. For example, analysis of stress and strain can be accurately predicted on the product because of the ability to specify exact material

Table 1.2. Parallel phases between geometric modeling and prototyping.

Geometric modeling	Prototyping
First phase: 2D wireframe	*First phase: Manual prototyping*
Started in mid-1960s	Traditional practice for many centuries
Few straight lines on display may be	Prototyping as a skilled crafts is
circuit path on a PCB	traditional and manual
plan view of a mechanical component	based on material of prototype
"Natural" drafting technique	"Natural" prototyping technique
Second phase: 3D curve and surface modeling	*Second phase: Soft or virtual prototyping*
Mid-1970s	Mid-1970s
Increasing complexity	Increasing complexity
Representing more information about	Virtual prototype can be stressed,
precise shape, size and surface	simulated and tested, with exact
contour of parts	mechanical and other properties.
Third phase: Solid modeling	*Third phase: Rapid prototyping*
Early 1980s	Mid-1980s
Edges, surfaces and holes are knitted	Benefit of a hard prototype made in a
together to form a cohesive whole	very short turnaround time is its main
Computer can determine the inside of	strong point (relies on CAD modeling)
an object from the outside. Perhaps,	Hard prototype can also be used for
more importantly, it can trace across	limited testing
the object and readily find all	Prototype can also assist in the
intersecting surfaces and edges	manufacturing of the products
No longer ambiguous but exact	

attributes and properties. With such tools on the computer, several iterations of designs can be easily carried out by changing the parameters of the computer models.

As products and their corresponding prototypes become twice as complex as before,[9] the time required to make the physical model increases tremendously to about 16 weeks. This is because the prototyping process still depends on craft-based techniques albeit higher precision machines, such as CNC, become available.

Even with the advent of RP in the third phase, there is still strong support for virtual prototyping. Lee[10] argued that there are still unavoidable limitations with RP. These include material limitations (either because of expense or through the use of materials dissimilar to that of the intended part), the inability to perform endless what-if scenarios and the likelihood that little or no reliable data can be gathered from the rapid prototype to perform finite element analysis (FEA). Specifically in the application of kinematic and dynamic analysis, he described a program which can assign physical properties of many different materials, such as steel, ice, plastic, clay or any custom material imaginable and perform kinematics and motion analysis as if a working prototype existed. Despite such strengths of virtual prototyping, there is one inherent weakness, that such soft prototypes cannot be tested for phenomena that is not anticipated or accounted for in the computer program. As such, there is no guarantee that the virtual prototype is 100% problem-free.

1.2.1.3. *Third Phase: RP*

RP of physical parts, otherwise known as solid free-form fabrication or desktop manufacturing or layer manufacturing technology, represents the third phase in the evolution of prototyping. The invention of this series of RP methodologies is described as a "watershed event"[11] because of the tremendous time savings, especially for complicated and difficult to produce models. Though parts (individual components) are relatively three times as complex as parts made in 1970s, the time required to make such a part now averages only three weeks.[9] Since 1988, more than 30 different RP techniques have emerged and commercialized.

1.3. FUNDAMENTALS OF RAPID PROTOTYPING

Regardless of the different techniques used in the RP systems developed, they adopt the same basic approach, which can be described as follows:

(1) A model or component is modeled on a computer-aided design–computer-aided manufacturing (CAD–CAM) system. The model which represents the physical part to be built must be represented as closed

surfaces which unambiguously define an enclosed volume. This means that the data must specify the inside, the outside and the boundary of the model. This requirement will become redundant if the modeling technique used is based on solid modeling. This is by virtue that a valid solid model will automatically have an enclosed volume. This requirement ensures that all horizontal cross-sections that are essential to RP are closed curves to create the solid object.

(2) The solid or surface model to be built is next converted into a format dubbed the "STL" (StereoLithography) file format which originates from 3D Systems. The STL file format approximates the surfaces of the model by polygons. Highly curved surfaces must employ many polygons, which mean that STL files for curved parts can be very large. However, there are some RP systems which also accept data in the initial graphics exchange specifications (IGES) format provided it is of the correct "flavor".

(3) A computer program analyzes an .STL file that defines the model to be fabricated and "slices" the model into cross-sections. The cross-sections are systematically recreated through the solidification of either liquids or powders and then combined to form a 3D model. Another possibility is that the cross-sections are already thin, solid laminations and these thin laminations are glued together with adhesives to form a 3D model. Other similar methods may also be employed to build the model.

Fundamentally, the development of RP can be seen in four primary areas. The RP wheel in Fig. 1.3 depicts these four key aspects of RP. They are: input, method, material and applications.

1.3.1. Input

Input refers to the electronic information required to describe the physical object with 3D data. There are two possible starting points — a computer model or a physical model. The computer model created by a CAD system can be either a surface model or a solid model. On the other hand, 3D data from the physical model is not at all straightforward. It requires data acquisition through a method known as reverse engineering. In reverse engineering, a wide range of equipment can be used, such as coordinate

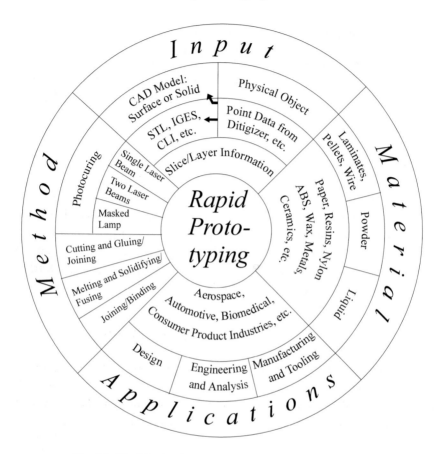

Fig. 1.3. The RP wheel depicting the four major aspects of RP.

measuring machine (CMM) or a laser digitizer, to capture data points of the physical model and "reconstruct" it in a CAD system.

1.3.2. Method

While there are currently more than 30 vendors for RP systems, the method employed by each vendor can be generally classified into the following categories: photocuring; cutting and gluing or joining; melting and solidifying or fusing; and joining or binding. Photocuring can be further divided into categories of single laser beam, double laser beams and masked lamp.

1.3.3. Material

The initial state of material can come in one of the following forms: solid, liquid or powder state. In solid state, it can come in various forms such as pellets, wire or laminates. The current range materials include paper, nylon, wax, resins, metals and ceramics.

1.3.4. Applications

Most of the RP parts are finished or touched up before they are used for their intended applications. Applications can be grouped into (1) design, (2) engineering analysis and planning and (3) manufacturing and tooling. A wide range of industries can benefit from RP and these include, but are not limited to, aerospace, automotive, biomedical, consumer, electrical and electronics products.

1.4. ADVANTAGES OF RAPID PROTOTYPING

Today's automated, toolless, patternless RP systems can directly produce functional parts in small production quantities. Parts produced in this way usually have an accuracy and surface finish inferior to those made by machining. However, some advanced systems are able to produce near tooling quality parts that are close to or are in the final shape. The parts produced, with appropriate post-processing, have material qualities and properties close to the final product. More fundamentally, the time to produce any part — once the design data are available — will be fast and can be in a matter of hours.

The benefits of RP systems are immense and can be broadly categorized into direct and indirect benefits.

1.4.1. Direct Benefits

The benefits to the company using RP systems are many. One would be the ability to experiment with physical objects of any complexity in a relatively short period of time. It is observed that over the last 35 years, products released in the market place have increased in complexity in

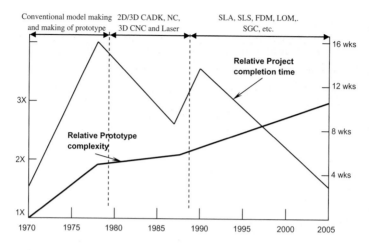

Fig. 1.4. Project time and product complexity in 25 years time frame.

shape and form.[9] For instance, compare the aesthetically beautiful car body of today with that of the 1970s. On a relative complexity scale of 1–3 as seen in Fig. 1.4, it is noted that from a base of 1 in 1970, this relative complexity index has increased to about 2 in 1980 and close to 3 in the 1990s. More interestingly and ironically, the relative project completion times have not been drastically increased. Initially, from a base of about four weeks project completion time in 1970, it increased to 16 weeks in 1980. However, with the use of CAD–CAM and CNC technologies, project completion time reduces to eight weeks. Eventually, RP systems allowed the project manager to further cut the completion time to three weeks in 1995.

To the individual in the company, the benefits and impact of using RP are varied. It depends on the role they play in the company. The full production of any product encompasses a wide spectrum of activities. Kochan and Chua[12] described the impact of RP technologies on the entire spectrum of product development and process realization. In Fig. 1.5, the activities required for full production in a conventional model are depicted at the top. The bottom of Fig. 1.5 shows the RP model. Depending on the size of production, savings on time and cost could range from 50% up to 90%.

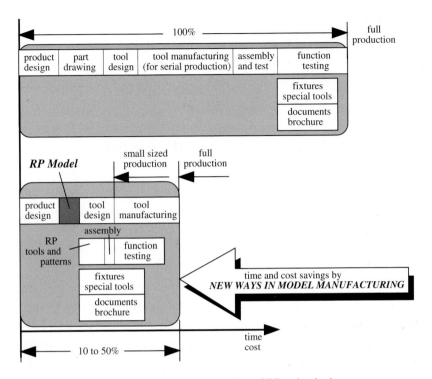

Fig. 1.5. Results of the integration of RP technologies.

1.4.1.1. *Benefits to Product Designers*

The product designers can increase part complexity with little significant effects on lead time and cost. More organic, sculptured shapes for functional or aesthetic reasons can be accommodated. They can optimize part design to meet customer requirements, with little restrictions by manufacturing. In addition, they can reduce parts count by combining features in single-piece parts that are previously made from several because of poor tool accessibility or the need to minimize machining and waste. With fewer parts, time spent on tolerance analysis, selecting fasteners, detailing screw holes and assembly drawings is greatly reduced.

There will also be fewer constraints in the form of parts design without regard to draft angles, parting lines or other such constraints. Parts which cannot easily be set up for machining, or having accurate, large thin walls, or not using stock shapes to minimize machining and waste can now be designed. They can minimize material and optimize strength/ weight ratios without regard to the cost of machining. Finally, they can minimize time-consuming discussions and evaluations of manufacturing possibilities.

1.4.1.2. *Benefits to the Tooling and Manufacturing Engineer*

The main savings are in costs. The manufacturing engineer can minimize design, manufacturing and verification of tooling. He can realize profit earlier on new products, since fixed costs are lower. He can also reduce parts count and, therefore, assembly, purchasing and inventory expenses.

The manufacturer can reduce labor content of manufacturing, since part-specific setting up and programming are eliminated, machining or casting labor is reduced and inspection and assembly are also consequently minimized as well. Reducing material waste, waste disposal costs, material transportation costs, inventory cost for raw stock and finished parts (make only as many as required, therefore, can reduce storage requirements) can contribute to low overheads. Fewer inventories are scrapped because of fewer design changes or reduced risks of disappointing sales.

In addition, the manufacturer can simplify purchasing since unit price is almost independent of quantity, therefore, only as many as are needed short-term need be ordered. Quotations vary little among supplies, since fabrication is automatic and standardized. One can purchase one general purpose machine rather than many special purpose machines and therefore, reduce capital equipment and maintenance expenses and need fewer specialized operators and training. A smaller production facility will also result in less effort in scheduling production. Furthermore, one can reduce inspection reject rate since the number of tight tolerances required when parts must mate can be reduced. One can avoid design misinterpretations (instead, "what you design is what you get"), quickly change design dimensions to deal with tighter tolerances and achieve higher part repeatability, since tool wear is eliminated. Lastly, one can reduce spare parts inventories (produce spare on demand, even for obsolete products).

1.4.2. Indirect Benefits

Outside the design and production departments, indirect benefits can also be derived. Marketing as well as the customers will also benefit from the utilization of RP technologies.

1.4.2.1. *Benefits to Marketing*

To the market, it presents new capabilities and opportunities. It can greatly reduce time-to-market, resulting in (1) reduced risk as there is no need to project customer needs and market dynamics several years into the future, (2) products which fit customer needs much more closely, (3) products offering the price/performance of the latest technology and (4) new products being test-marketed economically.

Marketing can also change production capacity according to the market demand, possibly in real time and with little impact on manufacturing. One can increase the diversity of product offerings and pursue market niches currently too small to justify due to tooling cost (including custom and semi-custom production). One can easily expand distribution and quickly enter foreign markets.

1.4.2.2. *Benefits to the Consumer*

The consumer can buy products which more closely suit individual needs and wants. Firstly, there is a much greater diversity of offerings to choose from. Secondly, one can buy (and even contribute to the design of) affordable built-to-order products. Furthermore, the consumer can buy products at lower prices, since the manufacturers' savings will ultimately be passed on.

1.5. COMMONLY USED TERMS

The number of terms used by the engineering communities around the world to describe this technology is still many. Perhaps this is due to the versatility and continued development of the technology. It certainly does not help as already there are so many buzz words used today. Worldwide, the most commonly used term is RP. The term is apt as the key benefit of RP is its *rapid* creation of a physical model. However, prototyping is slowly growing to include other areas. Soon, *rapid prototyping, tooling and manufacturing (RPTM)* should be used to include the utilization of the prototype as a master pattern for tooling and manufacturing.

Some of the less commonly used terms include *direct CAD manufacturing, desktop manufacturing* and *instant manufacturing*. The rationale behind these terms is also based on speed and ease, though not exactly direct or instant! *CAD oriented manufacturing* is another term and provides an insight into the issue of orientation, often a key factor influencing the output of a prototype made by RP methods.

Another group of terms emphasizes the unique characteristic of RP — layer by layer addition as opposed to traditional manufacturing methods such as machining which is material removal from a block. This group includes *layer manufacturing, material deposit manufacturing, material addition manufacturing* and *material incress manufacturing*.

There is yet another group which chooses to focus on the words "solid" and "free-form" — *solid free-form manufacturing* and *solid free-form fabrication*. *Solid* is used because while the initial state may be liquid, powder, individual pellets or laminates, the end result is a solid, 3D object, while *free-form* stresses on the ability of RP to build complex shapes with little constraint on its form.

1.6.   CLASSIFICATION OF RAPID PROTOTYPING SYSTEMS

While there are many ways in which one can classify the numerous RP systems in the market, one of the better ways is to classify RP systems broadly by the initial form of its material, i.e., the material that the prototype or part is built with. In this manner, all RP systems can be easily categorized into (1) liquid-based, (2) solid-based and (3) powder-based.

1.6.1. Liquid-Based

Liquid-based RP systems have initial form of their material in liquid state. Through a process commonly known as curing, the liquid is converted to the solid state. The following RP systems fall into this category:

(1) 3D Systems' stereolithography apparatus (SLA)
(2) Objet Geometries Ltd.'s Polyjet
(3) D-MEC's solid creation system (SCS)
(4) EnvisionTec's Perfactory
(5) Autostrade's E-Darts
(6) CMET's solid object ultraviolet–laser printer (SOUP)
(7) EnvisionTec's Bioplotter
(8) Rapid freeze prototyping
(9) Microfabrication
(10) Microfabrica®'s EFAB® Technology
(11) D-MEC's ACCULAS
(12) Two Laser Beams
(13) Cubital's solid ground curing (SGC)
(14) Teijin Seiki's soliform system
(15) Meiko's rapid prototyping system for the jewelry industry.

As is illustrated in the RP wheel in Fig. 1.3, three methods are possible under the *"photocuring"* method. The *single laser beam* method is a widely used method and includes many of the above RP systems (1), (3), (5), (6), (12), (14) and (15). Objet Geometries's Polyjet (2) uses a UV lamp for curing after desposition via jetting heads while Cubital's SGC (13) uses UV masked lamp. EnvisionTec's Perfactory (4) uses an imaging system called digital light processing (DLP) while D-MEC's ACCULAS (11) uses a different system called digital mirror device (DMD). The *two laser beam* (12) method is still not commercialized. EnvisionTec's Bioplotter (7) uses an extrusion method in a liquid medium. Rapid Freeze (8) involves the freezing of water droplets and deposits in a manner much like FDM (see below) to create the prototype. Microfabrica®'s EFAB® Technology (10) uses an electro-deposition method in a liquid medium. These will be described in detail in Chap. 3.

1.6.2. Solid-Based

Except for powder, solid-based RP systems are meant to encompass all forms of material in the solid state. In this context, the solid form can include the shape in the form of wires, rolls, laminates and pellets. The following RP systems fall into this definition:

(1) Stratasys' fused deposition modeling (FDM)
(2) Solidscape's benchtop system
(3) Cubic Technologies' laminated object manufacturing (LOM)
(4) 3D Systems' multi-jet modeling system (MJM)
(5) Solidimension's plastic sheet lamination (PSL)/3D System's invision LD sheet lamination
(6) Kira's paper lamination tech (PLT)
(7) CAM-LEM's CL 100
(8) Ennex Corporation's offset fabbers
(9) Shape deposition manufacturing process

Referring to the RP wheel in Fig. 1.3, two methods are possible for solid-based RP systems. RP systems (1), (2), (4), (7) and (9) belong to the *Melting and Solidifying or Fusing* method, while the *Cutting and Gluing or Joining* method is used for RP systems (3), (5), (6) and (8). The various RP systems will be described in detail in Chap. 4.

1.6.3. Powder-Based

In a strict sense, powder is by-and-large in the solid state. However, it is intentionally created as a category outside the solid-based RP systems to mean powder in grain-like form. The following RP systems fall into this definition:

(1) 3D systems' selective laser sintering (SLS)
(2) Z Corporation's three-dimensional printing (3DP)

(3) EOS's EOSINT systems

(4) Optomec's laser engineered net shaping (LENS)

(5) Arcam's electron beam melting (EBM)

(6) Concept Laser GmbH's LaserCUSING®

(7) MCP-HEK Tooling GmbH's Realizer II (SLM)

(8) Phenix Systems's PM series (LS)

(9) Sintermask Technologies AB's selective mask sintering (SMS)

(10) 3D-Micromac AG's microsintering

(11) Therics Inc.'s theriform technology

(12) The Ex One Company's ProMetal

(13) Voxeljet Technology GmbH'S VX system

(14) Soligen's direct shell production casting (DSPC)

(15) Fraunhofer's multiphase jet solidification (MJS)

(16) Aeromet Corporation's lasform technology

All the above RP systems employ the *joining/binding* method. The method of joining/binding differs for the above systems in that some employ a laser while others use a binder/glue to achieve the joining effect. Similarly, the above RP systems will be described in detail in Chap. 5.

REFERENCES

1. S. C. Wheelwright and K. B. Clark, *Revolutionizing Product Development: Quantum Leaps in Speed, Efficiency, and Quality* (The Free Press, New York, 1992).

2. K. T. Ulrich and S. D. Eppinger, *Product Design and Development*, 2nd edn. (McGraw-Hill, Boston, 2000).

3. A. S. Hornby and S. Wehmeier (eds.) *Oxford Advanced Learner's Dictionary of Current English*, 6th edn. (Oxford University Press, Oxford, 2000).

4. Y. Koren, *Computer Control of Manufacturing Systems* (McGraw-Hill, Singapore, 1983).

5. J. Hecht, *The Laser Guidebook*, 2nd edn. (McGraw-Hill, New York, 1992).

6. K. Taraman, *CAD/CAM: Meeting Today's Productivity Challenge* (Computer and Automated Systems Association of SME, Michigan, 1982).

7. C. K. Chua, Three-dimensional rapid prototyping technologies and key development areas, *Comput. Contr. Eng. J.* **5**(4): 200–206 (1994).

8. C. K. Chua, Solid modeling — A state-of-the-art report, *Manufacturing Equipment News*, September 1987, pp. 33–34.

9. J. Metelnick, How today's model/prototype shop helps designers use rapid prototyping to full advantage, Society of Manufacturing Engineers Technical Paper (1991), MS91–475.

10. G. Lee, Virtual prototyping on personal computers, *Mech. Eng.* **117**(7), 70–73 (1995).

11. D. Kochan, Solid freeform manufacturing — Possibilities and restrictions, *Comput. Ind.* **20**, 133–140 (1992).

12. D. Kochan and C. K. Chua. State-of-the-art and future trends in advanced rapid prototyping and manufacturing, *Int. J. Inform. Technol.* **1**(2), 173–184 (1995).

PROBLEMS

1. How would you define prototype in the context of the modern product development?
2. What are the three aspects of interest in describing a prototype? Describe them clearly.
3. What are the main roles and functions of prototypes? How do you think RP satisfy these roles?
4. Describe the historical development of RP and related technologies.
5. What are the three phases of prototyping? Contrasting these with those of geometric modeling, what similarities can be drawn?
6. Despite the increase in relative complexity of product's shape and form, project time has been kept relatively shorter. Why?
7. What are the fundamentals of RP?

8. What is the *RP wheel*? Describe its four primary aspects. Is the *wheel* a static representation of what is RP today? Why?

9. Describe the advantages of RP in terms of its beneficiaries such as the product designers, tool designer, manufacturing engineer, marketers, and consumers?

10. Many terms have been used to mean RP. Discuss three of such terms and explain why they have been used in place of RP.

11. Name three RP systems that are liquid-based.

12. How can the liquid form be converted to the solid form as in these liquid-based RP systems?

13. In what form of material can RP systems be classified as solid-based? Name three such systems.

14. What is the method used in powder-based RP systems?

Chapter 2
RAPID PROTOTYPING PROCESS CHAIN

2.1. FUNDAMENTAL AUTOMATED PROCESSES

There are three fundamental fabrication processes[1,2] as shown in Fig. 2.1. They are *subtractive, additive and formative* fabricators.

In a subtractive process, one starts with a single block of solid material larger than the final size of the desired object and portions of the material are removed until the desired shape is reached.

In contrast, an additive process is the exact reverse in that the end product is much larger than the material when it started. Materials are manipulated so that they are successively combined to form the desired object.

Lastly, the formative process is one where mechanical forces or restricting forms are applied on a material so as to form it into the desired shape.

There are many examples for each of these fundamental fabrication processes. Subtractive fabrication processes include most forms of machining processes — computer numerical control (CNC) or otherwise. These include milling, turning, drilling, planning, sawing, grinding, EDM, laser cutting, water-jet cutting and the likes. Most forms of rapid prototyping (RP) processes such as stereolithography apparatus (SLA) and selective laser sintering (SLS) fall into the additive fabrication processes category. Examples of formative fabrication processes are: bending, forging, electromagnetic forming and plastic injection molding. These include both bending of sheet materials and molding of molten or curable liquids. The examples given are not exhaustive but indicative of the range of processes in each category.

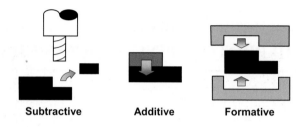

Subtractive　　　　**Additive**　　　　**Formative**

Fig. 2.1. Three types of fundamental fabrication processes.

Hybrid machines combining two or more fabrication processes are also possible. For example, in progressive press-working, it is common to see a hybrid of subtractive (as in blanking or punching) and formative (as in bending and forming) processes.

2.2. PROCESS CHAIN

As described in Sec. 1.3, all RP techniques adopt the same basic approach. As such all RP systems generally have a similar sort of process chain. Such a generalized process chain is shown in Fig. 2.2.[3] There are a total of five steps in the chain and these are: 3D modeling, data conversion and transmission, checking and preparing, building and postprocessing. Depending on the quality of the model and part in steps 3 and 5, respectively, the process may be iterated until a satisfactory model or part is achieved.

However, like other fabrication processes, process planning is important before the RP commences. In process planning, the steps of RP process chain are listed. The first step is 3D geometric modeling. In this instance, the requirement would be a computer and a CAD modeling system. The factors and parameters which influence the performance of each operation are examined and decided upon. For example, if SLA is used to build the part, the orientation of part is an important factor which would, amongst others, influence the quality of the part and the speed of the process. Thus, an operation sheet used in this manner requires proper documentation and sound guidelines. Good documentation, such as a process logbook, allows

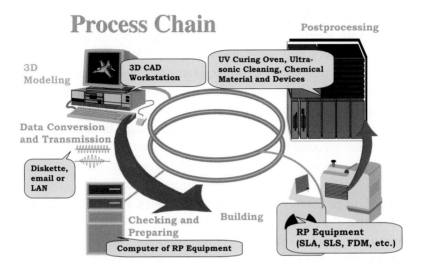

Fig. 2.2. Process chain of rapid prototyping process.

future examination and evaluation and subsequent improvements can be implemented to process planning. The five steps are discussed in the following sections.

2.2.1. 3D Modeling

Advanced 3D computer-aided design (CAD) modeling is a general prerequisite in RP processes and, usually is the most time-consuming part of the entire process chain. It is most important that such 3D geometric models can be shared by the entire design team for many different purposes, such as interference studies, stress analyses, FEM analysis, detail design and drafting, planning for manufacturing, including NC programming, etc. Many computer-aided design–computer-aided manufacturing (CAD–CAM) systems now have a 3D geometrical modeler facility with these special purpose modules.

There are two common misconceptions amongst new users of RP. Firstly, unlike NC programming, RP requires a closed volume of the model, whether the basic elements are surfaces or solids. This confusion arises because new users are usually acquainted with the use of NC programming where a single surface or even a line can be an NC element. Secondly, new users also usually assume *what you see is what you get*. These two misconceptions often lead to the user under specifying process parameters to the RP systems, resulting in poor performance and nonoptimal utilization of the system. Examples of considerations that have to be taken into account include orientation of part, need for supports, difficult-to-build part structure such as thin walls, small slots or holes and overhanging elements. Therefore, RP users have to learn and gain experience from working on the RP system. The problem is usually more complex than one can imagine because there are many different RP machines which have different requirements and capabilities. For example, while SLA requires supports, SLS does not and Solid Ground Curing (SGC) works most economically if many parts are nested together and processed simultaneously (see Secs. 3.1, 5.1 and 3.12.2).

2.2.2. Data Conversion and Transmission

The solid or surface model to be built is next converted into a format dubbed as the .STL (stereolithography) file format. This format originates from 3D Systems which pioneers the STL system. The STL file format approximates the surfaces of the model using tiny triangles. Highly curved surfaces must employ many more triangles, which mean that STL files for curved parts can be very large. The STL file format will be discussed in detail in Chap. 6.

Almost, if not all, major CAD–CAM vendors supply the CAD–STL interface. Since 1990, all major CAD–CAM vendors have developed and integrated this interface into their systems.

This conversion step is probably the simplest and shortest of the entire process chain. However, for highly complex model coupled with an extremely low performance workstation or PC, the conversion can take several hours. Otherwise, the conversion to STL file should take only several minutes. Where necessary, supports are also converted to a separate STL file. Supports can alternatively be created or modified in the next step by third party software which allows verification and modifications of models and supports.

The transmission step is also fairly straightforward. The purpose of this step is to transfer the STL files which reside in the CAD computer to the computer of the RP system. It is typical that the CAD computer and the RP system are situated in different locations. The CAD computer, being a design tool, is typically located in a design office. The RP system, on the other hand, is a process or production machine and is usually located in the shop floor. More recently however, there are several RP systems, usually the Concept Modelers or sometimes called 3D printers, which can be located in the design office. Data transmission via agreed data formats such as STL or IGES may be carried out through a diskette, electronic mail (email) or local area network (LAN). No validation of the quality of the STL files is carried out at this stage.

2.2.3. Checking and Preparing

The computer term, *garbage in garbage out*, is also applicable to RP. Many first time users are frustrated at this step to discover that their STL files are faulty. However, more often than not, it is due to both the errors of CAD models and the nonrobustness of the CAD–STL interface. Unfortunately, today's CAD models, whose quality are dependent on the CAD systems, human operators and postprocesses, are still afflicted with a wide spectrum of problems, including the generation of unwanted shell-punctures (i.e., holes, gaps, cracks, etc.). These problems, if not rectified, will result in the frequent failure of applications downstream. These problems are discussed in detail in the first few sections of Chap. 6.

Much of the CAD model errors are corrected by human operators assisted by specialized software such as Magics, a software developed by Materialise, N.V., Belgium.[4] Magics software enables users to import a wide variety of CAD formats and to export STL files ready for RP, tooling and manufacturing. Its applications include repairing and optimizing 3D models; analyzing parts; making process-related design changes on the STL files; documentation; and production planning. The manual repair process is, however, still very tedious and time consuming especially considering the great number of geometric entities (e.g., triangular facets) encountered in a CAD model. The types of errors and its possible solutions are discussed in Chap. 6.

Once the STL files are verified to be error-free, the RP system's computer analyzes the STL files that define the model to be fabricated and slices the model into cross sections. The cross sections are systematically recreated through the solidification of liquids or binding of powders, or fusing of solids, to form a 3D model.

In SLA, for example, each output file is sliced into cross sections, between 0.12 (minimum) and 0.50 mm (maximum) thickness. Generally, the model is sliced into the thinnest layer (approximately 0.12 mm) as they have to be very accurate. The supports can be created using coarser settings. An internal cross hatch structure is generated between the inner and the outer surface boundaries of the part. This serves to hold up the walls and entrap liquid that is later solidified in the presence of UV light.

Preparing building parameters for positioning and stepwise manufacturing in the light of many available possibilities can be difficult if not accompanied by proper documentation. These possibilities include determination of the geometrical objects, the building orientation, spatial assortments, arrangement with other parts, necessary support structures and slice parameters. They also include the determination of technological parameters such as cure depth, laser power and other physical parameters as in the case of SLA. It means that user-friendly software for ease of use and handling, user support in terms of user manuals, dialogue mode and on-line graphical aids will be very helpful to users of the RP system.

Many vendors are continually working to improve their systems on this aspect. For example, 3D Systems' Buildstation 5.5 software[5] enables users to simplify the process of setting parameters for the SLA. In early SLA systems, parameters (such as the location in the 250×250 mm box and the various cure depths) had to be set manually. This was very tedious for there could be up to 12 parameters to be keyed in. These parameters are shown in Table 2.1.

However, the job is now made simpler with the introduction of default values that can be altered to other specific values. These values can be easily retrieved for use in other models. This software also allows the user

Table 2.1. Parameters used in the SLA process.

1. X–Y shrink
2. Z shrink
3. Number of copies
4. Multi-part spacing
5. Range manager (add, delete, etc.)
6. Recoating
7. Slice output scale
8. Resolution
9. Layer thickness
10. X–Y hatch-spacing or 60/120 hatch spacing
11. Skin fill spacing (X, Y)
12. Minimum hatch intersecting angle

to orientate and move the model such that the whole model is in the positive axis' region (the SLA uses only positive numbers for calculations). Thus the original CAD design model can also be in "negative" regions when converting to STL format.

2.2.4.   Building

For most RP systems, this step is fully automated. Thus, it is usual for operators to leave the machine on to build a part overnight. The building process may take up to several hours depending on the size and number of parts required. The number of identical parts that can be built is subject to the overall build size constrained by the build volume of the RP system. Most RP systems come with user alert systems that inform the users remotely via electronic communication, e.g., cellular phone, once the building of the part is complete.

2.2.5.   Postprocessing

The final task in the process chain is the postprocessing task. At this stage, generally some manual operations are necessary. As a result, the danger of damaging a part is particularly high. Therefore, the operator for this last process step has a high responsibility for the successful process realization.

Table 2.2. Essential postprocessing tasks for different RP processes.

Postprocessing tasks	Rapid prototyping technologies			
	SLS[a]	SLA[b]	FDM[c]	LOM[d]
1. Cleaning	✓	✓	✗	✓
2. Postcuring	✗	✓	✗	✗
3. Finishing	✓	✓	✓	✓

✓ = is required; ✗ = not required.
[a] SLS — selective laser sintering.
[b] SLA — stereolithography apparatus.
[c] FDM — fused deposition modeling.
[d] LOM — layered object manufacturing.

The necessary postprocessing tasks for some major RP systems are shown in Table 2.2.

The cleaning task refers to the removal of excess parts which may have remained on the part. Thus, for SLA parts, this refers to excess resin residing in entrapped portion such as blind hole of part, as well as the removal of supports. Similarly, for SLS parts, the excess powder has to be removed. Likewise for LOM, pieces of excess wood-like blocks of paper which acted as supports have to be removed.

As shown in Table 2.2, the SLA postprocessing procedures require the highest number of tasks. More importantly, for safety reasons, specific recommendations for postprocessing tasks have to be prepared, especially for cleaning of SLA parts. It was reported by Peiffer that accuracy is related to post-treatment process.[6] Specifically, Peiffer referred to the swelling of SLA-built parts with the use of cleaning solvents. Parts are typically cleaned with solvent to remove unreacted photo-sensitive resin. Depending upon the "build style" and the extent of crosslinking in the resin, the part can be distorted during the cleaning process. This effect was particularly pronounced with the more open "build styles" and aggressive solvents. With the "build styles" approaching a solid fill and more solvent-resistant materials, damage with the cleaning solvent can be minimized. With newer cleaning solvents, like tripropylene glycol monomethyl ether (TPM) introduced by 3D Systems, part damage due to the cleaning solvent can be reduced or even eliminated.[6]

For reasons to be discussed in Chap. 3, SLA parts are built with pockets of liquid embedded within the part. Therefore, postcuring is required. All other nonliquid RP methods do not need to undergo this task.

Finishing refers to the secondary processes such as sanding and painting used primarily to improve surface finish or aesthetic appearance of the part. It also includes additional machining processes such as drilling, tapping and milling to add necessary features to the parts.

REFERENCES

1. M. Burns, Research notes. *Rapid Prototyping Rep.* **4**(3), 3–6 (1994).

2. M. Burns, *Automated Fabrication* (PTR Prentice Hall, New Jersey, 1993).

3. D. Kochan and C. K. Chua, Solid freeform manufacturing — assessments and improvements at the entire process chain, *Proc. Int. Dedicated Conf. Rapid Prototyping for the Automotive Industries, ISATA94*, 31 October–4 November 1994, Aachen, Germany.

4. N. V. Materialise, Magics 3.01 Software for the RP&M Professional. [Online]. Available: http://www.materialise.com/materialise/view/en/92074-Magics.html, September, 2008.

5. 3D Systems, SLA System Software, [On-line]. Available: http://www.3dsystems.com/products/software/3d_lightyear/index.asp, September, 2008.

6. R. W. Peiffer, The laser stereolithography process — photosensitive materials and accuracy, *Proc. First Int. User Congress on Solid Freeform Manufacturing*, 28–30 October 1993, Germany.

PROBLEMS

1. What are the three types of automated fabricators? Describe them and give two examples each.
2. Each one of the following manufacturing processes/methods in Table 2.3 belongs to one of the three basic types of fabricators. Tick [✓]

under the column if you think it belongs to that category. If you think that it is a hybrid machine, you may tick [✓] more than 1 category.

S/No	Manufacturing process	Subtractive	Additive	Formative
1	Pressworking			
2	*SLS			
3	Plastic Injection Molding			
4	CNC Nibbling			
5	*CNC CMM			
6	*LOM			

*For a list of abbreviations used, please refer to the front part of the book.

3. Describe the five steps involved in a general RP process chain. Which steps do you think are likely to be iterated?

4. After 3D geometric modeling, a user can either make a part through NC programming or through rapid prototyping. What are the basic differences between NC programming and RP in terms of the CAD model?

5. STL files are problematic. Is it fair to make this a statement? Discuss.

6. Preparing for building appears to be fairly sophisticated. In the case of SLA, what are some of the considerations and parameters involved?

7. Distinguish between cleaning, postcuring and finishing which are the various tasks of postprocessing. Name two RP processes that do not require postcuring and one that does not require cleaning.

8. Which step in the entire process chain is, in your opinion, the shortest? Most tedious? Most automated? Support your choice.

Chapter 3
LIQUID-BASED RAPID PROTOTYPING SYSTEMS

Most liquid-based rapid prototyping (RP) systems build parts in a vat of photo-curable liquid resin, an organic resin that cures or solidifies under the effect of exposure to light, usually in the UV range. The light cures the resin near the surface, forming a thin hardened layer. Once the complete layer of the part is formed, it is lowered by an elevation control system to allow the next layer of resin to be coated and similarly formed over it. This continues until the entire part is complete. The vat can then be drained and the part removed for further processing, if necessary. There are variations to this technique by the various vendors and they are dependent on the type of light or laser, method of scanning or exposure, type of liquid resin and type of elevation and optical system used.

3.1. 3D SYSTEMS' STEREOLITHOGRAPHY APPARATUS (SLA)

3.1.1. Company

3D Systems was founded in 1986 by inventor Charles W. Hull and entrepreneur Raymond S. Freed. Amongst all the commercial RP systems, the Stereolithography Apparatus, or SLA® as it is commonly called, is the pioneer with its first commercial system marketed in 1988. The company has grown significantly through increased sales and acquisitions, most notably of EOS GmbH's Stereolithography business in 1997 and DTM Corp., the maker of the Selective Laser Sintering (SLS) system in 2001. By 2007, 3D Systems is a global company that delivers advanced rapid prototyping solutions to every major market around the world. It has a global portfolio of nearly 400 US and foreign patents, with additional patents filed or pending in the US and several other major industrialized countries. 3D Systems Inc. is headquartered in 333 Three D Systems Circle Rock Hill, SC 29730, USA.

3.1.2.　Products

3D Systems produces a wide range of RP machines to cater to various part sizes and throughput. There are several models available, including those in the series of Viper SLA, Viper HA SLA, SLA 5000, SLA 7000 and Dual-Vat Viper™ Pro SLA system. The Viper SLA is an economical and versatile starter system that uses Nd:YVO_4 laser. The Viper system has two different built-in "modes", which are standard mode and HR (High resolution) mode, in a single system. Standard mode utilizes a beam diameter of 0.254 ± 0.0254 mm (0.010 ± 0.001 in.) and HR mode has a beam diameter of 0.0705 ± 0.0127 mm (0.003 ± 0.0005 in.). Depending on the size of the design part, the function enables the user to choose the appropriate mode to obtain high quality surfaces. HR mode is specially used for ultra small parts where parts are built with smooth surface finish, excellent optical clarity, high accuracy and thin straight vertical walls. It is ideal for a myriad of solid imaging applications, from rapid modeling and prototyping, to injection melding and investment casting. Viper HA SLA system shares the same functionality as Viper SLA system and it has additional hearing aid specific enhancements. Single-Vat Viper HA SLA system can be upgraded to dual-vat to produce two different color hearing aid shells in a single build.

For larger build envelopes, the SLA 5000, SLA 7000 and Dual-Vat Viper PRO SLA (see Fig. 3.1) are available. These three machines use the same laser as the Viper SLA system (solid-state Nd:YVO_4). The SLA 7000 can build parts up to four times faster than the SLA 5000 with the capacity of building thinner layers (minimum layer thickness 0.025 mm) for finer surface finish. Its fast speed is determined by its dual spot laser's ability. For the dual spot laser technology, a smaller beam spot is used for the border for accuracy, whereas the bigger beam spot is used for internal cross-hatching for the increased speed.

The largest and the top of the series is the Viper™ Pro SLA® system. It has a new configuration called "dual Resin Delivery Module (RDM)"

Fig. 3.1. 3D Systems' Viper™ Pro SLA® system (courtesy 3D Systems).

which allows the customer to build parts from two different materials simultaneously while utilizing a single proprietary imaging and scanning module. RDM contains a resin management system which will automatically detect and cross-check for the proper resin type in the system. Additionally a new feature is integrated in the system to filter and recirculate the material to extend the use of materials. With the two inbuilt chambers, the size of parts can be manufactured up to a volume of 737 × 635 × 533 mm (29 × 25 × 21 in.) while having a similar capability as the Viper™ SLA system in producing high resolution products. Specifications of these machines are summarized in Tables 3.1(a) and 3.1(b).

The Zephyr™ system was introduced in 1996 as a product enhancement in all the SLA systems.[1] The Zephyr™ system eliminates the need for the traditional "deep dip" in which a part is dunked into the resin vat after each layer and then raised to within one layer's depth of the top of the vat. With the deep dip, a wiper blade sweeps across the surface of the vat to remove excess resin before scanning the next layer. The Zephyr™ system has a vacuum blade that picks up resin from the side of the vat and applies a thin layer of resin as it sweeps across the part. This speeds up the build process by reducing the time required between layers and greatly reduces the problems involved when building parts with trapped volumes.

Table 3.1. Summary specifications of (a) the Viper and Viper HA SLA machines and (b) the rest of SLA machines (source from 3D Systems).

	a	
	Model	
	Viper SLA	Viper HA SLA
System characteristics		
Description	A dual-resolution, constant power, longer-life laser.	A dual-resolution system with hearing aid specific enhancements.
Vat capacity		
Maximum build envelope, mm (in.)	$250 \times 250 \times 250$ $(10 \times 10 \times 10)$	
Volume		
Volume, l (US gal)	32.2 (8.5l)	
Laser		
Type	Solid-state ($Nd:YVO_4$)	
Wavelength, nm	354.7	
Power at vat @ h	@ 7,500/h 100 mW	
Warranty, h	7,500	
Optical and scanning		
Dual spot	Yes	
Beam diameter: standard mode, mm (in.)	$0.25 +/-0.025 (0.010 \pm 0.001)$	
Beam diameter: HR mode, mm (in.)	$0.075 +/-0.015 (0.0030 \pm 0.0006)$	
Recoating system		
	Zephyr	
Features		
Interchangeable vat	Yes	
SmartSweep	No	
Auto resin refill	No	
Software		
Control software	Buildstation™	Buildstation™
Operating systems	Windows NT (4.0 with Service Pack 3.0 or higher)	Windows XP Professional

(Continued)

Table 3.1. (*Continued*)

	b		
		Model	
	SLA 5000	SLA 7000	Viper™ Pro SLA® system

Warranty

1 yr from installation date

System characteristics

	SLA 5000	SLA 7000	Viper™ Pro SLA® system
Description	A large-frame system with three times the build volume of SLA 3500.	A supercharged large-frame system two times faster than SLA 5000 with the capability of building thinner layers for finer surface finish.	An outstanding system with excellent build speed and adjustable beam size. Integrated configuration to build a part with two different types of materials.

Vat capacity

	SLA 5000	SLA 7000	Viper™ Pro SLA® system
Maximum build envelope, mm (in.)	$508 \times 508 \times 584$ $(20 \times 20 \times 23)$	$508 \times 508 \times 600$ $(20 \times 20 \times 23.6)$	$737 \times 635 \times 533$ $(29 \times 25 \times 21)$

Volume

	SLA 5000	SLA 7000	Viper™ Pro SLA® system
Volume, l (US gal)	253.6 (67)		935 (247)

Laser

	SLA 5000	SLA 7000	Viper™ Pro SLA® system
Type		Solid-state (Nd:YVO$_4$)	
Wavelength, nm		354.7	
Power at vat @	@ 5,000/h 216 mW	@ 5,000/h 800 mW	@ 5,000/h 1000 mW
Warranty, h		5,000	

Optical and scanning

	SLA 5000	SLA 7000	Viper™ Pro SLA® system
Dual spot	No	Yes	
Beam diameter; border @ l/e^2, mm (in.)	0.25 ± 0.025 (0.010 ± 0.001)	0.25 ± 0.025 (0.010 ± 0.001)	Nominal 0.13 mm (0.005)
Beam diameter; specialty spot @ l/e^2, mm (in.)	Nil	Nil	Nominal 0.3 mm (0.012 in.)

(*Continued*)

Table 3.1. (*Continued*)

	b	
	Model	
SLA 5000	SLA 7000	Viper™ Pro SLA® system
Beam diameter; hatch @ *l/e²*, mm (in.)	0.25 ± 0.025 mm (0.010 ± 0.001 in.)	0.7615 ± 0.0765 mm (0.03 ± 0.003 in.)	Nominal 0.76 mm (0.030 in.)

Recoating system
Zephyr

Features

Interchangeable vat	Yes
SmartSweep	Yes
Auto resin refill	Yes

Software

| Control software | Buildstation™ | Buildstation™ | 3DView™, 3DManage™, 3DPrint™ |
| Operating system | Windows NT | Windows NT 4.0 | Windows XP Professional (SP 2) |

Warranty
1 yr from installation date

All these machines rely on photo-curable liquid resins as the material for building. There are several grades of resins available and usage is dependent on the laser on the machine and the mechanical requirements of the part. Specific details on the correct type of resins to be used are available from the manufacturer. The other main consumable used by these machines is the cleaning solvent which is required to clean the part of any residual resin after the building of the part is completed on the machine.

3.1.3.   Process

3D Systems' stereolithography process creates 3D plastic objects directly from computer-aided design (CAD) data. The process begins with the vat

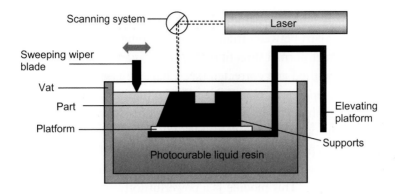

Fig. 3.2. Schematic of SLA Process.

filled with the photo-curable liquid resin and the elevator table set just below the surface of the liquid resin (see Fig. 3.2). The operator loads a 3D CAD solid model file into the system. Supports are designed to stabilize the part during building. The translator converts the CAD data into an STL file. The control unit slices the model and supports into a series of cross sections from 0.025 to 0.5 mm (0.001–0.020 in.) thick. The computer-controlled optical scanning system then directs and focuses the laser beam so that it solidifies a 2D cross section corresponding to the slice on the surface of the photo-curable liquid resin to a depth greater than one layer thickness. The elevator table then drops enough to cover the solid polymer with another layer of the liquid resin. A leveling wiper or vacuum blade (for Zephyr™ recoating system) moves across the surfaces to recoat the next layer of resin on the surface. The laser then draws the next layer. This process continues building the part from bottom up, until the system completes the part. The part is then raised out of the vat and cleaned of excess polymer.

The main components of the SLA system are a control computer, a control panel, a laser, an optical system and a process chamber. The workstation software used by the SLA system, known as 3D Lightyear exploits the full power of the Windows NT operating system and delivers far richer functionality than the UNIX-based Maestro software. Maestro includes the following software modules[2]:

(1) *3dverify™ module.* This module can be accessed to confirm the integrity and/or provide limited repair to stereolithography (STL) files

before part building without having to return to the original CAD software. Gaps between triangles, overlapping or redundant triangles and incorrect normal directions are some examples of the flaws that can be identified and corrected (see Chap. 6).

(2) *View*^{TM} *module.* This module can display the STL files and slice file (SLI) in graphical form. The viewing function is used for visual inspection and the orientation of these files so as to achieve optimal building.

(3) *MERGE module.* By using MERGE, several SLI files can be merged into a group which can be used together in the future process.

(4) *Vista*^{TM} *module.* This module is a powerful software tool that automatically generates support structures for the part files. The support structure is an integral part to successful part building, as it helps to anchor parts to the platform when the part is free floating or there is an overhang.

(5) *Part Manager*^{TM} *module.* This software module is the first stage of preparing a part for building. It utilizes a spreadsheet format into which the STL file is loaded and set-up with the appropriate build and recoat style parameters.

(6) *Slice*^{TM} *module.* This is the second stage of preparing a part for building. It converts the spreadsheet information from the *Part Manager*^{TM} module to a model of 3D cross sections or layers.

(7) *Converge*^{TM} *module.* This is the third and last stage of preparing a part for building. This module creates the final build files used by the SLA.

3.1.4. Principle

The SLA process is based fundamentally on the following principles[3]:

(1) Parts are built from a photo-curable liquid resin that cures when exposed to a laser beam (basically, undergoing the photo-polymerization process) which scans across the surface of the resin.

(2) The building is done layer by layer, each layer being scanned by the optical scanning system and controlled by an elevation mechanism which lowers at the completion of each layer.

These two principles will be discussed briefly in this section to lay the foundation for the understanding of RP processes. They are mostly applicable to the liquid-based RP systems described in this chapter. This first principle deals mostly with photo-curable liquid resins, which are essentially photo-polymers and the photo-polymerization process. The second principle deals mainly with CAD data, the laser and the control of the optical scanning system as well as the elevation mechanism.

3.1.4.1. *Photo-Polymers*

There are many types of liquid photopolymers that can be solidified by exposure to electro-magnetic radiation, including wavelengths in the gamma rays, X-rays, UV and visible range, or electron-beam (EB).[4,5] The vast majority of photo-polymers used in commercial RP systems, including 3D Systems' SLA machines are curable in the UV range. UV-curable photo-polymers are resins which are formulated from photo-initiators and reactive liquid monomers. There are a large variety of them and some may contain fillers and other chemical modifiers to meet specified chemical and mechanical requirements.[6] The process through which photo-polymers are cured is referred to as the photo-polymerization process.

3.1.4.2. *Photo-Polymerization*

Loosely defined, polymerization is the process of linking small molecules (known as monomers) into chain-like larger molecules (known as polymers). When the chain-like polymers are linked further to one another, a cross-linked polymer is said to be formed. Photo-polymerization is polymerization initiated by a photo-chemical process whereby the starting point is usually the induction of energy from an appropriate radiation source.[7]

Polymerization of photo-polymers is normally an energetically favorable or exothermic reaction. However, in most cases, the formulation of photo-polymer can be stabilized to remain unreacted at ambient temperature. A catalyst is required for polymerization to take place at a reasonable rate. This catalyst is usually a free radical which may be generated either thermally or photo-chemically. The source of a photo-chemically generated radical is a photo-initiator, which reacts with an actinic photon to produce the radicals that catalyze the polymerization process.

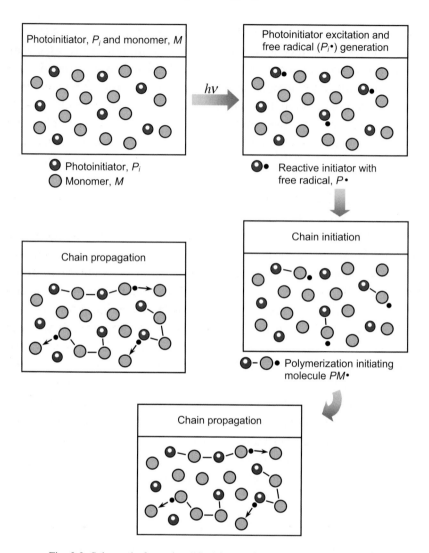

Fig. 3.3. Schematic for a simplified free-radical photo-polymerization.

The free-radical photo-polymerization process is schematically presented in Fig. 3.3.[8] Photo-initiator molecules, P_i, which are mixed with the
monomers, M, are exposed to a UV source of actinic photons, with energy
of $h\nu$, where h is the Planck constant and ν is the frequency of the radiation. The photo-initiators absorb some of the photons and are in an excited

state. Some of these are converted into reactive initiator molecules, P•, after undergoing several complex chemical energy transformation steps. These molecules then react with a monomer molecule to form a polymerization initiating molecule, PM•. This is the chain initiation step. Once activated, additional monomer molecules go on to react in the chain propagation step, forming longer molecules, PMMM• until a chain inhibition process terminates the polymerization reaction. The longer the reaction is sustained, the higher will be the molecular weight of the resulting polymer. Also, if the monomer molecules have three or more reactive chemical groups, the resulting polymer will be cross-linked and this will generate an insoluble continuous network of molecules.

During polymerization, it is important that the polymers are sufficiently cross-linked so that the polymerized molecules do not re-dissolve back into the liquid monomers. The photo-polymerized molecules must also possess sufficient strength to remain structurally sound while the cured resin is subjected to various forces during recoating.

While free-radical photo-polymerization is well-established and yields polymers that are acrylate-based, there is another newer 'chemistry' known as cationic photo-polymerization.[9] It relies on cationic initiators, usually iodinium or sulfonium salts, to start polymerization. Commercially available cationic monomers include epoxies, the most versatile of cationally polymerizable monomers and vinylethers. Cationic resins are attractive as prototype materials as they have better physical and mechanical properties. However, the process may require higher exposure time or a higher power laser.

3.1.4.3. *Layering Technology, Laser and Laser Scanning*

Almost all RP systems use layering technology in the creation of prototype parts. The basic principle is the availability of computer software to slice a CAD model into layers and reproduce it in an "output" device like a laser scanning system. The layer thickness is controlled by a precision elevation mechanism. It will correspond directly to the slice thickness of the computer model and the cured thickness of the resin. The limiting aspect of the RP system tends to be the curing thickness rather than the resolution of the elevation mechanism.

The important component of the building process is the laser and its optical scanning system. The key to the strength of the SLA is its ability to rapidly direct focused radiation of appropriate power and wavelength onto the surface of the liquid photo-polymer resin, forming patterns of solidified photo-polymer according to the cross-sectional data generated by the computer.[10] In the SLA, a laser beam with a specified power and wavelength is sent through a beam expanding telescope to fill the optical aperture of a pair of cross axis, galvanometer driven and beam scanning mirrors. These form the optical scanning system of the SLA. The beam comes to a focus on the surface of a liquid photo-polymer, curing a pre-determined depth of the resin after a controlled time of exposure (inversely proportional to the laser scanning speed).

The solidification of the liquid resin depends on the energy per unit area (or "exposure") deposited during the motion of the focused spot on the surface of the photo-polymer. There is a threshold exposure that must be exceeded for the photo-polymer to solidify.

To maintain accuracy and consistency during part building using the SLA, the cure depth and the cured line width must be controlled. As such, accurate exposure and focused spot size become essential.

Parameters which influence performance and functionality of the parts are physical and chemical properties of resin, speed and resolution of the optical scanning system, the power, wavelength and type of the laser used, the spot size of the laser, the recoating system and the post-curing process.

3.1.5.   Strengths and Weaknesses

The main strengths of the SLA are:

(1) *Round the clock operation.* The SLA can be used continuously and unattended round the clock.
(2) *Build volumes.* The different SLA machines have build volumes ranging from small ($250 \times 250 \times 250$ mm) to large ($737 \times 635 \times 533$ mm) to suit the needs of different users.
(3) *Good accuracy.* The SLA has good accuracy and can thus be used for many application areas.

(4) *Surface finish.* The SLA can obtain one of the best surface finishes amongst RP technologies.
(5) *Wide range of materials.* There is a wide range of materials, from general-purpose materials to specialty materials for specific applications.

The main weaknesses of the SLA are:

(1) *Requires support structures.* Structures that have overhangs and undercuts must have supports that are designed and fabricated together with the main structure.
(2) *Requires post-processing.* Post-processing includes removal of supports and other unwanted materials, which is tedious, time-consuming and can damage the model.
(3) *Requires post-curing.* Post-curing may be needed to cure the object completely and ensure the integrity of the structure.

3.1.6. Applications

The SLA technology provides manufacturers cost justifiable methods for reducing time to market, lowering product development costs, gaining greater control of their design process and improving product design. The range of applications includes:

(1) Models for conceptualization, packaging and presentation.
(2) Prototypes for design, analysis, verification and functional testing.
(3) Parts for prototype tooling and low volume production tooling.
(4) Patterns for investment casting, sand casting and molding.
(5) Tools for fixture and tooling design and production tooling.

Software developed to support these applications includes QuickCast™, a software tool which is used in the investment casting industry. QuickCast™ enables highly accurate resin patterns that are specifically used as an expendable pattern to form a ceramic mould to be created. The expendable pattern is subsequently burnt out. The standard process uses an expendable wax pattern which must be cast in a tool. QuickCast™

eliminates the need for the tooling use to make the expendable patterns. QuickCast™ produces parts which have a hard thin outer shell and contain a honeycomb like structure inside, allowing the pattern to collapse when heated instead of expanding, which would crack the shell.

3.1.7.   Examples

3.1.7.1.   *INCS Prototyping and Manufacturing Services Make Japan a Model for the World Market*

INCS is one of the largest manufacturing and prototyping process bureaus in Japan. It offers innovative 3D prototyping services to reduce the product life cycle and to produce the prototype within the tight lead time. Some of INCS's customers include leading automotive and electronics manufacturers and they require INCS to provide highly complex and durable prototypes (an example is shown Fig. 3.4).

INCS' founder, Yamada saw the great potential of the SLA from 3D Systems for the Japanese market and started INCS in 1990. It began with the SLA 250 and in 1993 INCS obtained the distributor rights from 3D Systems for Japan. Subsequently in 1997, INCS increased the number of systems to 10 and installed more than 30 systems by 2007. With all the different range of systems, INCS is able to handle prototypes of different sizes, providing high accuracy and tight tolerances. INCS is able to deliver products within an average of four days with the systems working round the clock.

Fig. 3.4. Prototype made by INCS (courtesy 3D Systems).

3.1.7.2. *3D Systems Helps Walter Reed Army Medical Centre Rebuild Lives*

Walter Reed Army Medical Centre is one support unit which cannot be removed from action. Walter Reed provides continuous health care and services to soldiers, their families and a large population of military retirees. The key operation of the Medical Centre is to support and treat the military personnel in surgical, accidental or combat trauma.

Walter Reed 3D Medical Applications Centre plays an important role in providing 3D models in assisting surgeons for pre-surgical and post-surgical planning. In 2001, the SLA 7000 system was installed in the Medical Centre. Besides allowing the physicians to physically examine and identify problems from the actual 3D model, the SLA 7000 system has dual-spot technology and high throughput to produce highly complex 3D models. With the combination of computed tomography (CT) or magnetic resonance imaging (MRI) together with third-party software, the 2D images are transformed into accurate 3D rendering files. The data generated is then transferred to the SLA 7000 system. Models can also be multi-colored to specifically indicate different regions to plan an operation and for other activities, such as biomedical education. In months after the installation of SLA 7000, the fabricated models have been involved in nearly 90 cases. Not only have the 3D models provided great opportunity for the physicians to plan surgery, the models have also effectively shortened the operating time and cutting cost.

Having a prototype also means laying a communication bridge between the patient and the surgeon as the patient is able to understand and discuss the available options with the surgeon. A typical example is shown in Fig. 3.5.

3.1.7.3. *DePuy Speeds Down the Information Highway*

The demand for RP models for medical application is growing rapidly and fabricating models eases the working procedure for surgeons to identify the problems which eventually reduces the risk of the operation. DePuy, one of the world's largest suppliers of orthopedic appliances and surgical

Fig. 3.5. Medical model prototype (courtesy 3D Systems).

tooling, has long learnt about the advantages of building models by stereolithography. Time is always the crucial parameter when it comes to medical services, not only the total fabrication time but also the time taken for the transferring of the models' data. Together with stereolithography and the internet, things can be made possible. Depuy Inc. has a total of three SLA systems: one in Leeds and the other two in Warsaw, Indiana. Transferring SLA files from Leeds to Warsaw can be very efficient and cost-saving as the process of transmitting data in intergraphy's engineering modeling system, (EMS) or STL is fast and the downloading of a file can be completed in a short time, depending on its size.

With the reduction in development time, new appliances and tooling can be created and this benefits the surgeons and patients. Figure 3.6 shows an example. Over the years, DePuy has always updated itself with the latest technologies to provide the best services to its existing and future customers.

3.1.7.4. *Stereolithography Makes a Strategic Difference at Xerox by Helping Designers Muffle Copier Noise*

Mufflers are what copiers have in common with auto mobile vehicles. To filter the copier noise, the air turbulence must be tightly controlled. Stereolithography was the key drive to generating a new copier's muffler assembly design at Xerox's RP laboratories. To conduct a simulation, the centre muffler components were built into two pieces using Stereolithography and then assembled together (see Fig. 3.7). The components that were built using Stereolithography were then used for silicone

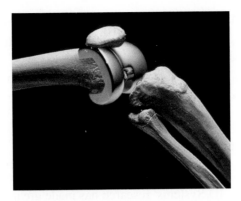

Fig. 3.6. Bone joint made using SLA system and QuickCast™ (courtesy 3D Systems).

Fig. 3.7. Muffle copier noise using SLA (courtesy 3D Systems).

molding. Through this process, Xerox was able to optimize the budget for design as well as meet the limitation of the tight schedule.

Stereolithography has helped Xerox to hold firm in their position in a number of areas:

(1) confirmation in assemblies with the manufactured parts,
(2) rapid, easy and low cost method for testing new design ideas,
(3) money saving before embarking on tooling investing and
(4) usage of silicone molding.

In the fabrication process, the design drawings were first produced by a Xerox project group which used Intergraph wireframes to represent the assembly. The file was then converted to solids using Intergraph's EMS. At times, when a large portion of a component was to be constructed in one SLA system, the other portion was constructed on another SLA system to shorten the time taken. With SLA systems, two machines can be integrated together to provide greater versatility.

3.1.8.   Research and Development

To stand as a leading company in the competitive market and to improve customers' needs, 3D Systems' research has made great improvement not only in developing new materials and applications, but also in software applications.

3D Systems and Ciba-Geigy Ltd are in a joint research and development program continually working on new resins which have better mechanical properties, are faster and easier to process and are able to withstand higher temperatures.[11]

One of the other most important areas of research is in rapid tooling, i.e., the realization of prototype moulds and ultimately production tooling inserts.[12] 3D Systems is involved in 15 cooperative rapid tooling partnerships with various industrial, university and government agencies. Methods studied, tested or being evaluated include those used for soft and hard tooling. With the purchase of 3D Keltool in September 1996, 3D Systems now has a means for users to go from a CAD model, to an SL master to Keltool cores and cavities capable of producing in excess of one million injection molded parts in a wide range of engineering thermoplastics such as polypropylene, nylon, ABS, polyethylene and polycarbonate.

3.2.   OBJET GEOMETRIES LTD.'S POLYJET

3.2.1. Company

Objet was founded in 1998 and has established itself as the leading platform for high-resolution three-dimensional printing (3DP). Objet also has proven installations worldwide where 3D modeling can be created in office environment.[13] Using its patented and market-proven PolyJet™

inkjet-head technology, it is able to print out the most complex 3D models with exceptionally high quality. PolyJet-based systems are used in hundreds of manufacturing sites across the world and across a wide spectrum of industries: automotive, electronics, toy, consumer goods, medical footwear and more. It has been awarded more than 40 patents with additional patents filed or pending internationally. Objet Geometries Ltd is currently headquartered at 2 Holzman St. Science Park P.O. Box 2496, Rehovot 76124, Israel.

3.2.2. Products

Objet's current line of PolyJet-based systems, the Eden™ family, is a group of four machines that can deliver high-resolution prototypes within an office environment.[14] The Eden™ family consists of the Eden 500V™, Eden 350™/350V™, Eden 260™ and Eden 250™, giving options to the users in terms of build size, productivity and budget requirements. For economical and effective small models, both Eden 250™ and Eden 260™ are able to fit in a small office. Eden 250™ features of two printing modes, high quality (HQ) and high speed (HS), for user to choose from in order to produce high quality prototype. Eden 260™ consists of 8 units of single head replacement (SHR) to jet identical amounts of resin compared to Eden 250™ resulting in better and more even surface finish. Eden 350™/350V™ are the medium build professional machines in the Eden series which features printing modes (HQ and HS) and higher material capacity. The Eden 500V™ (see Fig. 3.8) is the largest build system with a build volume of $490 \times 390 \times 200$ mm. It has the best features including dual printing modes, 8 units of SHR and an automatic function to switch between cartridges. Specifications of the Eden™ family of machines are summarized in Table 3.2.

The Eden™ systems utilize Objet FullCure® materials and Objet Studio™ software to provide a complete 3DP solution for any RP application. Objet systems provide a range of different materials for user to choose from, depending on the required properties. All Eden™ systems are able to print high accuracy ultra-thin 16 μm layers, producing models with exceptionally fine details and ultra-smooth surfaces. The Eden™ family works on the same principle where the jetting head lays both the

Fig. 3.8. Eden 500V™ (courtesy Objet Geometries Ltd).

Fullcure M (model material) and Fullcure S (support material) on the build tray. At the same instance, the UV light integrated with the jetting head cures the already just-laid FullCure® materials, virtually laying and curing the model in a single process.

3.2.3. Process

The Eden™ family of all PolyJet systems undergoes the same simple 3DP process. Objet PolyJet™ process creates 3D objects with the use of Objet Studio software. The designer loads the 3D CAD solid model file into the system which is compatible with Windows XP and Windows 2000. The Objet Studio will convert the CAD data into an STL or SLC file. The designer will also have to set the orientation arrangement of the designed part on the build tray.

Before the actual building process commences, the designer has to ensure that the build tray and the two types of material cartridges are inserted into the machine. The two types of material cartridges consist of the part material and the supporting material. When the procedure is done, the jetting heads, based on Objet's patent Polyjet inkjet technology, will move along the *x*-axis and lay the first layer of material onto the tray (see Fig. 3.9). Depending on the size of the part, the jetting head will

Table 3.2. Summary specifications of Eden™ series (source from Objet Geometric Ltd).

	Models			
	Eden 500V™	Eden 350/350V™	Eden 260™	Eden 250™
Tray Size (X × Y × Z), mm	500 × 400 × 200	350 × 350 × 200	260 × 260 × 200	250 × 250 × 200
Net build size (X × Y × Z), mm	490 × 390 × 200	340 × 340 × 200	258 × 250 × 205	250 × 250 × 200
Print resolution				
X axis	600 dpi: 42 μm	600 dpi: 42 μm	600 dpi: 42 μm	600 dpi: 42 μm
Y axis	600 dpi: 42 μm	600 dpi: 42 μm	300 dpi: 84 μm	300 dpi: 84 μm
Z axis	1,600 dpi: 16 μm	1,600 dpi: 16 μm	1,600 dpi: 16 μm	1,600 dpi: 16 μm
Accuracy, mm	0.1–0.3 typical	0.1–0.3 typical	0.1–0.3 typical	0.1–0.2 typical
Materials cartridges	Sealed 4 × 3.6 kg	Sealed 2 × 3.6 kg/ 4 × 3.6 kg	Sealed 2 × 2 kg cartridges	Sealed 2 × 2 kg cartridges
Tango net build (mm)	49 × 39 × 20	49 × 39 × 20	31 × 25 × 10/ 34 × 34 × 20	Not available
Materials supported	FullCure®720 VeroWhite VeroBlue VeroBlack TangoBlack TangoGray FullCure®705 Support	FullCure®720 VeroWhite VeroBlue VeroBlack TangoBlack TangoGray FullCure®705 Support	FullCure®720 VeroWhite VeroBlue VeroBlack FullCure®705 Support	FullCure®720 VeroWhite VeroBlue VeroBlack FullCure®705 Support

(Continued)

Table 3.2. (*Continued*)

	Eden 500V™	Eden 350/350V™	Eden 260™	Eden 250™
			Models	
Input format	STL and SLC File	STL and SLC File	STL and SLC File	STL and SLC File
Software			Objet Studio™	
Jetting heads	SHR (single head replacement), 8 units	SHR (single head replacement), 8 units	SHR (single head replacement), 8 units	SHR (single head replacement), 4 units
Machine dimension (W × D × H), mm	1,320 × 990 × 1,200	1,320 × 990 × 1,200	870 × 740 × 1,200	870 × 740 × 1,200
Weight, kg	410	410	280	280
Operational environment	Temperature 18–25°C relative humidity 30–70%	Temperature 18–25°C relative humidity 30–70%	Temperature 18–25°C relative humidity 30–70%	Temperature 18–25°C relative humidity 30–70%

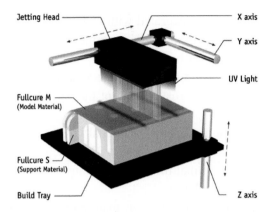

Fig. 3.9. Schematic of the Objet Polyjet process (courtesy Objet Geometries Ltd).

move on the y-axis and move on to the x-axis to lay the next layer after the first layer is completed. During the printing process, the jetting head will release the actual amounts of part material and support material. The materials will be immediately cured by the UV light from the jetting head. Whenever the materials are about to be used up, the material cartridges can be easily replaced without interrupting the fabrication process. Once the jetting head cures the first 2D cross section, the build tray will drop by one layer thickness of 16 µm. The jetting head will repeat the process continuously until the system completes the part. The part will be raised up and can be taken out for post-processing. The support material can then be removed easily by the water jet and the part is complete.

3.2.4. Principle

Objet's PolyJet™ technology creates high quality models directly from the computerized 3D files. Complex parts are produced with the combination of Objet's Studio software and the jetting head.

The process is based on the following principles:

(1) Jetting heads release the required amount of material which shares the same method as the normal inkjet printing method. At the same time when the material is printed on the tray, the material is cured by

the UV light which is integrated with the jetting head. Parts are built layer by layer, from a liquid photo-polymer where a similar polymerization process as described in Sec. 3.1.4 takes place.

(2) Jetting heads are moved only along the *xy*-axes and each slice of the building process is the cross section of the parts arranged in the software.

(3) With the completion of a cross-sectional layer, the build tray will be lowered for the next layer to be laid. The *z*-height of the elevator is leveled accurately so that the corresponding cross-sectional data can be calculated for that layer.

(4) Both the part material and support material will be fully cured when they are exposed to the UV light and most importantly the nontoxic support material can be removed easily by the water jet.

3.2.5. Strengths and Weaknesses

The Eden™ system has the following strengths[15]:

(1) *High quality.* The PolyJet™ can build layers as thin as 16 μm in thickness with accurate details depending on the geometry, part orientation and print size.

(2) *High accuracy.* Precise jetting and build material properties enable fine details and thin walls (600 μm or less depending on the geometry and materials).

(3) *Fast process speed.* Certain RP systems require draining, resin stripping, polishing and others whereas Eden™ systems only require an easy wash of the support material which is a key strength.

(4) *Smooth surface finish.* The models built have smooth surface and fine details without any post-processing.

(5) *Wide range of materials.* Objet has a range of materials suited for different specifications, ranging from tough acrylic-based polymer, to polypropylene-like plastics (Duruswhite) to the rubber-like Tango materials.

(6) *Easy usage.* The Eden™ family utilizes a cartridge system for easy replacement of build and support materials. Material cartridges provide

an easy method for insertion without having any risk of contact with the materials.

(7) *SHR technology.* The Eden™ machines' nozzles consist of heads and nozzles. With Single Head Replacement (SHR) these individual nozzles can be replaced instead of replacing the whole unit whenever the need arises.

(8) *Safe and clean process.* Users are not exposed to the liquid resin throughout the modeling process and the photo-polymer support is nontoxic. Eden™ systems can be installed in the office environment without increasing the noise level.

The Eden™ system has the following weaknesses:

(1) *Post-processing.* A water jet is required to wash away the support material used in PolyJet™, meaning that water supply must be nearby. This is somewhat a let-down to the claim that the machine is suitable for an office environment. In cases where the parts built are small, thin or delicate, the water jet can damage these parts, so care in post-processing must be exercised.

(2) *Wastage.* The support material which is washed away with water cannot be reused, meaning additional costs are added to the support material.

3.2.6. Applications

The applications of Objet's systems can be divided into different areas:

(1) *General applications.* Models created by Objet's systems can be used for conceptual design presentation, design proofing, engineering testing, integration and fitting, functional analysis, exhibitions and pre-production sales, market research and inter-professional communication.

(2) *Tooling and casting applications.* Parts can also be created for investment casting, direct tooling and rapid, tool-free manufacturing of plastic parts. Also they can be used to create silicon molding, aluminum epoxy moulds, VLT Molding (alternative rubber mould) and vacuum forming.

(3) *Medical imaging.* Diagnostic, surgical,[16] operation and reconstruction planning and custom prosthesis design. Parts built by PolyJet™ have outstanding detail and fine features which can make the medical problems more visible for analysis and surgery simulation. Due to its fast building time, prototype models are always built for trauma or tumors. Most importantly, it reduces the surgical risks and provides a communication bridge for the patients.

(4) *Jewelry industry.* Presentation of concept design, actual display, design proof and fitting. Pre-market survey and market research can be conducted using these models.

(5) *Packing.* Vacuum forming is an easy method to produce inexpensive parts and it requires a very short time for the part to be formed.

3.2.7.   Examples

3.2.7.1.   *Adidas–Salomon AG Uses Objet's PolyJet™ to Produce Prototypes*

Adidas–Salomon AG is always among the top athletic footwear manufacturers worldwide. To keep up with the highly competitive market, product enhancements and timely product presentations to the market are very crucial. The search for systems to produce high quality models ended with the presence of PolyJet™ inkjet technology. Physical models are directly produced from STL files in a short period of time. Objet Geometric RP systems' assistance has brought Adidas–Salomon AG closer to their company's vision of streamlined digital process for sharing among all business units. Physical models have proven to Adidas–Salomon AG that they would greatly benefit using the Objet machines where parts can be used for design verification, development review and production tooling in a short development time.

Objet Geometries' PolyJet™ has given Adidas–Salomon AG the flexibility to collaborate with each factory in Asia on specific product enhancements while reducing the time necessary. With Objet Geometric Ltd sales and training services, Adidas–Salomon AG is now working with Objet to search for ways to apply this technology into other aspects of their products (Fig. 3.10).

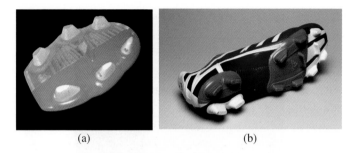

(a)   (b)

Fig. 3.10. High quality finishing part that (a) appear close to a real part and (b) fitted onto a real soccer boot (courtesy Objet Geometric Ltd).

3.2.7.2. *Realizing Efficiency and Reducing Cost in the Design and Development Process Through the Installation of an Eden Machine at MegaHouse*

Tokyo-based MegaHouse develops and manufactures high-precision figures and toy foods targeted at both adults and children. Because the company serves a wide range of customers and its products often comprise many finely detailed parts and usually have short life cycles, MegaHouse works hard to streamline development.

MegaHouse considered its RP options and decided that installing an RP system onsite was the best way to reduce prototyping costs, increase access to prototypes, improve design quality and reduce development time. It selected Objet's Eden 260™ because the 3D printing system enabled molding with high precision, supporting their need to verify mechanisms and because it was easy to use. This enabled MegaHouse to create molds with nonexpert staff. The strength of Objet's FullCure® materials was another advantage of the Eden 260™ as it allowed MegaHouse designers to test the models without fear of damaging them.

The Eden 260™ brings benefits that went far beyond what MegaHouse had expected. It makes it possible to easily obtain a high-precision prototype in the early stages of conceptual design. This means that design work can now proceed without the delays MegaHouse used to incur while waiting for the service bureau, thus considerably reducing design time.

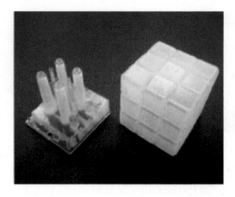

Fig. 3.11. Prototype of an illuminated cube created on the Eden 260™ (courtesy Objet Geometric Ltd).

Also, the Eden 260™ enables design defects to be discovered before tooling, avoiding expensive tool delays due to changes that needed to be made. The ability to handle virtually all prototyping in-house has created significant cost savings for MegaHouse. With the Eden260™ produced models, product strength and parts fit can be verified during development before tooling is committed.

As it is so easy and cost effective to make models that can be tested using the Eden 260™, it become possible to allow everyone in the design and development team to have a hand with them, improving understanding and enabling more effective product development (Fig. 3.11).

Even the sales department has benefited from the installation of the Eden260™, as presentation models are now used in sales activities right from the early stages — an important competitive advantage for MegaHouse Corporation.

3.2.8. Research and Development

Objet is doing research in developing faster processing, higher performance, higher resolution graphics and smoother and more accurate details. One way of improving the accuracy of the Polyjet procedure can be done by optimizing the scaling factor.[17] Objet Geometries Ltd has generated a new solution for fast and cost-effective production of high quality hearing

Fig. 3.12. Prototype of hearing aid (courtesy Objet Geometric Ltd).

aids (see Fig. 3.12). With the combination of easy operation and high speed capabilities, Objet Hearing Aid Solution reduces the hearing aid makers' time and cost to market.

In software development, Materialise NV, world leader in software development for RP industry and Objet Geometries Ltd have cooperated to develop a customized version of Materialise's Magics Software to be tuned for Objet's 3DP. This agreement will definitely benefit both parties in satisfying customers and increasing their competitive advantage in the RP industry. Fabrication materials are increasing the end users' ability to select mechanical properties for specific applications. With the new rubber mold materials, molds can be made by spin casting where parts can be cast within hours.

3.3. D-MEC'S SOLID CREATION SYSTEM (SCS)

3.3.1. Company

The SCS has been jointly developed by Sony Corporation, JSR Corporation and D-MEC Ltd. The software and hardware have been created by Sony Corporation, the UV curable resin by JSR Corporation and the forming and applied technology by D-MEC Ltd, a company that was established in 1990. The address of the D-MEC Ltd is Hamarikyu Park Side Place, 5-6-10, Tsukiji, Chuo-ku, Tokyo 104-0045, Japan.

3.3.2. Products

Based on the principle of laser cured polymer using layer manufacturing, D-MEC Ltd is the first company to offer a 0.5 m cubic tank size, amongst the largest in the market, with a scanning speed of up to 5 m/s. In cooperation with JSR which developed the resin used in the system, the Solid Creator was well received in Japan, especially among auto makers and electronics industries. D-MEC has five models: SCS-1000HD, SCS-6000, SCS-8100, SCS-8100D and SCS-9000. Each model has its own scan speed in producing different quality of parts and SCS-9000 has the highest scan speed of 20 m/s. SCS-1000HD is the smallest size model which uses He–Cd laser producing the smallest size of at least 0.3mm. SCS-8100 (see Fig. 3.13) is a high speed modeling system which has advance features like improved radiation effect, dual-beam function and newly developed self-running tank which has been adopted in the system for easy replacement of resins.

The specifications of D-MEC's available models are summarized in Table 3.3.

3.3.3. Process

The SCS creates the 3D model by laser curing polymer layer by layer. Its process comprises five steps: generating the CAD model, slicing the CAD

Fig. 3.13. Solid creation system SCS-8100 (courtesy D-MEC Corporation).

Table 3.3. Solid creation systems model specifications (courtesy D-MEC Corporation).

	Model				
	SCS-1000HD	SCS-6000	SCS-8100	SCS-8100D	SCS-9000
Description	High precision, ultraline modeling machine	Small-size modeling laser	Large-size modeling machine	Large-size modeling machine	Large-size modeling machine
Laser type, power, and frequency	He–Cd laser		Solid-state semi-conductor laser		
		200 mW, 25 kHz	1,000 mW, 60 kHz	1,000 mW, 60 kHz × 2	1,000 mW, 60 kHz (first in Japan)
Modulator	AOM (acusto-optic element)			Built-in	AOM (acusto-optic element)
Deflection equipment	Galvanometer mirror system (includes sweep-defocus correction function)				
Model creation range (mm)	300 × 300 × 270	300 × 300 × 250	610 × 610 × 500	610 × 610 × 500	1,000 × 800 × 500
Spot size, mm (automatic adjustment)	0.05–0.3	0.075–0.4	0.15–0.4	0.15–0.4	0.2–0.4
Scan speed, m/s	Max. 2 m/s	Max. 7.5 m/s	Max. 10 m/s	Max. 10 m/s	Max. 20 m/s
Layer thickness, mm	0.02–0.15	0.05–0.2	0.1–0.2	0.1–0.2	0.1–0.4
Tank volume (exchangeable)	45 L	55 L	280 L	280 L	840 L
Power source	100 V, 20 A	100 V, 30 A	100 VAC, 40 A	100 VAC, 45 A	100 V, 35 A
Coolant			Not needed		
Main unit dimensions (W × D × H) (mm)	1,425 × 1,100 × 1,590	1,425 × 1,115 × 1,610	1,940 × 1,150 × 1,990	1,940 × 1,150 × 1,990	2,340 × 1,640 × 2,760
Main unit weight (including resin)	710 kg	710 kg	1,700 kg	1,750 kg	2,920 kg

model and transferring data, scanning the resin surface, lowering the elevator, completion of prototype and post-processing.

First, a CAD model, usually a solid model, is created in a commercial CAD system, like CATIA or Pro/Engineer. Three-dimensional CAD data of the part from the CAD system are converted to the sliced cross-sectional data which the SCS will use in creating the solid. This is the slicing process. Editing may be necessary if the slicing is not carried out properly. Both the slicing and editing processes can be done either ON- or OFF-line. Consequently, the section data is passed to the laser controller for the UV curing process.

The ultraviolet laser then scans the resin surface in the tank to draw the cross-sectional shape based on the data. The area of the resin surface which is hit by the laser beam is cured, changing from liquid to solid on the elevator. This process is similar to the one that is illustrated in Fig. 3.2. The elevator than descends to allow the next solid layer to be created by the same process. This is repeated continuously to laminate the necessary number of thin cross-sectional layers to form the 3D part. Finally, when the model is complete, the elevator is raised and the model is lifted out before post-curing treatment is applied.

The main hardware of the SCS includes:

(1) the Sony NEWS UNIX workstation,
(2) the main machine controller (VME based, MTOS — multi-tasking operating system),
(3) the galvanometer mirror and its controller,
(4) the optical system including the laser, the lens and the acoustic optical modulator (AOM).
(5) the photo-polymer tank and
(6) the elevator mechanism.

The main software of the SCS consists of two parts: the slice data generator which uses inputs from various CAD systems and creates the cross-sectional slices and the editing software for slice data which includes the automatic support generation software. The software used to control the SCS also uses principles of man–machine interfaces.

3.3.4. Principle

D-MEC's SCS is based on the principle of polymer curing by exposure to ultraviolet light and manufacturing by layering. The basic principles and techniques used are similar to that described in Sec. 3.1.4.

In the process, parameters which affect performance and functionality of the machine are scanning pitch, step period, step size, scanner delay, jump size, jump delay, scanning pattern and resin's properties.

The required software for SCS models are:

- Magics — SCS models use the Magics series, produced by Materialise Corporation, as data processing software. Magics offers functions such as STL measurement and correction, support creation and output in slicing data format.
- Solidware — Solidware is a software which enables user to browse slice data and to detect an error layer automatically. Solidware provides options such as adding or combining or deleting lines when errors are detected during the processing of slice data.

3.3.5. Strengths and Weaknesses

SCS has the following strengths:

(1) *Large build volume.* The tank size is among the largest in the market and large prototypes (especially large full-scale prototypes) can be produced.
(2) *Accurate.* High accuracy (0.04 mm repeatability) models may be produced.
(3) *Wide variety of resins with special characteristic.* D-MEC offers a wide variety of generally used resins and they include epoxy resins, heat resistant resins, ABS-like resins and elastomer-like resins.

It has the following weaknesses:

(1) *Requires support structures.* Structures that have overhangs and undercuts must have supports that are designed and fabricated with the main structure.

(2) *Requires post-processing.* Post-processing includes removal of supports and other unwanted materials, which is tedious, time-consuming and can damage the model.

(3) *Requires post-curing.* Post-curing may be needed to cure the object completely and ensure the integrity of the structure.

3.3.6. Applications

The general application areas are given as follows:

(1) Mock-up in product design.
(2) Design study, medical models analysis and sales sample of new products.
(3) Use as parts without need of modification in small lot production.
(4) Simplified mold tool and master for investment casting and other similar processes.

D-MEC Corporation has used SCS mainly to design and prototype its own products like the drum base for video tape recorders. By using SCS, D-MEC is able to ascertain the optimum design for its products. Important functional tests can be carried out on the prototype as well.

When Solid Creator is used for medical purposes, surgical operation time can be significantly shortened by checking the defective or diseased areas using models reconstructed on the system from other frequently used computer images such as from CT-scans.

3.3.7. Others

Solid Creator machines have been installed throughout the world, though mainly in Japan and Taiwan. Besides these, D-MEC provides comprehensive bureau services to many industries in Japan.

3.4. ENVISIONTEC'S PERFACTORY

3.4.1. Company

The company was founded when it was a young start-up as Envision Technologies GmbH in August 1999 and corporatized as Envisiontec

GmbH in 2002. Envisiontec provides RP systems and solutions to serve customers in a wide variety of applications. Envisiontec even deals with software and materials development to increase productivity and cost effectiveness. Currently, it has over 20 patents and patents application pending worldwide. The company's address is Brüsseler Straße 51, 45968 Gladbeck, Germany.

3.4.2. Product

Perfactory®[18] (see Fig. 3.14) is the RP system built by Envisiontec and it is a versatile system suitable for an office environment. Perfactory® undergoes the basic process by means of converting a CAD file to STL format and then transferring the STL data to the system to build the model. Resins are cured by photo-polymerization but Perfactory® uses a different approach in curing the resins. The photo-polymerization process is created by an image projection technology called digital light processing technology (DLP™) from Texas Instrument and it requires mask projection to cure the resin layer by layer. The standard system alone can achieve resolutions between 148 and 93 μm but with Voxel, resolution can

Fig. 3.14. Perfactory® SXGA+ standard (courtesy Envisiontec GmbH).

be adjusted to between 50 and 150 µm. Additional components or devices such as the mini multi lens system and the enhanced resolution module (ERM) enable designers and manufacturers to build smaller figures which required high surface quality. Perfactory® mini is able to create parts of a higher quality of 32 µm in solution and voxel thickness adjusted to between 25 and 50 µm.

The specifications of Perfactory® systems are shown in Tables 3.4(a) and 3.4(b).

3.4.3. Process

To build the part in a layer-by-layer manner, Perfactory® undergoes a simple process (see Fig. 3.15) where the 3D model of a solid model is first created with a commercial CAD system. For medical applications, data acquired with MRT or CT systems can be processed directly. The 3D data model in the STL format acquired is sliced within the software and each sliced layer is converted into a bitmap file with which the mask image is generated. The bitmap image consists of black and white where white represents the material and black represents the void. When the image is projected onto the resin with DLP™, the illuminated white portion will cure the resin while the black areas will not.

With the embedded operating system in Perfactory®, it can operate independently and is monitored by the device driver software installed in the PC. The software provides two types of mode which are auto and expert mode. The auto mode allows direct conversion of 3D-CAD data to STL format and other required set-ups are programmed automatically. The expert mode is specially programmed for advanced users to offer them with choice of manual set-up according to their experience, needs and preferences.

Unlike almost all other RP systems that build the model from bottom up, Perfactory® builds the model top down (see Fig. 3.16). The build or carrier is first immersed into a shallow trough of acrylate-based photo-polymer resin sitting on a transparent contact window. The mask is projected from below the build area onto the resin to cure it. Once the resin is cured, the build platform is raised a single layer, the thickness being dependent on the voxel thickness. While the

Table 3.4. (a) Perfactory® SXGA⁺ W/ERM mini multi lens system specifications and (b) Perfactory® SXGA⁺ standard and standard UV systems specifications (courtesy Envisiontec GmbH).

	a		
System	Perfactory® SXGA⁺ W/ERM mini multi lens		
Lens system, focal length, mm (in.)	60 (2.3)	75 (3)	85 (3.3)
Build envelope XYZ, mm, (in.)	84 × 63 × 230 (3.3 × 2.5 × 9)	59 × 44 × 230 (2.3 × 1.7 × 9)	45 × 34 × 230 (1.77 × 1.3 × 9)
Dynamic voxel thickness Z, μm (in.)		15–50 (0.005–0.0019)	
Voxel size XY W/ERM, μm (in.)	30 (0.0011)	21 (0.0008)	16 (0.0006)
Resolution SXGA⁺ W/ERM		2800 × 2100	

	b		
	Perfactory® SXGA⁺		
System	Standard ZOOM with integrated ERM	Standard UV system	
Lens system, focal length, mm	25–45	Fixed focal UV lens	
Build envelope XYZ (mm)	120 × 90 × 230 to 190 × 142 × 230	100 × 75 × 230	140 × 105 × 230	175 × 131 × 230
Native voxel size XY, μm	86–136	71	100	125
ERM voxel size XY, μm	43–68	35	50	62
Dynamic voxel thickness Z, μm	25–150	25–150		
Resolution SXGA⁺	1400 × 1050	1400 × 1050		
		Resin		
Group		Acrylate		
Color		Red-brown (not transparent)		
Light source		High pressure mercury vapor lamp		

+ Double Perfactory, ® Resolution.

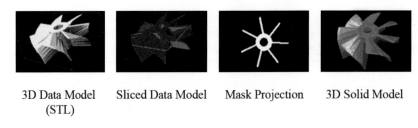

3D Data Model   Sliced Data Model   Mask Projection   3D Solid Model
   (STL)

Fig. 3.15. Perfactory® process (courtesy Envisiontec GmbH).

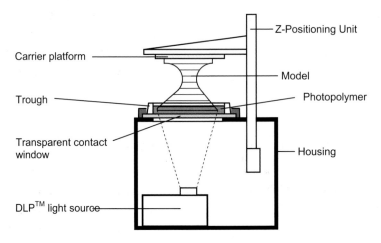

Fig. 3.16. Schematic of the Perfactory® build process with DLP™ technology (adapted from Envisiontec GmbH).

platform is raised, it peels the model away from the transparent contact window, thus allowing fresh resin to flow in through capillary action. The next layer is then built in a similar manner. The whole cycle takes about 25 s without the need for planarization or leveling for each layer.

The two key differences of the Perfactory® with other RP systems are the use of mask projection for photo-polymerization and the part is moved upwards with each completed cured layer instead of moving downwards in other systems. Once the model is built, the user simply has to peel the model off from the platform as the model is stuck to the carrier platform during the entire process.

3.4.4. Principle

Perfactory® uses the basic principles of stereolithography by undergoing the following:

(1) Parts are built from acrylate photo-polymer and the user is able to select different material properties with different materials colors. Resins are cured when exposed to a mask projection image using digital light processing (DLP™) technology from Texas Instrument.
(2) Every completion of cured layer is moved away from the build trough containing the resin vertically upwards by a precision linear drive. This is due to the projection system integrated at the bottom of the RP system. Also, the fabricated part does not require any support and removing the model from the transparent platform is easy.

3.4.4.1. *Digital Light Processing Technology (DLP™)*

DLP™ is a projection technology invented by Dr. Larry Hornbeck and Dr. William E. "Ed" Nelson of Texas Instruments in 1987. In DLP™ technology, the key device is the digital micromirror device (DMD), the producer of image. DMD is an optical semi-conductor and each DMD chip has hundreds of thousands of mirrors arranged in a rectangular array on its surface to steer the photons with great accuracy. This means that each mirror is represented as one pixel in a projected image and therefore the resolution of an image depends on the number of mirrors.

The mirrors in the DMD are made of aluminum and are 16 μm in size. Each individual mirror is connected to two support posts where it can be rotated ±10°–12° of an ON or OFF state. In the ON state, the light source is reflected from the mirrors into the lens making pixels on the screen. For the OFF state, the reflected light is redirected to the other direction allowing the pixels to appear in a dark tone. During each mask projection, the cross section of each layer is projected by mirrors in the ON state and resins are cured by the visible light projected from below the transparent contact window.

Every single mirror is mounted on a yoke by compliant torsion hinges with its axle fixed on both ends and able to twist in the middle. Due to its extremely small scale, damages hardly occur and the DMD structure is able to absorb shock and vibration thus providing high stability.

Position control of each mirror is done by two pairs of electrodes as positioning significantly affects the overall image of the cross-sectional layer. One pair is connected to the yoke while the other is connected to the mirror and every pair has an electrode on each side of the hinge. Most of the time, equal bias charges are applied to both sides to hold the mirror firmly in its current position. In order to move the mirrors, the required state has to load into a static random access memory (SRAM) cell connecting to the electrodes and mirror. Once all the SRAM cells have been loaded, the bias voltage is removed, allowing every individual movement of the mirrors through released charges from the SRAM cell. When the bias is restored, all of the mirrors will be held in their current position to wait for the next loading into the cell. Bias voltage enables instant removal from the DMD chip so that every mirror can be moved together providing more accurate timing while requiring a lower amount of voltage for the addressing of the pixel.

3.4.5.   Strengths and Weaknesses

The main strengths of Perfactory® systems are:

(1) *High building speed.* The use of a mask image directly exposed to the resin enables a part to be built at approximately 10 mm height per hour at 100 μm pixel height. This speed is independent of part size and geometry and is one of the fastest systems in the RP market.

(2) *Office friendly process.* Perfactory® allows operation in an office environment as its foot print is under 0.3 sq m. The curing of the photo-polymer does not use UV light and there is no need for special facility. It operates with low noise and zero dust emissions.

(3) *Small quantity of resin during build.* The shallow trough of resin means that the amount of material in use at any one time is small (about 200 ml). This means that should a number of different resins are required, they can be swapped out easily with minimal wastage.

(4) *No wiper or leveler.* When the carrier platform is raised with the model, there is no need for planarization or leveling. This eliminates

the possibility of causing problems to the stability of the parts during the build, e.g., a wiper damaging a small detail on the part during the wiping action.

(5) *Less shrinkage.* Due to immediate curing of a controlled layer (based on the voxel thickness) there is less shrinkage during the process.

(6) *Safe supply cartridges.* Liquid photo-polymers are packed in cartridges and this reduces the risk of users coming into contact with them.

(7) *Additional components.* Perfactory® is able to build even higher quality tough parts with the use of additional components which can be integrated into the system.

The main weaknesses of Perfactory® systems are:

(1) *Limited building volumes.* Structures are built from the bottom of the build chamber and stuck to the carrier platform; this limits the size of the build.

(2) *Peeling of completed part.* The user has to peel the completed model from the build platform as the model is built on a movable carrier platform which moves vertically upwards. Care has to be taken so as not to damage the model during the peeling process.

(3) *Requires post-processing.* After the model is complete, cleaning and post-processing, sometimes including post-curing are required.

3.4.6. Applications

The application areas of the Perfactory® systems include the following:

(1) Concept design models for design verification, visualization, marketing and commercial presentation purposes.

(2) Working models for assembly purpose, simple functional tests and for conducting experiments.

(3) Master models and patterns for simple molding and investment casting purposes.

(4) Building and limited production of completely finished parts.

(5) Medical[19] and dental applications. Creating exact physical models of patient's anatomy from CT and MRI scans.

3.4.7. Others

EnvisionTec's main offices are located in Germany, North America and United Kingdom. Perfactory® systems have been sold worldwide, especially in Europe and the US.

3.5. AUTOSTRADE'S E-DARTS

3.5.1. Company

Autostrade Co., Ltd, the Japanese manufacturer of the E-DARTS system that is based on stereolithography, was founded in 1984. Besides the E-DARTS system, the company is also involved with developing CAE and similar software tools for its E-Darts systems. It also developed a parallel virtual machines (PVM) system, a software package that permits a heterogeneous collection of UNIX and Windows computers that are connected together in a network to be used as a single large parallel computer. Autostrade Co., Ltd is located at 13-54 Ueno-machi, Oita-City, Oita 870-0832, Japan. Its first machine was sold in 1998.

3.5.2. Products

The E-DARTS system uses the stereolithography method of hardening liquid resin by a beam of laser light. The set-up of the laser system is found under the resin tank. The laser beam is directed into the resin which is in a container with a clear bottom plate. This system uses an acrylic photo-polymer resin to produce models. Figure 3.17 shows the E-DARTS system while Table 3.5 lists its specifications.

3.5.3. Process

The software comprises two components. Firstly, the STL file editing software changes the CAD data into slice data. Secondly, the controller software drives the hardware. Both of these components use Windows Operating System.

Fig. 3.17. E-DARTS system (courtesy Autostrade Co., Ltd).

Table 3.5. Specifications of E-DARTS machine.

Model	E-DARTS
Laser type	Semiconductor
Laser power, mW	30
Spot size, mm	0.1
XY sweep speed, m/s	0.03
Elevator vertical resolution, mm	0.05
Work volume, XYZ, mm	$200 \times 200 \times 200$
Maximum part weight, kg	No count
Minimum layer thickness, mm	0.05
Size of unit, XYZ, mm	$430 \times 500 \times 515$
Data control unit	Win 98/ME
Overall system weight, kg	21
Power supply, V	12

Like many other RP process, the E-DARTS system builds the model in a layer-by-layer manner. Similar to the Envisiontec Perfactory® method, the modeling platform is linked to the Z-position elevator over the build chamber. The build or carrier is first immersed into a shallow trough of acrylate-based photo-polymer resin sitting on a transparent contact window. The semiconductor laser mounted on an *x–y* table cures a single layer of resin through a plotting action that is similar to that of the

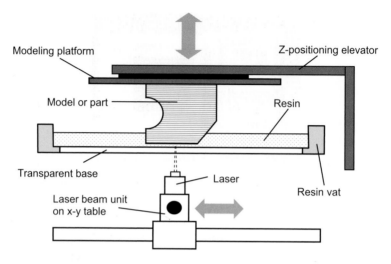

Fig. 3.18. E-DARTS process.

3D Systems. The main difference here is that the resin is cured through a transparent window from the bottom of the model. Once the resin is cured, the build platform is raised a single layer upwards and fresh resin to flow in between the model and the transparent table is ready for the next layer. The cycle repeats in a similar manner until the whole model is complete. The whole cycle takes about 25 s without the need for planarization or leveling for each layer (see Fig. 3.18).

3.5.4. Principle

The model is formed by liquid resin which is cured by laser light beamed from below the resin chamber. While 3D Systems' SLA uses a platform which is dipped into the resin tank when one layer is completed, E-DARTS system uses another method. The platform or modeling base is raised upwards each time a layer is completed. Otherwise, the E-DARTS is based on the laser lithography technology which is similar to that described in Sec. 3.1.4. Parameters which influence performance and functionality are the control and precision of the x–y table, laser spot diameter, slicing thickness and resin properties.

3.5.5. Strengths and Weaknesses

The E-DARTS system has the following strengths:

(1) *Low price.* The price of the E-Darts system is 2,980,000 yen (US$ 27,000), which has one of the more affordable prices amongst most RP systems.
(2) *Low operating cost.* The introduction of an improved fluid surface regulation system has decreased the necessary volume of resin, which actually eliminates the need to stock the resin. This in turn reduces the running cost of the system.
(3) *Compact size.* The overall system size is $430 \times 500 \times 515$ mm, requiring only a relatively small foot-print for the machine.
(4) *Portable.* The overall weight of the system is less than 25 kg, allowing the system to be easily transported by one person.
(5) *Ease of installation.* The E-DARTS is designed compact and light weight and this enables the system to be set up by one person. The set-up time for the system is less than 1 h.

The E-DARTS system has the following weaknesses:

(1) *Requires support structures.* Structures that have overhangs and undercuts must have supports that are designed and fabricated together with the main structure.
(2) *Requires post-processing.* Post-processing includes removal of supports and other unwanted materials, which is tedious, time-consuming and can damage the model.
(3) *Requires post-curing.* Post-curing may be needed to cure the object completely and ensure the integrity of the structure.

3.5.6. Applications

The application areas of the E-DARTS include the following:

(1) Concept design models for design verification, visualization, marketing and commercial presentation purposes.

(2) Working models for assembly purpose, simple functional tests and conducting experiments.
(3) Master models and patterns for simple molding and investment casting purposes.

3.6.   CMET'S SOLID OBJECT ULTRAVIOLET–LASER PRINTER

3.6.1.   Company

CMET (Computer Modeling and Engineering Technology) Inc. was established in November 1990 with Mitsubishi Corporation as the major share holder and two other substantial shareholders, NTT Data Communication Systems and Asahi Denak Kogyo K.K. However, in 2001, CMET Inc. merged with Opto-Image Company, Teijin Seiki Co., Ltd. The company's address is CMET Inc., Sumitomo Fudosan Shin-yokohama Building, 2-5-5 Shin-yokohama Kohoku-ku Yokohama, Kanagawa 222-0033, Japan.

3.6.2.   Products

CMET has two models of machines: Rapid Meister 6000II and Rapid Meister 3000. Each of them serves individual designers and manufacturers with different parameters. Rapid Meister 6000II has a higher scanner system and a different re-coater. Figure 3.19 shows a Rapid Meister

Fig. 3.19. Rapid Meister 6000II machine (courtesy CMET Inc).

Table 3.6. Specifications of the RM 3000 and RM 6000II (courtesy CMET Inc).

	Model	
	Rapid Meister 3000	Rapid Meister 6000 II
Laser type	LD Laser 400 mW 25 kHz	LD Laser 800 mW, 60 kHz
Scan mode	Digital scanner system	Digital scanner system
Max. scanning speed, m/s	12	22
Beam diameter	Variable system	
Maximum modeling size $X \times Y \times Z$ (mm)	$300 \times 300 \times 250$	$610 \times 610 \times 500$
Minimum build layer, μm	50	
Vat	Interchangeable	
Recoater	Braid recoater	Multipurpose high speed (MH) recoater
Power	AC 100 V single phase, 30 A	AC 100 V, single phase × two circuits (main part 30 A, heater part 10 A)
Dimension $X \times Y \times Z$ (mm)	$1{,}430 \times 1{,}045 \times 1{,}575$	$1{,}020 \times 2{,}045 \times 2{,}050$
Weight, kg	400 (without resin)	1,400 (without resin)

6000II machine. The digital scanner system uses ultra high speed scan to create parts in a short time. Scanning speeds range from 12 to 22 m/s. The central working section provides a high position precision suited for very small models like connectors and solid structure models like components. A summary table of the machines available is given in Table 3.6.

3.6.3. Process

CMET's Rapid Meister machines have a process that contains three main steps:

(1) *Creating a 3D model with CAD system*: The 3D model, usually a solid model, of the part is created with a commercial CAD system. 3D data of the part are then generated.
(2) *Processing data with the software*: Often data from CAD system are not sufficiently error-free to be used by the RP system directly. Rapid Meister's software is able to edit CAD data, repair its defects, like

gaps or overlaps (see Chap. 6), slice the model into cross sections and finally generate the corresponding Rapid Meister machine data.

(3) *Making model with Rapid Meister unit*: The laser scans the resin, solidifying it according to the cross-sectional data from the software. The elevator lowers by a single layer and the liquid covers the top of the part which is recoated and prepared for the curing of the next layer. This is repeated until the whole part is created.

Rapid Meister's hardware contains: a communication controller, a laser controller, a shutter/filter controller, a scanner controller (galvanometer mirror unit) and an elevator controller. The Rapid Meister software is a real-time, multi-user and multi-machine control software. It has functions such as simulation, a convenient editor for editing and error repair, 3D off-set, loop scan for filling area between the outlines, selectable structures and automatic support generation.

3.6.4. Principle

The Rapid Meister system is based on the laser lithography technology which is similar to that described in Sec. 3.1.4. The one major difference is in the optical scanning system. The main trade-off is in scanning speed and consequently, the building speed. Parameters which influence per-formance and functionality are galvanometer mirror precision for the galvanometer mirror machine, laser spot diameter, slicing thickness and resin properties.

3.6.5. Strengths and Weaknesses

The main strengths of Rapid Meister systems are:

(1) *New recoating system.* The new recoating system provides a more accurate Z-layer and shorter production time.
(2) *High scanning speed.* It has scanning speeds of up to 20 m/s.
(3) *Variety of resins.* Rapid Meister has a variety of materials with different properties for users to select from (ABS performance resin, high precision resin, rubber like resin, etc.).

The main weaknesses of Rapid Meister systems are:

(1) *Requires support structures.* Structures that have overhangs and undercuts must have supports that are designed and fabricated together with the main structure.
(2) *Requires post-processing.* Post-processing includes removal of supports and other unwanted materials, which is tedious, time-consuming and can damage the model.
(3) *Requires post-curing.* Post-curing may be needed to cure the object completely and ensure the integrity of the structure.

3.6.6. Applications

The application areas of the Rapid Meister systems include the following:

(1) Concept models for design verification, visualization and commercial presentation purposes.
(2) Working models for form fitting and simple functional tests.
(3) Master models and patterns for silicon molding, lostwax, investment casting and sand casting.
(4) Building and limited production of completely finished parts.
(5) 3D stereolithography copy of existing component, parts of or the full product.

3.6.7. Research and Development

CMET focuses its research and development on a new recoating system. The new recoating system can make Z-layers with high accuracy even when a small Z slice pitch (0.05 mm) is used. As the down-up process of the Z table is not productive, the "no-scan" time can be shortened by more than 50%. More functions will be added into CMET's software. CMET is also conducting the following experiments for improving accuracy:

(1) Investigation on relationship between process parameters and curl distortion: This experiment is done by changing laser power, scan speed, fixing other parameters like Z-layer pitch and beam diameter.

(2) Investigation on effects of scan pattern: This experiment aims to compare building results between using and not using scanning patterns.
(3) Comparative testing of resins' properties on distortion.

3.7. ENVISIONTEC'S BIOPLOTTER

3.7.1. Company

EnvisionTec GmbH is the same company that has developed the Perfactory® System discussed in Sec. 3.4. Details of the company can be found in Sec. 3.4.1.

3.7.2. Product

Medical applications have been gaining a foothold in RP and Bioplotter[20,21] has become one of the solutions for computer-aided tissue engineering. Bioplotter technology was first invented in 1999 at Frieburg Materials Research Centre for producing scaffold structures from various biochemical materials and even living cells. The EnvisionTec Bioplotter (see Fig. 3.20) is being sold in Europe since 2002.

Bioplotter undergoes a simple process of CAD data handling, dispensing material and finally solidification of material to create the structures. The system is based on 3D dispenser which allows only a

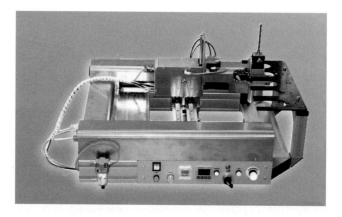

Fig. 3.20. EnvisionTec Bioplotter (courtesy EnvisionTec GmbH).

Table 3.7. Bioplotter system specifications (courtesy Envisiontec GmbH).

Model	Bioplotter
Machine size $L \times W \times H$, mm	$600 \times 500 \times 500$
Working area $X \times Y \times Z$, mm	$200 \times 200 \times 130$
Resolution in $X/Y/Z$, μm	50
Dispensing heads	1 (2nd optional)
Dispensing process	Pressure over time
Dispensing velocity $X/Y/Z$, mm/s	1–30
Cartridges	— at room temp: 30 cc standard PE;
	— with heating: 20 cc glass syringe
Dispensing needle	Standard, diameter 0.1–2 mm
Cartridge heating	Room temp. — 210°C
Dispensing basin	DURAN case
	— 200 cc, $\varnothing = 80$, $H = 50$;
	— 400 cc, $\varnothing = 100$, $H = 60$
Basin heating	Room temp. — 100°C
Machine control	Via PC incl. CAD/CAM software with integrated device driver

certain amount of materials to be dispensed into a plotting medium. Specific 2½D computer-aided design–computer-aided manufacturing (CAD–CAM) software is used to handle CAD data and machine/process control. The specifications of the Bioplotter are summarized in Table 3.7.

3.7.3. Process

The Bioplotter process contains the following three main steps:

(1) *3D data handling*. Digital data models of the design structure need to be generated which is the first condition for building 3D scaffolds. The 3D scaffolds can be designed on any commercial 3D CAD software or data can be generated from data acquired through MRI, CT or X-ray. Additional software like VoXim from IVS-solutions can also be used for volume data for 3D image processing and then exported in data formats like DXF and CLI.

2½D CAD–CAM software PrimCam® is the main program used to import data formats specially programmed for the Bioplotter system. Data formats are transferred to Primus data to modify single layers of the imported data or to design another structure design for plotting process.

(2) *Control process*: PrimCam® software is designed especially for the Bioplotter system allowing users to calibrate and design machine and process control. Users can input machine set-up and process parameters either manually or automatically. Two key functions of the software enable the selection of the dispensing path in the object layers and the simulation of the dispensing process, which can then be checked before the commencement of the process.

(3) *Dispensing process*: Liquids or solutions are dispensed to build the model in layers and all the plotting materials are first stored in a cartridge (see Fig. 3.21). Plotting material is then later forced to extrude through a small dispensing needle of a diameter close to 80 microns

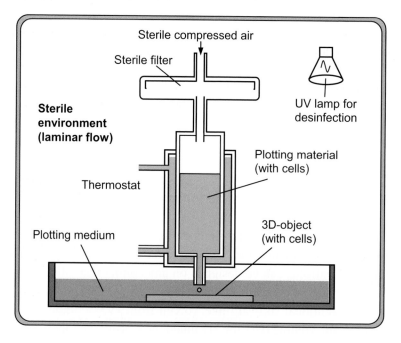

Fig. 3.21. Schematic process diagram (adapted from Envisiontec GmbH).

into the plotting medium. The purpose of this second medium is to cause the solidification of the first solution through certain reactions when both materials are in contact. The solidification of material depends on the material, medium and the temperature control creating the precipitation reaction, phase transition or chemical reaction[22] (see Fig. 3.22). As mentioned before about the solidification by heating, storage cartridges can be heated up to 230°C while the build platform can be heated up to 100°C.

Secondly, the plotting medium acts as a supporting force to create buoyancy to prevent structure from collapsing. Normally when materials are dispensed without a supporting force, gravity will cause the complex scaffold design to collapse.

3.7.4. Principle

The design of the Bioplotter system is created for the purpose of tissue engineering applications and its approach in terms of the process will be different even though the object is created layer by layer. The process of bioplotter system is based on the following principles:

(1) Under normal condition dispensing material, complex structures will collapse due to gravity (see Fig. 3.23). With the Bioplotter, the compensation method is to dispense the plotting material into

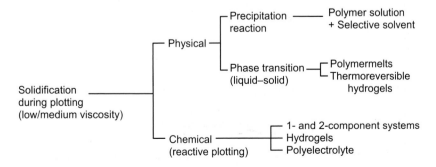

Fig. 3.22. Material reaction flow diagram.

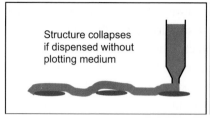 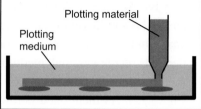

Fig. 3.23. Diagram for compensation due to buoyancy force (adapted from Envisiontec).

a matched plotting medium. This can be done based on the buoyancy force where the densities of both plotting material and medium are similar and create the required gravity force to prevent the collapse of the structure. Additionally with the use of the medium, support structures are not required in the process which means no waste or unwanted support will be built.

(2) The production or dispensing process can also be carried out in a sterile environment when living cells are to be incorporated into the plotting material.

3.7.5.   Strengths and Weaknesses

The main strengths of Bioplotter systems are:

(1) *Broad range of biomaterials.* Materials that can be processed on the Bioplotter include natural and thermoplastic polymers, reactive resins and living cells.
(2) *Layer modification.* User is able to modify process parameters within a specific layer.
(3) *No support required.* Bioplotter consists of plotting medium which provides the up-trust and thus acts as a support during the process.

The main weaknesses of Bioplotter systems are:

(1) *Sterile environment.* The system requires a sterile environment as materials include biomaterial, biochemicals and living cells.

(2) *Temperature control.* Control of temperature is necessary to ensure solidification between plotting material and medium by specific reaction.

3.7.6. Applications

The application areas of the Bioplotter system include the following:

(1) Scaffolds models for design verification, visualization, experimental and fabrication purposes.
(2) Testing and experimental purposes.
(3) Medical purposes like creating implantations for human and generation of living tissues.
(4) Fabrication of biodegradable materials for tissue engineering and drug delivery purposes.

3D Bioplotter is designed with unique capabilities differentiating it from other commercial systems based on its ability to plot biological cells. Due to this capability, it is mainly meant for medical applications, surgical and biomedical research.

3.7.7. Others

University of Freiburg and Envisiontec are partners in developing and marketing Bioplotter. Envisiontec technology supports users by providing the required information and materials for the 3D plotting process. 3D plotting technology is still under development to bring biomedical research to a higher level in the future.

3.8. RAPID FREEZE PROTOTYPING (RFP)

3.8.1. Introduction

Most of the existing RP processes are relatively expensive and many of them generate substances such as smoke, dust, hazardous chemicals, etc., which are harmful to human health and the environment. Continuing innovation is essential in order to create new RP processes that are fast, clean and of low-cost.

Dr. Ming Leu of the University of Missouri-Rolla, has been developing a novel, environmentally benign RP process that uses cheap and clean materials and can achieve good layer binding strength, fine build resolution and fast build speed. They have invented such a process, called Rapid Freeze Prototyping (RFP)[23]; it makes 3D ice parts layer-by-layer by freezing water droplets.

3.8.2. Objectives

The fundamental study of this process has three objectives:

(1) Developing a good understanding of the physics of this process, including the heat transfer and flow behavior of the deposited material in forming an ice part.
(2) Developing a part building strategy to minimize part build time, while maintaining the quality and stability of the build process.
(3) Investigating the possibility of investment casting application using the ice parts generated by this process.

3.8.3. Process

The experimental RP system[24] in Fig. 3.24 consists of the following mechanical mechanism: (a) a 3D positioning subsystem; (b) a material depositing subsystem (see Fig. 3.25); (c) a freezing chamber; and (d) an electronic control device. Tedious modeling and analysis efforts[25] have been carried out to understand and improve the behavior of material solidification and rate of deposition during the ice-part building process. The process starts with software which receives STL file from CAD software and generates the sliced layers of contour information under a CLI file. Together with a CLI file, process parameters like nozzle transverse speed, environment temperature and fluid viscosity are then further processed to generate the NC code. The NC code is sent to the experimental system to control the fabrication of the ice parts.

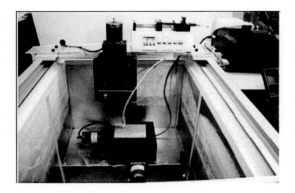

Fig. 3.24. Experimental system of RFP.

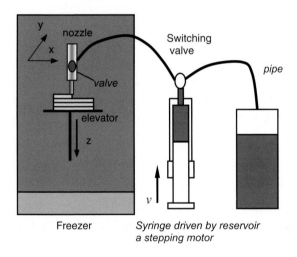

Fig. 3.25. Building environment and water extrusion subsystem.

The 3D positioning system consists of an *XY* table and a *Z* elevator and it is seated in the freezer at a low temperature of −20°C or lower. The depositing nozzle is mounted on the *Z* elevator whereby water droplets are extracted from the nozzle onto the substrate. Once the first layer solidifies, the nozzle will eventually level up and continue depositing for the next layer.

During the fabrication process, building and support solutions are used. The building solution will be water and the support solution is the

Fig. 3.26. An ice part before removal of support material and after removal.

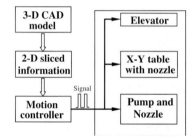

Fig. 3.27. Control system schematics

eutectic sugar solution ($C_6H_{12}O_6$–H_2O) of a melting temperature −5.6°C.[26] Due to the difference in each solution's melting temperature, the built part is then placed in an environment between 0°C and the melting point of the support material to melt the support material (see Fig. 3.26). Experiments have also been conducted where small parts are fabricated with radius of 0.205 mm and thin wall of 0.48 mm in thickness.[27] Figure 3.27 shows the control system schematics for the system.

3.8.4. Principles

The rapid freezing prototyping system is based on the principle of using a water freezing method to build an icy cold prototype. Besides being environmental friendly, the water solution can also be easily recycled. Before the frozen part can be built, an STL file from a CAD design is required.

The software will convert the STL file to a CLI file and finally to NC codes. NC codes together with other process parameters are then transferred to a motion control card to begin fabrication.

The system consists of three major hardware equipments: a positioning subsystem, a water ejecting subsystem and an electronic control device. Each of these holds an important role during the building process. Firstly, the *XY* table is placed in the freezer under a low temperature in order for the extracted water-droplets from the nozzle to freeze in a short time. The nozzle is attached at the Z-elevator and it will move in an upwards direction after every completed layer. For the first layer of the ice part, only water droplets are deposited. Every water-droplet does not solidify immediately upon deposition as the water-droplets are spread and form a continuous water line. The water line will then be frozen by convection through the cold environment and conduction from the previous frozen layer rapidly.

During each experimental process for building an ice part, process parameters set in the water ejection subsystem and electronic control device are important as they affect the overall fabrication accuracy. The key parameters include ambient the build temperature, scan speed, water feed rate. The ratio of material flow to *XY* movement speed is kept at the best value to prevent discontinuity in the freezing strands. Each layer thickness and smoothness is determined by the adjustment of nozzle scanning speed and water feed rate rather than the mechanical mechanism.[28]

3.8.5.   Strengths and Weaknesses

RFP has the following strengths:

(1) *Low running cost.* The RFP process is cheaper and cleaner than all the other RP processes. The energy utilization of RFP is low compared to other RP processes such as laser stereolithography or selective laser sintering.
(2) *Good accuracy.* RFP can build accurate ice parts with excellent surface finish. It is easy to remove the RFP made ice part in a mold making process, by simply heating the mold to melt the ice part.

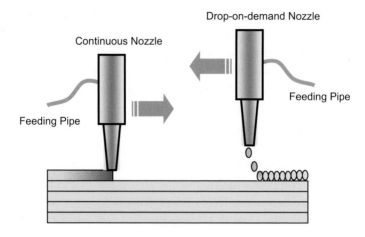

Fig. 3.28. Two material extrusion method.

(3) *Good building speed.* The build speed of RFP can be significantly faster than other RP processes, because a part can be built by first depositing water droplets to form a waterline and then filling in the enclosed interior with a water stream (see Fig. 3.28). This is possible due to the low viscosity of water. It is easy to build color and transparent parts with the RFP process.

(4) *Environmentally friendly materials.* Both the building material and support material are clean and nontoxic when handled.

(5) *Suitability in investment casting.* Eliminating problems created from traditional wax patterns in investment casing like pattern expansion and shelling cracking during pattern removal.

On the other hand, RFP has also the following weaknesses:

(1) *Requires cold environment.* The prototype of RFP is made of ice and hence it cannot maintain its original shape and form at room temperature.

(2) *Needs additional processing.* The prototype made with RFP cannot be used directly but has to be subsequently cast into a mold and so on and this increases the production cost and time.

(3) *Repeatability.* Due to the nature of water, the part built in one run may differ from the next one. The composition of water is also hard to control and determine unless tests are carried out.
(4) *Post-processing.* The final part is required to be placed in a lower temperature environment to remove the support material.

3.8.6. Potential Applications

(1) *Part visualization.* Parts can be built for the purpose of visualization. Examples can be seen in Fig. 3.29.
(2) *Ice sculpture fabrication.* One application of RFP is making ice sculptures for entertainment purposes. Imagine how much more fun a dinner party can provide if there are colorful ice sculptures that can be prescribed one day before and they vary from table to table. This is not an unlikely scenario because RFP can potentially achieve a very fast build speed by depositing water droplets only for the part boundary and filling in the interior with a water stream in the ice sculpture fabrication process. Fig. 3.30(a) shows the CAD model of an ice sculpture, while Fig. 3.30(b) shows the part made by RFP.
(3) *Silicon molding.* The experiments on UV silicone molding have shown that it is feasible to make silicone molds with ice patterns

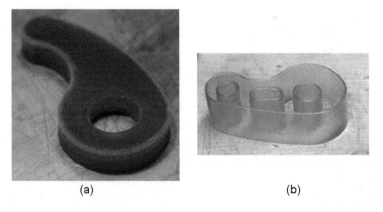

(a)   (b)

Fig. 3.29. Examples of visualization parts: (a) solid link rod (10 mm height) and (b) contour of a link rod.

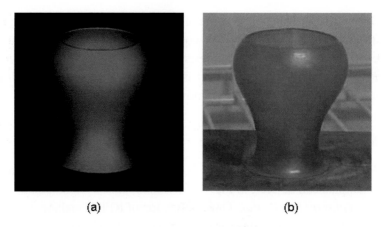

(a)   (b)

Fig. 3.30. Ice sculpture: (a) the CAD model and (b) the RFP fabricated part.

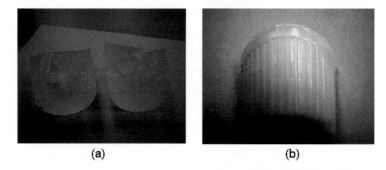

(a)   (b)

Fig. 3.31. (a) UV silicone mold made by ice pattern and urethane part made by UV silicone mold.

(see Fig. 3.31) and further make metal parts from the resulting silicone moulds (see Fig. 3.32). The key advantage of using ice patterns instead of plastic or wax patterns is that ice patterns are easier to remove (without pattern expansion) and no demolding step is needed before injecting urethane or plastic parts. This property can avoid demolding accuracy loss and can allow more complex molds to be made without the time-consuming and experience-dependent design of demolding lines.

(4) *Investment casting.* A promising industrial application of RFP technology is investment casting. DURAMAX recently developed the

| (a) | (b) | (c) | (d) |

Fig. 3.32: (a) Making of the ice part by RFP, (b) UV silicone mold made by ice part, (c) urethane part made by UV silicone mold and (d) metal part made by UV silicone mold.

Fig. 3.33. Metal model made by investment casting with ice pattern.

freeze cast process (FCP), a technology of investment casting with ice patterns made by molding. The company has demonstrated several advantages of this process over the competing wax investment and other casting processes, including low cost (35–65% reduction), high quality, fine surface finish, no shell cracking, easy process operation and faster run cycles. Additionally, there is no smoke and smell in investment casting with ice patterns. Apparently there is a finding that the alcohol base binder can damage the surface of the ice pattern due to the solubility of ice and alcohol binder. Research has been going on for the search of an interface agent to solve the problem.[29] Figure 3.33 shows a metal model made by investment casting using ice patterns from RFP.

3.9. MICROFABRICATION

In microfabrication, many methods have been developed and improved over the last few years whereby miniature objects are built. For example, Koji Ikuta and his group in Kyushu Institute of Technology in Japan worked on a process of microfabrication called "integrated hardened polymer stereolithography" (IH process) to build prototype venous valve and 3D microintegrated fluid system (MIFS).[30,31] These are very small structures, e.g., the valve's inner diameter is 80 μm. Using the IH process, they have also built a functioning microelectrostatic actuator approximately 700 μm tall and 120 μm wide, consisting of two adjacent bars. When charged with electricity, these bars close a 50-μm gap. Such a switch is used for small mechanical devices such as the microvenous valve.

Recently, they have improved the process, naming it "Super IH process".[32] The unique feature of this process is that liquid UV (ultra violet) polymer can be solidified at a pinpoint position in 3D space by optimizing the apparatus and focusing the laser beam. This pinpoint exposure allows the 3D microstructure to be made without any supporting parts or sacrificial layers. Hence, freely movable micro mechanisms such as gear rotators and free connecting chains can be made easily. Since the total fabrication time of the super IH process is extremely fast, it is easily applied to mass production of 3D microstructures.

Similar microlithography works can be found in the prototyping of multidirectional inclined structures,[33] flexible microactuator (FMA),[34] and piezoelectric micropump and microchannels.[35]

Another method known as the LIGA technique is a new method for microstructure fabrication.[36] LIGA (Lithographie, Galvanoformung und Abformung — in German) is developed in Germany and it means lithography, galvanoforming and plastic molding. Synchrotron radiation X-ray lithography is used due to its good parallelism, high radiation intensity and broad spectral range. The template having a thickness of several hundred microns can be fabricated with aspect ratio up to 100 and a 2D structure with submicron deviation is obtained. This structure can then be transformed into metal or plastic microstructural product by galvanoforming and plastic molding.

The LIGA technique has been successfully used in making engine, turbine, pump, electric motor, gear, connector and valve.[36] In the construction of milli-actuators, Lehr explained that the LIGA technique offers a large variety of materials to fabricate components with submicron tolerances so as to meet specific actuator functions.[37]

X-ray LIGA relies on synchrotron radiation to obtain necessary X-ray fluxes and uses X-ray proximity printing.[38] It is being used in micromechanics (micromotors, microsensors, spinnerets, etc.), microoptics, microhydrodynamics (fluidic devices), microbiology, medicine, biology and chemistry for microchemical reactors. It is comparable to microelectromechanical systems (MEMS) technology, offering a larger, nonsilicon choice of materials and better inherent precision. Inherent advantages are its extreme precision, depth of field and very low intrinsic surface roughness. However, the quality of fabricated structures often depends on secondary effects during exposure and effects like resist adhesion.

Deep X-ray lithography (DXRL) is able to create tall microstructures with heights ranging from 100 to 1,000 μm and aspect ratios of 10–50.[39] DXRL provides lithographic precision of placement, small feature sizes in the micrometer range and submicrometer details over the entire height of the structures. Figure 3.34 shows an SEM picture of a gear train assembly.

UV-LIGA relies on thick UV resists as an alternative for projects requiring less precision.[38] Modulating the spectral properties of synchrotron

Fig. 3.34. SEM picture of a gear train assembly (courtesy Louisiana State University).

radiation, different regimes of X-ray lithography lead to (a) the mass-fabrication of classical nanostructures, (b) the fabrication of high aspect ratio nanostructures (HARNST), (c) the fabrication of high aspect ratio microstructures (HARMST) and (d) the fabrication of high aspect ratio centimeter structures (HARCST).

Microstereolithography (MSL) is very similar to stereolithography. The manufacture of 3D microobjects by using a stereolithographic technique needs first a strong correction in the process control, in order to have an accuracy of less than 10 μm in the three directions of space.[40] This method differs from the stereolithographic method in that the focus point of the laser beam remains fixed on the surface of the resin, while an *x–y* positioning stage moves the resin reactor in which the object is made. However, the reactor must be translated very slowly to ensure that the surface of the liquid resin is stable during polymerization. As such, the outer size of the microstructure has to be limited unless a long manufacturing time is allowed. The fabrication of 20 μm thick ceramic microcomponents has been achieved with this method.

The apparatus consists of a He–Cd laser with acoustic-optic shutter controlled by the computer as shown in Fig. 3.35.[41] The laser beam is then deflected by two computer-controlled low inertia galvanometric

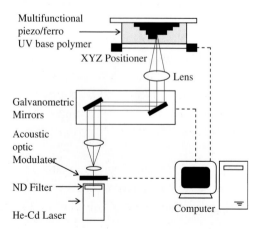

Fig. 3.35. Schematic diagram of the microstereolithography unit.

mirrors with the aid of focusing lens on the open surface of the polymer containing photo-initiators. An *XYZ* positioner moves the reactor containing the polymer and the laser beam is focused on the layer to be solidified.

Multifunctional smart materials involve the integration of polymers and nanoceramic particles by chemical bonding as side groups on a polymer backbone. The concept is to design a backbone with functional groups that will serve as anchor points for the metal oxides. The nanoparticles such as PZT, PLZT, etc. must have active surfaces or functional groups that can bond with the polymer chain. The nanoparticles provide the piezoelectric function in the polymer and the backbone provides mechanical strength and structural integrity, electrical conductivity, etc. The multifunctionality of these polymers provides a large-scale strain under electric field and thus can be used as actuators for MEMS based devices such as micropumps.

Functional and structural ceramic materials possess unique properties such as high temperature/chemical resistance, low thermal conductivity, ferroelectricity, piezoelectricity, etc. Three-dimensional ceramic microstructures are of special interest in applications such as microengines and microfluidics. The fabrication of ceramic microstructures differs from that of polymeric MSL. In ceramic MSL, the homogeneous ceramic suspension is prepared. Submicron ceramic powders are mixed with monomer, photo-initiator, dispersant, dilutents, etc. by ball milling for several hours. The prepared ceramic suspension is then put into the vat and ready for MSL based on the CAD design. After MSL, the green body ceramic microparts are then obtained. To obtain the dense microceramic parts, the green body is next put into the furnace to burn out the polymer binders and further sintered at a high temperature furnace. The binder burnout and the sintering temperature vary with different polymer and ceramics. After sintering, ceramic microstructures are ready for assembly and application.

MSL can be very useful for building microparts in micromechanics, microbiotics (microactuators) and microfluidics.[41] Current lithographic processes mentioned previously have the limitation that complex structures cannot be made easily. Thus, MSL can be used for more complex geometries.

3.10. MICROFABRICA®'S EFAB® TECHNOLOGY

3.10.1. Company

Microfabrica®, formerly known as MEMGen Corporation, is a private company which was founded in 1999.[42] It started off collaborating with Defense Advanced Research Projects Agency (DARPA) at University of Southern California (USC). Microfabrica is the first company with EFAB® technology that can build 3D complex microdevices out of metal with high precision. Microfabrica® builds the designs provided by the customers and provides foundry services. Microfabrica® is located at 7911 Haskell Ave. Van Nuys, California, USA.

3.10.2. Product

EFAB® technology was developed by Microfabrica® and brings microdevice manufacturing to a higher level and great benefits of a wide range of applications into the market. With EFAB® technology, greater opportunities are created to provide the market with highly complex microdevices (see Fig. 3.36). Microfabrica® is an application-specific microdevice supplier based on the customer's specific requirements and conditions to build custom components. Microfabrica® will either work together with the customers or the customers will provide their completed design themselves and fabricated parts will be manufactured solely by Microfabrica®, with quality checks done by the customers before shipment.

3.10.3. Process

The process of EFAB® technology has similarities with the combination of semi-conductor manufacturing and RP layered manufacturing. The designer

Fig. 3.36. Fabricated biomedical devices (courtesy Microfabrica®).

has to create the generated idea in CAD first in order to build the 3D microdevice. Once the CAD design is ready, fabrication can then commence.

Completed parts are manufactured by the electrodeposition of selective metals layer by layer. Each layer is made up of two different selective metals and the first layer is layered on alumina substrates. Basically, every individual layer consists of three basic steps which are repeated many times until the desired complex device is completed. These three basic steps are: patterned layer deposition, blanket layer deposition and planarization.

For patterned layer deposition, the first metal that is layered on the substrate is known as the sacrificial metal. The sacrificial metal is deposited in a cross-sectional pattern of the device which is to be fabricated. After the sacrificial metal is deposited, the second step called the blanket layer deposition is applied. The second material called the structural material (e.g., a Nickel–Cobalt alloy) is electroplated on the substrate and filled in the empty areas from the previous patterned layer completely. Finally, the two materials are planarized to a high precision thickness to remove the excess materials deposited. Each set of the steps is repeated until all the sliced cross-section of the 3D design have been built. Once all of the layers are deposited, the sacrificial material is removed by a release etchant, presenting the desired device. The overall fabrication process is shown in Fig. 3.37.

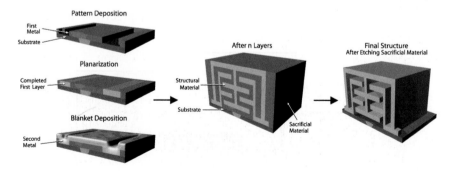

Fig. 3.37. Process diagram (courtesy Microfabrica).

3.10.4.   Principle

EFAB® Technology is an additive microfabrication process which has the following fundamental principles:

(1) Parts are built from the CAD design done by the designer. The final structure is formed from layers and layers of metal of high precision thickness.
(2) Fabrication consists of two different materials which are the structure material and the sacrificial material. The sacrificial material has an important role to guide the structure material into the right position and to support the next layer preventing it from collapse. The structural material is deposited on the sacrificial material by electroplating to fully cover the previous layer completely. Electroplating will leave excess material or leftovers, creating uneven surfaces. Planarization is required to maintain the accuracy of the layer thickness.
(3) Each layer presents a cross-sectional image of the designed parts and the sacrificial material can be easily etched away for the completion of the entire process.

3.10.4.1.   *Electrodeposition*

Electroposition is an electroplating process (see Fig. 3.38) and it involves the coating or adding of layers of metal onto an electrically conductive material. This method has advantages over other additive process in achieving a desirable layer with high precision thickness.

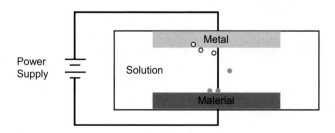

Fig. 3.38. Electroplating diagram.

The electroplating process also offers corrosion protection, wear resistance, improvement of aesthetic qualities and other desirable properties on the substrate.

Electroplating requires an electrical current to deposit metal onto the required material. In the process of electroplating, the material to be plated is connected to the negative terminal, the cathode, while the metal which is to be plated onto the material is connected to the positive terminal, the anode. Both of the components have to be immersed into a solution (the electrolyte) or it has to be supplied through the small gap between the anode and the cathode. The ions in the electrolyte act as a connector between the metal and material allowing the flow of the electricity.

Both ends of the terminal are connected to a power supply, commonly a rectifier. When the power supply is switched on, the ions from the metal will lose their charge and move across the electrolyte, plating the cathode. The desired thickness of the metal plating depends on the duration and the selection of areas where electroplating is not required.

3.10.5. Strengths and Weaknesses

EFAB® Technology has the following strengths:

(1) *Fast production.* Capable of building different microdevices on the same substrate.
(2) *Fast development cycle.* Easy and fast development time compared to traditional microelectro mechanical systems (MEMS) processes.
(3) *High complexity parts.* High complexity design from 3D data can be built from EFAB® Technology.

EFAB® Technology has the following weaknesses:

(1) *Requires post-processing.* Parts are required etching to remove sacrificial material.
(2) *Limited model size.* Restricted to devices smaller than several millimeters and larger than 20 μm.

3.10.6.   Applications

Microfabrica® has been working closely with their customers from a wide range of markets and industries. Devices are manufactured according to the customers' designs and the applications listed below are built with EFAB® technology.

(1) RF devices — switching, relays, inductors, transmission lines, filters[43,44] and transformers.
(2) Biomedical devices — 3D pumps, values, surgical tools, drug delivery, air flow sensors, etc.
(3) Chemical analysis — microchemical reactors, heat exchangers, filter, microspectrometers etc.
(4) Information technology — hard disk drive head petitioners, industrial print heads.
(5) Automotive — fuel injectors, tire pressure sensors, active suspension, navigation gyroscopes, etc.

3.11.   D-MEC'S ACCULAS

3.11.1.   Company

ACCULAS is developed through a collaborative effort between Laser Solutions Co., Ltd, JSR Corporation and D-MEC Ltd. D-MEC has taken the role of technology development and marketing to introduce the first practical machine in the manufacturing of microdevices. Laser Solutions Co., Ltd is responsible in manufacturing ACCULAS and JSR produces the specific resin for ACCULAS. D-MEC Ltd. was established in 1990 and is also the manufacturer of the SCS (see Sec. 3.3). The address of the company is JSR Building, 2-11-24 Tsukiji, Chuo-ku, Tokyo 104-0045, Japan.

3.11.2.   Product

ACCULAS is the first commercially released RP system to fabricate small precision parts in microns. As a new microstereolithography machine, ACCULAS has brought more possible applications into different areas in

Fig. 3.39. The microstereolithography machine ACCULAS (courtesy D-MEC).

the biomedical field, chemistry field, engineering field and many others. The fabrication undergoes the stereolithography process and a photo-polymer resin is cured layer-by-layer to form the product. However, the ACCULAS system does not use the conventional method of using gal-vanometers and laser to cure the photo-polymer resin. Instead, a DMD is used to project the image onto the photo-polymer for solidification. ACCULAS has its own software to convert the 3D CAD data into sliced data before manipulating the mirrors of DMD. ACCULAS, built with spe-cial techniques, is able to produce slice thickness ranging from 2 to 10 µm. The fabricated model can have a resolution of up to 1.7 µm in the X–Y direction and 5 µm in the Z direction. The ACCULAS machine is shown in Fig. 3.39.

The specifications for ACCULAS are summarized in Table 3.8.

3.11.3. Process

The process of the ACCULAS system in building 3D microstructures has similarities with the other liquid-based RP systems available com-mercially. A 3D CAD design has to be designed first and the data is then exported in the STL format. To complete the data preparation, the

Table 3.8. ACCULAS specifications (courtesy D-MEC).

Model	ACCULAS
Light source	LD (405 nm)
Image modulation	Digital mirror device (DMD)
Exposure resolution, μm	1
Molding range, mm	$150 \times 150 \times 50$
Maximum model size, mm	50×50
Lamination layer pitch, μm	5–10 (Machine accuracy: 2 μm)
Resin	Special high-resolution resin
Data interface	Dedicated I/F software "Viola" (plugged in magics)
Power supply	100 VAC, 2 KVA
Outside dimensions (body), mm	1,000 (W) $\times$ 1,000 (D) $\times$ 1,000 (H)
Weight (body), kg	Approx. 600

STL data is sliced by Magics RP software and eventually the contour data is converted to bitmap data. Materialise has collaborated with D-MEC for the past several years, compiling Magics RP software with the RP machines to create solutions based on their expertise. The conversion is programmed by the customized software from D-MEC called VIOLA.

After the data handling stage is completed, the next key component for the next stage of process is the DMD. The DMD is managed by the bitmap data and it has a size of 2×2 cm comprising of 1 million mirrors. Each individual mirror acts independently as it represents a pixel in the resolution of the image. The size of the mirror is about 14×14 μm and is able to project an image of 1.7×1.7 μm which a typical laser beam is unable to match (usually the spot size is about 100–200 μm). Every projection creates each desired layer to form the final device. ACCULAS is integrated with a special technique to control the curing depth of the photo-polymerizing resin between 2 and 10 μm.

3.11.4.   Principle

ACCULAS is based on the principles of the photo-polymerization process described in Sec. 3.1.4 and the description of the DMD device is similar to that found in Sec. 3.4.4.1. The DMD device is used instead of the laser system because of the limitation of the laser resolution in producing microstructures. DMD, which is made up of millions of mirrors microns in size, is able to produce highly precise microdevices from the multi-layered cured resin. ACCULAS comprises two software, Magics RP software to convert STL data into slice files and VIOLA, used for conversion of sliced files to bitmap files.

3.11.5. Strengths and Weaknesses

ACCULAS as a new microstereolithography system enabling high quality 3D microstructures to be produced has the following strengths:

(1) *High resolution.* Structures and parts built can have resolutions of up to 1 μm. Thus very fine details can be achieved.
(2) *Fast development cycle.* 3D microstructures are built within a short processing time compared to conventional MEMS processes.
(3) *No support required.* Parts are built directly onto the Alumina substrates and no additional supports are required.
(4) *Master model for production.* Models built can be used as master patterns for electroform coating and for injection molding.

The weaknesses of ACCULAS are:

(1) *Requires post-processing.* Parts are required to undergo etching to remove sacrificial material.
(2) *Limited variety of resins.* The materials for users to select from are limited to those that are supplied by D-MEC Ltd.

3.11.6. Applications

The demand for ACCULAS will grow because of its ability to produce 3D microstructures for research, prototype production and the manufacturing of microdevices. ACCULAS covers the following applications:

(1) Master molds for mass production and injection molding.
(2) MEMS research applicable for testing.
(3) Biomedical applications.

(4) Engineering applications like micromachining to produce microgears or micromotors.
(5) Optical electronics field.

3.12. OTHER NOTABLE LIQUID-BASED RP SYSTEMS

There are several other commercial and noncommercial liquid-based RP methods that are similar in terms of technologies, principles and applications to those presented in the previous sections, but with some interesting variations. Some of these RP systems have not made it commercially in the market while others are no longer available due to the severe competition in the RP market. However their technologies are of significant interests to the RP community.

3.12.1. Two Laser Beams

Unlike many of the earlier methods which use a single laser beam, or the DLP method (EnvisionTec GmbH), another liquid-based RP method is to employ two laser beams. First conceptualized by Wyn Kelly Swainson in 1967, the method involves penetrating a vat of photo-curable resin with two lasers.[45] As opposed to SLA and other similar methods, this method no longer works only on the top surface of a vat of resin. Instead, the work volume resides anywhere within the vat.

In the two-laser-beams method, the two lasers are focused to intersect at a desired point in the vat. The principle of this method is a two-step variation of ordinary photo-polymerization: a single photon initiates curing and two photons of different frequencies are required to initiate polymerization. Therefore, the point of intersection of the two lasers is the desired point of curing. The numerous points of intersection will collectively form the 3D part.

In this method, it is assumed that one laser excites molecules all along its path in the vat. While most of these molecules will return back to their unexcited state after a short time, those also in the path of the second laser will be further excited to initiate the polymerization process. On this basis, the resin cures at the point of intersection of the two laser beams, while the rest lying along either of the two paths will remain liquid.

Unfortunately, after more than three decades of work on this method, the research work have not been realized into a commercial system, despite the continuation of Swainson's work by Formigraphic Engine Co. in collaboration with Battell Development Corporation, USA and another independent research group, French National Center for Scientific Research (CNRS), France.

Three major problems need to be resolved before the process is deemed to deliver parts of usable size and resolution. The first two problems relate to the control of the laser beams so that they intersect precisely at the desired 3D point in the vat. Firstly, the focusing problem exists because each laser beam undergoes numerous refractions as its ray moves from the source in air through the liquid resin. Nonuniformities in the refractive index of the resin are difficult to control and monitor because of the following reasons:

(1) Different progress of curing can cause nonuniformities.
(2) Inherent nonuniformities in the resin itself.
(3) Change of temperature in the vat.

Assuming that the resin can be completely homogenous and temperature changes are minimized so that its effects are insignificant, the second problem to contend with is how two laser beams moving at high speed can be focused to a very small and accurate point of intersection. The problem is compounded by the high energies required of the two lasers, which means that the two lasers will have short focal lengths. Thus, the distance is limited and in turn, this limits the size of the part that can be made by this method.

Thirdly, it is not always the case that molecules lying along the paths of either laser beams will return to the liquid state. This means unwanted resins may be formed along the path of either laser beams. Consequently, the cured resins may pose further problems to the rest of the process.

3.12.2. Cubital's Solid Ground Curing (SGC)

3.12.2.1. *Product*

The SGC System, produced by Cubital Ltd and Israelis' company, is one of the more sophisticated high-end systems providing wider range and

Table 3.9. Cubital Inc.'s Solider (source from Cubital Inc.).

Model	Solider
Irradiation medium	High power UV lamp
XY resolution (mm)	Better than 0.1
Surface definition (mm)	0.15
Elevator vertical resolution (mm)	0.1–0.2
Minimum feature size (mm)	0.4 (horizontal, X–Y)
	0.15 (vertical, Z)
	–0.4 (horizontal, X–Y)
	0.15 (vertical, Z)
Work volume, XYZ (mm × mm × mm)	$350 \times 350 \times 350$–$500 \times 350 \times 500$
Production rate (cm³/h)	550–1,311
Minimum layer thickness (mm)	0.06
Dimensional accuracy	0.1%
Size of unit, XYZ (m × m × m)	$1.8 \times 4.2 \times 2.9$–$1.8 \times 4.2 \times 2.9$
Data control unit	Data front end (DFE) workstation

options for various modeling demands. Table 3.9 summarizes the specifications of the SGC system.

Cubital's system uses several kinds of resins, including liquid resin and cured resin as materials to create parts, water soluble wax as a support material and ionographic solid toner for creating an erasable image of the cross section on a glass mask.

3.12.2.2. *Process*

The Cubital's SGC process comprises three main steps: data preparation, mask generation and model making.[46]

(1) *Data preparation.* In this first step, the CAD model is prepared and the cross sections are generated digitally and transferred to the mask generator. The software used, Cubital's Solider DFE (Data Front End) software, is a CAD application package that processes data prior to sending them to the Cubital Solider system. DFE can search and correct flaws in the CAD files and render files on-screen for visualization purposes. Solider DFE accepts CAD files mainly in the STL format.

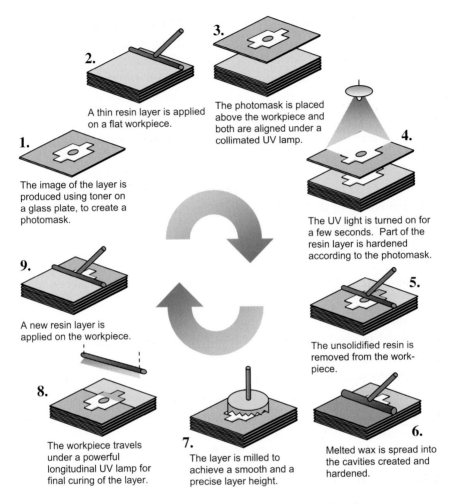

2. A thin resin layer is applied on a flat workpiece.

3. The photomask is placed above the workpiece and both are aligned under a collimated UV lamp.

1. The image of the layer is produced using toner on a glass plate, to create a photomask.

4. The UV light is turned on for a few seconds. Part of the resin layer is hardened according to the photomask.

9. A new resin layer is applied on the workpiece.

5. The unsolidified resin is removed from the workpiece.

8. The workpiece travels under a powerful longitudinal UV lamp for final curing of the layer.

7. The layer is milled to achieve a smooth and a precise layer height.

6. Melted wax is spread into the cavities created and hardened.

Fig. 3.40. Solid ground curing process (courtesy Cubital Ltd).

(2) *Mask generation.* After data are received, the mask plate is charged through an "image-wise" ionographic process (see item 1, Fig. 3.40). The charged image is then developed with electrostatic toner.

(3) *Model making.* In this step, a thin layer of photo-polymer resin is spread on the work surface (see item 2, Fig. 3.40). The photo mask from the mask generator is placed in close proximity above the work piece and aligned under a collimated UV lamp (item 3). The UV light

is turned on for a few seconds (item 4). The part of resin layer which is exposed to the UV light through the photo mask is hardened. Note that the layers laid down for exposure to the lamp are actually thicker than the desired thickness. This is to allow for the final milling process. The unsolidified resin is then collected from the work piece (item 5) by vacuum suction. Melted wax is then spread into the cavities (item 6). Consequently, the wax in the cavities is cooled to produce a wholly solid layer. The layer is then milled to its exact thickness, producing a flat surface ready to receive the next layer (item 7). In the SGC 5600, an additional step (item 8) is provided for final curing of the layer whereby the workpiece travels under a powerful longitudinal UV lamp. The cycle repeats itself until the final layer is completed.

The main components of the Solider system are:

(1) Data front end (DFE) workstation.
(2) Model production machine (MPM): It includes the process engine, operator's console and vacuum generator.
(3) Automatic Dewaxing machine (optional).

3.12.2.3. *Principle*

Cubital's RP technology creates highly physical models directly from computerized 3D data files. Parts of any geometric complexity can be produced without tools, dies or molds by Cubital's RP technology. The process is based on the following principles:

(1) Parts are built, layer by layer, from a liquid photo-polymer resin that solidifies when exposed to UV light. The photo-polymerization process is similar to that described in Sec. 3.1.4, except that the irradiation source is a high power collimated UV lamp and the image of the layer is generated by masked illumination instead of optical scanning of a laser beam. The mask is created from the CAD data input and "printed" on a transparent substrate (the mask plate) by a nonimpact

ionographic printing process, a process similar to the Xerography process used in photo-copiers and laser printers.[47] The image is formed by depositing black powder, a toner which adheres to the substrate electrostatically. This is used to mask the uniform illumination of the UV lamp. After exposure, the electrostatic toner is removed from the substrate for reuse and the pattern for the next layer is similarly "printed" on the substrate.

(2) Multiple parts may be processed and built in parallel by grouping them into batches (runs) using Cubital's proprietary software.

(3) Each layer of a multiple layer run contains cross-sectional slices of one or many parts. Therefore, all slices in one layer are created simultaneously. Layers are created thicker than desired. This is to allow the layer to be milled precisely to its exact thickness, thus giving overall control of the vertical accuracy. This step also produces a roughened surface of cured photo-polymer, assisting adhesion of the next layer to it. The next layer is then built immediately on the top of the created layer.

(4) The process is self-supporting and does not require the addition of external support structures to emerging parts since continuous structural support for the parts is provided by the use of wax, acting as a solid support material.

3.12.2.4. *Strengths and Weaknesses*

The Solider system has the following strengths:

(1) *Parallel processing*. The process is based on instant, simultaneous curing of the whole cross-sectional layer area (rather than point-by-point curing). It has high speed throughput that is about eight times faster than its competitors. Its production costs can be 25–50% lower. It is a time and cost saving process.

(2) *Self-supporting*. It is user friendly, fast and simple to use. It has a solid modeling environment with unlimited geometry. The solid wax supports the part in all dimensions and therefore support structure is not required.

(3) *Fault tolerance.* It has good fault tolerances. Removable trays allow job changing during a run and layers are erasable.

(4) *Unique part properties.* The part that the Solider system produces is reliable, accurate, sturdy, machinable and can be mechanically finished.

(5) *CAD to RP software.* Cubital's RP software, DFE, processes solid model CAD files before they are transferred to the machine.

(6) *Minimum shrinkage effect.* This is due to the full curing of every layer.

(7) *High structural strength and stability.* This is due to the curing process that minimizes the development of internal stresses in the structure. As a result, they are much less brittle.

(8) *No hazardous odors generated.* The resin stays in liquid state for a very short time and the uncured liquid is wiped off immediately. Thus safety is considerably higher.

The Solider system has the following weaknesses:

(1) *Requires large physical space.* The size of the system is much larger than other systems with similar build volume size.

(2) *Wax gets stuck in corners and crevices.* It is difficult to remove wax from parts with intricate geometry. Thus, some wax may be left behind.

(3) *Waste material produced.* The milling process creates shavings, which have to be cleaned from the machine.

(4) *Noisy.* The Solider system generates a high level of noise as compared to other systems.

3.12.2.5. *Applications*

The applications of Cubital's system can be divided into four areas:

(1) *General applications.* Conceptual design presentation, design proofing, engineering testing, integration and fitting, functional analysis, exhibitions and pre-production sales, market research and inter-professional communication.

(2) *Tooling and casting applications.* Investment casting, sand casting and rapid, tool-free manufacturing of plastic parts.

Fig. 3.41. Cubital prototypes and metal prototypes created from them (courtesy Cubital America Inc.).

(3) *Mold and tooling.* Silicon rubber tooling, epoxy tooling, spray metal tooling, acrylic tooling and plaster mold casting.
(4) *Medical imaging.* Diagnostic, surgical, operation and reconstruction planning and custom prosthesis design.

3.12.2.6. *Examples: New Cubital-Based Solicast Process Offers Inexpensive Metal Prototypes in Two Weeks*

Schneider prototyping GmbH created a metal prototype (see Fig. 3.41) for investment cast directly from CAD files in two weeks with SoliCast.[48] Starting with a CAD file, Schneider can deliver a metal prototype in two weeks, whereas similar processes with other RP methods take 4–6 weeks and conventional methods based on computer numerical control (CNC) prototyping can take 10–16 weeks. In general, Schneider's prototypes can be as much as 50% cheaper than those produced by other RP methods.

3.12.3.   Teijin Seiki's Soliform System

3.12.3.1. *Product*

Teijin Seiki Co., Ltd produces two main series (250 and 500 series) of the Soliform system. The Soliform 250 series machine is shown in Fig. 3.42.

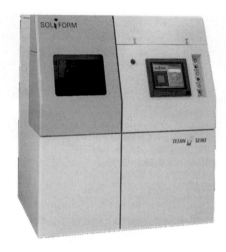

Fig. 3.42. The Soliform 250B solid forming system (courtesy Teijin Seiki Co., Ltd).

The specifications of the machines are summarized in Table 3.10. The materials used by Soliform are primarily SOMOS photo-polymers supplied by Du Pont and TSR resins, its own developed resin.

3.12.3.2. *Process*

The process of the Soliform system comprises the following steps: concept design, CAD design, data conversion, solid forming and plastic model.

(1) *Concept design*: Product design engineers create the concept design. This may or may not necessarily be done on a computer.
(2) *CAD design*: The 3D CAD model of the concept design is created in the SUN-workstation.
(3) *Data conversion*: The 3D CAD data is transferred to the Soliform software to be tessellated and converted into the STL file.
(4) *Solid forming*: This is the part building step. A solid model is formed by the ultraviolet laser layer by layer in the machine.
(5) *Plastic model*: This is the postprocessing step where the completed plastic part is cured in an oven to be further hardened.

Table 3.10. System specifications of Soliform (courtesy Teijin Seiki Co., Ltd).

	Model			
	SOLIFORM 250B	SOLIFORM 250EP	SOLIFORM 500C	SOLIFORM 500EP
Laser type	Solid state		Solid state	
Laser power (mW)	200	1000	400	1000
Scanning system	Digital scanner mirror		Digital scanner mirror	
Maximum scanning speed (m/s)	12 (max)		24 (max)	
Beam diameter	Fixed: 0.2 mm		Fixed: 0.2 mm	
	Variable: 0.1–0.8 mm (Option)		Variable: 0.2–0.8 mm (Option)	
Max. build envelope, *xyz* (mm)	250 × 250 × 250		500 × 500 × 500	
Min. build layer (mm)	0.05		0.05	
Vat	Interchangeable		Interchangeable	
Recoating system	Dip coater system		Dip coater system	
Resin level controller	Counter volume system		Counter volume system	
Power supply	100 VAC, single phase, 30 A		200 VAC, three phase, 30 A	
Size of unit, *xyz* (mm)	1430 × 1045 × 1575		1850 × 1100 × 2280	
Weight (kg)	400		1200	

The Soliform system contains the following hardware: an SUN-EWS workstation, an argon ion laser, a controller, a scanner to control the laser trace of scanning and a tank which contains the photo-polymer resin.

3.12.3.3. *Principle*

The Soliform system creates models from photo-curable resins based essentially on the principles described in Sec. 3.1.4. The resin developed by Teijin is an acrylic–urethane resin with a viscosity of 40,000 centapoise and a flexural modulus of 52.3 MPa compared to 9.6 MPa for a grade used to produce conventional prototype models.

Parameters which influence performance and functionality are generally similar to those described in Sec. 3.1.4, but for Soliform, the properties of the resin and the accuracy of the laser beam are considered more significant.

3.12.3.4. *Strengths and Weaknesses*

The Soliform system has several technological strengths:

(1) *Fast and accurate scanning.* Its maximum scanning speed of 24 m/s is faster than all other RP systems. Its accurate scan system is controlled by digital encoder servomotor technique.
(2) *Good accuracy.* It has exposure control technology for producing highly accurate parts.
(3) *Photo-resins.* It has a relatively wide range of acrylic–urethane resins for various applications.

The Soliform system has the following weaknesses:

(1) *Requires support structures.* Structures that have overhangs and undercuts must have supports that are designed and fabricated together with the main structure.
(2) *Requires post-processing.* Post-processing includes removal of supports and other unwanted materials, which is tedious, time-consuming and can damage the model.
(3) *Requires post-curing.* Post-curing may be needed to cure the object completely and ensure the integrity of the structure.

3.12.3.5. *Applications*

The Soliform system has been used in many areas, such as injection modeling (low cost die, see Fig. 3.44), vacuum molding, casting and lost wax molding.

(1) *Injection molding (low cost die).* The Soliform can be used to make injection molding or low cost die based on a process described in Fig. 3.43. Comparing this method with the conventional processes, this process has shorter development and execution time. The created

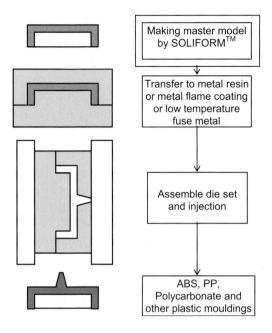

Making master model by SOLIFORM™

Transfer to metal resin or metal flame coating or low temperature fuse metal

Assemble die set and injection

ABS, PP, Polycarbonate and other plastic mouldings

Fig. 3.43. Creating injection molding parts (adapted from Teijin Seiki Co., Ltd).

Fig. 3.44. RP model (left), the ABS injection molding (center) and the low cost metal die (right) (courtesy Teijin Seiki Co., Ltd).

CAD data can be used for mass-production die and no machining is necessary. Photo-polymers that can be used include SOMOS-2100 and SOMOS-5100. An example of the product molded in ABS is shown together with the injection mold and the SOLIFORM pattern in Fig. 3.44.

(2) *Vacuum molding.* Vacuum molding molds are made by the Soliform system using a process similar to that illustrated in Fig. 3.52. The photo-polymers that can be used include SOMOS-2100, SOMOS-3100 and SOMOS-5100.

(3) *Casting.* The process of making casting part form is again similar to the process illustrated in Fig. 3.52 with the exception of using casting sand instead of the injection mold. The photo-polymer recommended is SOMOS-3100.

(4) *Lost wax molding.* Again this process is similar to the one described in Fig. 3.52. The main difference is that the wax pattern is made and later burnt out before the metal part is molded afterwards. The recommended photo-polymer is SOMOS-4100.

(5) *Making injection and vacuum molding tools directly.* The other application of SOLIFORM is the process of making the tools for injection molding and vacuum molding directly. From the CAD data of the part, the CAD model of the mold inserts are created. These new CAD data are then transferred the usual way to SOLIFORM to create the inserts. The inserts are then cleaned and post-cured before the gates and ejector pin holes are machined. They are then mounted on the mold sets and subsequently on the injection molding machine to mould the parts. Figure 3.45 describes schematically the process flow.

Figures 3.46 and 3.47 show the injection mold cavity insert and the molding for a handy telephone, respectively.

A process similar to that described in Fig. 3.45 can also be used for creating molds and dies for vacuum molding. Figure 3.48 shows a mouse cover that is molded using vacuum molding from tools created by SOLIFORM.

3.12.4. Meiko's RP System for the Jewelry Industry

3.12.4.1. *Company*

Meiko RPS Co., Ltd, a company making machines for analyzing components of gases such as CO_2, NO_x and O_2, factory automation units such as loaders and unloaders for assembly lines and two arm

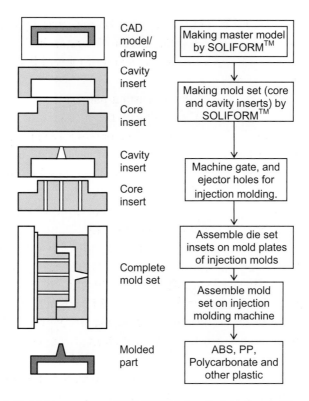

	CAD model/ drawing	Making master model by SOLIFORM™
	Cavity insert	Making mold set (core and cavity inserts) by SOLIFORM™
	Core insert	
	Cavity insert	Machine gate, and ejector holes for injection molding.
	Core insert	
	Complete mold set	Assemble die set insets on mold plates of injection molds
		Assemble mold set on injection molding machine
	Molded part	ABS, PP, Polycarbonate and other plastic

Fig. 3.45. Making process of SOLIFORM directly injection molding dies.

Fig. 3.46. Core and cavity mold inserts on molding machine of the handy telephone (courtesy Teijin Seiki Co., Ltd).

Fig. 3.47. Injection molded part from the SOLIFORM molds of the handy telephone (courtesy Teijin Seiki Co., Ltd).

Fig. 3.48. Direct vacuum molded mouse cover (courtesy Teijin Seiki Co., Ltd).

robots, was founded in June 1962. The system for the jewelry industry was developed together with Yamanashiken-Industrial Technical Center in the early 1990s. Meiko RPS Co., Ltd was established in 2000 selling 3D software and RP systems. The company's address is Meiko RPS Co., Ltd., 2F, 51-3 Oyamakanaicho, Itabashi-ku, Tokyo 173-0024, Japan.

3.12.4.2. *Product*

Meiko's LC-510 is an optical modeling system that specializes in building prototypes for small models such as jewelry. LCV-700 (see Fig. 3.49) and LCV-810 both can provide a model with a maximum size of $60 \times 60 \times 60$ mm. The two main differences between the systems are that LCV-810 has a scanning speed of four times higher than LCV-700 but LCV-700 can provide a better surface finish than LCV-810 due the difference in spot

Fig. 3.49. LCV-700 RP system (courtesy Meiko RPS Co., Ltd).

Table 3.11. Specifications of all Mekio's systems (courtesy Meiko RPS Co., Ltd).

	Model		
	LC-510	LCV-700	LCV-800
Laser type	He–Cd	Semiconductor	
Laser power, mW	5	30	60
Spot size, mm	0.08	0.04	0.05
Method of laser operation		X–Y plotter	
Scanning Speed (Max.), mm/min	1,000	500	2000
Resolution in X–Y direction, mm		0.01	
Max work volume, XYZ, mm	$100 \times 100 \times 60$	$60 \times 60 \times 60$	
Slice thickness, mm		0.2–0.01	
Operating system		Windows 98/Me/2000/XP/NT	
Power supply	100 VAC 15 A		100 VAC 10 A
CAD format		JSD, DXF, STL	

diameter. The specifications of the LC-510, LCV-700 and LCV-800 are summarized in Table 3.11.

Parts of different properties can be fabricated due to the types of photo-curable resins offered by Meiko Co., Ltd. The photo-curable resins are in translucent yellowish, greenish and bluish-violet colors. The consumable components in the systems are the laser tube and the photo-curable resin. The 3D CAD software, which is named JCAD3/Takumi, is the customized software for 3D printing.

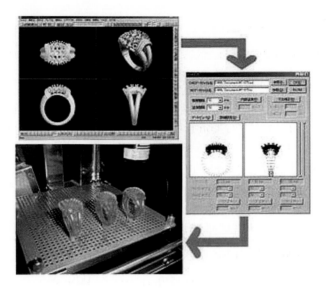

Fig. 3.50. Modeling jewelry on JCAD3/Takumi and MEIKO (courtesy Meiko Co., Ltd).

3.12.4.3. *Process*

The process building models by the system MEIKO is illustrated in Fig. 3.50.

(1) *Designing a model using the exclusive 3D CAD software*: A model is created in the personal computer using JCAD3/Takumi, specifically created for jewelry design in Meiko.

(2) *Creating NC data from the CAD data directly by the exclusive computer-aided manufacturing (CAM) module*: From the CAD data, NC data is generated using the CAM module. This is based on the standard software and methods used in CNC machines.

(3) *Forwarding the NC data to the LC-510 machine*: The NC data and codes are transferred to the NC controller in the LC-510 machine via RS-232C cable. This transfer is similar to those used in CNC machines.

(4) *Building the model by the LC-510*: From the downloaded NC data, the LC-510 optical modeling system will create the jewelry prototype using the laser and the photo-curable resin layer-by-layer.

3.12.4.4. *Principle*

The fundamental principle behind the method is the laser solidification process of photo-curable resins. Similar to other liquid-based systems, its principles are like those described in Sec. 3.1.4. The main difference is in the controller of the scanning system. The system MEIKO uses an *X–Y* (plotter) system with NC controller instead of the galvanometer mirror scanning system.

Parameters which influence performance and functionality of the system are the properties of resin, the diameter of the beam spot and the *XY* resolution of the machine.

3.12.4.5. *Strengths and Weaknesses*

The strengths of the product and process are as follows:

(1) *Good accuracy.* Highly accurate modeling that is primarily targeted for jewelry.
(2) *Low cost.* The cost of prototyping with the system is relatively inexpensive.
(3) *Cost saving.* Lead time is cut and cost is saved by labor saving.
(4) *Exclusive CAM software.* The CAM software is for jewelry and small parts.
(5) *Ease of manufacturing complicated parts.* Intricate geometries can be easily made using JCAD3/Takumi and LC-510 in a single run.

The weaknesses of the product and process are as follows:

(1) *Requires support structures.* Structures that have overhangs and under-cuts must have supports that are designed and fabricated together with the main structure.
(2) *Requires post-processing.* Post-processing includes removal of supports and other unwanted materials, which is tedious, time-consuming and can damage the model.
(3) *Require post-curing.* Post-curing may be needed to cure the object completely and ensure the integrity of the structure.

3.12.4.6. *Applications*

The general application areas of the system MEIKO are as follows:

(1) Production of jewelry.
(2) Production of other small models (such as hearing-aids and a part of spectacle frames).
(3) Production of machine parts.
(4) Production of medical/dental parts.

Users of the system MEIKO produce resin models using the exclusive resin UVM-8001 as its material as shown in Fig. 3.51.

There are two ways to create actual products from these resin models as follows:

(1) In the traditional way (see Fig. 3.52), rubber models are made and wax models are fabricated (see Fig. 3.53). Then using the lost wax method, actual products can be manufactured (see Fig. 3.54).
(2) These resin models can also be used as casting patterns, similar to that of the wax models.

3.12.4.7. *Research and Development*

Meiko Co., Ltd researches on the other new machine (including the resin) and the new edition of the JCAD3/Takumi, cooperating with

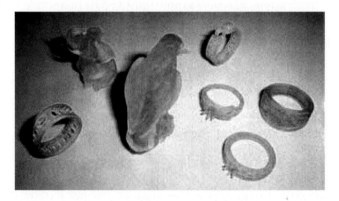

Fig. 3.51. Resin models made by the system MEIKO (courtesy Meiko Co., Ltd).

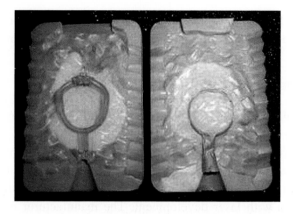

Fig. 3.52. Making rubber mold from resin model rings (courtesy Meiko Co., Ltd).

Fig. 3.53. Resin ring and metal casted ring models (courtesy Meiko Co., Ltd).

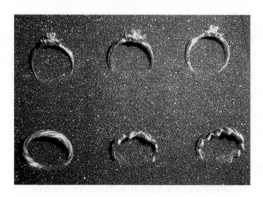

Fig. 3.54. Actual precious metal rings from casted from originals of resin ring models (courtesy Meiko Co., Ltd).

Professor S. Furukawa at Yamanashi University. Meiko will invest more manpower to produce more machines in order to widen its market.

3.12.5. Others

Other similar liquid-based systems are summarized as follows:

(1) *SLP (solid laser-diode plotter).* The SLP series of resin-based RP machines is a joint effort of Denken Engineering and Autostrade Co., Ltd, assisted by Nippon Kayaku with resin development and by Shimadzu with laser development. The manufacturer of the SLP is Denken Engineering Co, Ltd, 2-1-40, Sekiden-machi Oita City, Oita Prefecture 870, Japan.

(2) *COLAMM (computer-operated, laser-active modeling machine).* Opposed to most methods where parts are built on a descending platform, the COLAMM is built on an ascending platform.[45] The first (top) layer of the part is scanned by a laser placed below, through a transparent window plate. The layer is built on the platform and since it is upside down, the part is suspended. Next, the platform is raised by the programmed layer thickness to allow a new layer of resin to flow onto the window for scanning. The manufacturer is Mitsui Zosen Corporation of Japan.

(3) *LMS (layer modeling system).* The manufacturer is Fockele und Schwarze Stereolithographietechnik GmbH (F&S) of Borchen-Alfen and its address is Alter Kirchweg 34, W-4799 Borchen-Alfen, Germany. A unique feature of the LMS is its low-cost, nonoptical resin level control system (LCS). The LCS does not contact the resin and therefore allows vats to be changed without system recalibration.

(4) *Light Sculpting.* Not sold commercially, the light sculpting device is offered as a bureau service by light sculpting of Milwaukee, Wisconsin, USA [E20]. The light sculpting technique uses a descending platform like SLA and many others and irradiates the resin surface with a masked lamp as in Cubital's SGC. However, the unique feature of this technique is that the resin is cured in contact with a plate of transparent material on which the mask rests. The resulting close

proximity of the mask to the resin surface has good potential of ensuring accurate replication of high-resolution patterns.

REFERENCES

1. Rapid Prototyping Report. 3D Systems Introduces Upgraded SLA-250 with Zephyr Recoating, **6**(4), CAD/CAM Publishing Inc., April, 3.

2. 3D Systems Product brochure. SLA Series, 1999.

3. P. F. Jacobs, *Rapid Prototyping and Manufacturing, Fundamentals of Stereolithography* (Society of Manufacturing Engineers, 1992), Chapter 1, pp. 11–18.

4. J. Wilson, *Radiation Chemistry of Monomers, Polymers, and Plastics* (Marcel Dekker, NY, 1974).

5. K. Lawson, UV/EB Curing in North America, *Proceedings of the International UV/EB Processing Conference*, Florida, USA, 1–5 May, 1 (1994).

6. A. Reiser, *Photosensitive Polymers* (John Wiley, NY, 1989).

7. P. F. Jacobs, *Rapid Prototyping and Manufacturing, Fundamentals of Stereolithography* (Society of Manufacturing Engineers, 1992), Chapter 2, pp. 25–32.

8. P. F. Jacobs, *Stereolithography and other RP&M Technologies* (Society of Manufacturing Engineers, 1996), Chapter 2, pp. 29–35.

9. P. F. Jacobs, *Rapid Prototyping and Manufacturing, Fundamentals of Stereolithography* (Society of Manufacturing Engineers, 1992), Chapter 2, pp. 53–56.

10. P. F. Jacobs, *Rapid Prototyping and Manufacturing, Fundamentals of Stereolithography* (Society of Manufacturing Engineers, 1992), Chapter 3, pp. 60–78.

11. Rapid Prototyping Report. Ciba Introduces Fast Polyurethanes for Part Duplication, **6**(3), CAD/CAM Publishing Inc., March 6.

12. P. F. Jacobs, Insight: Moving toward rapid tooling, *The Edge*, **IV**(3), 6–7 (1995).

13. Anon, Rapid success for PolyJet, *Profess. Eng.* **19**, 43–43 (2006).

14. B. Vaupotič, M. Brezočnik and J. Balič, Use of PolyJet technology in manufacture of new product, *J. Achievements Mater. Manufact. Eng.* **18**, 319–322 (September 2006).

15. M. Durham, Rapid prototyping — Stereolithography, selective laser sintering, and polyjet, *Adv. Mater. Process.* **161**: 40–42, January 2003.

16. Y. L. Cheng and S. J. Chen, Manufacturing of cardiac models through rapid prototpying technology for surgery planning, *Mater. Sci. Forum*, **505–507**, 1063–1068 (2006).

17. T. Brajlih, I. Drstvensek, M. Kovacic and J. Balic, Optimizing scale factors of the PolyJet™ rapid prototyping procedure by generic programming, *J. Achievement in Materials and Manufacturing Engineering*, **16**, 101–106 (2006).

18. J. Hendrik, Perfactory® — A Rapid Prototyping system on the way to the "personal factory" for the end user (Envision Technologies GmbH).

19. J. Stampfl, R. Cano Vives, S. Seidler, R. Liska, F. Schwager, H. Gruber, A. Wöβ and P. Fratzl, *Proc. 1st Int. Conf. Advanced Research in Virtual and Rapid Prototyping*, 1–4 October 2003, Leiria, Portugal, pp. 659–666.

20 R. Landers, H. John and R. Mülhaupt, Scaffolds for tissue engineering applications fabricated by 3D plotting, Freiburger Materialforschungszentrum and Institut für makromolekulare Chemie of the Albert-Lubwigs-Universität, Freiburg, Germany (Envisiont Technologies, Marl, Germany).

21. L. Moroni, J. de Wijn and C. A. van Blitterswijk, 3D pltted scaffolds for tissue engineering: dynamical mechanical analysis, *Eur. Cells Mater.* **7** (Suppl. 1), 68 (2004).

22. R. Mülhaupt, R. Landers and Y. Thomann, Biofunctional processing: scaffold design, fabrication and surface modification, *Eur. Cells Mater.* **6** (Suppl. 1), 12 (2003).

23. W. Zhang, M. C. Leu, Z. Ji and Y. Yan, Rapid freezing prototyping with water, *Mater. Des.* **20**(June), 139–145 (1999).

24. M. C. Leu, W. Zhang and G. Sui, An experimental and analytical study of ice part fabrication with rapid freeze prototyping, *Ann. CIRP* **49**, 147–150 (2000).

25. G. Sui and M. C. Leu, Thermal analysis of ice wall built by rapid freeze prototyping, *ASME J. Manufact. Sci. Eng.* **125**(Nov), 824–834 (2003).

26. F. D. Bryant and M. C. Leu, Study on incorporating support material in rapid freeze prototyping, *Proc. Solid Freeform Fabrication Symp.* (Aug), 2–4 (2004).

27. M. C. Leu, Q. Liu and F. D. Bryant, Study of part geometric features and support materials in rapid freeze prototyping, *Ann. CIRP* **52**, 185–188 (2003).

28 G. Sui and M. C. Leu, Investigation of layer thickness and surface roughness in rapid freeze prototyping, *ASME J. Manufact. Sci. Eng.* **125**(Aug), 556–563 (2003).

29. Q. Liu and M. C. Leu, Investigation of interface agent for investment casting with ice patterns, *ASME J. Manufact. Sci. Eng.* **128**(May), 554–562 (2006).

30. Rapid prototyping report. Microfabrication, **4**(7), CAD/CAM Publishing Inc., July 4–5 (1994).

31. K. Ikuta, K. Hirowatari and T. Ogata, Three dimensional micro integrated fluid systems (MIFS) fabricated by stereolithography, *Proc. IEEE Micro Electro Mechanical Systems 1994*, Oiso, Japan (1994).

32. K. Ikuta, S. Maruo and S. Kojima, New micro stereo lithography for freely movable 3D micro structure — super IH process with submicron resolution, *Proc. IEEE Micro Electro Mechanical Systems 1998*, Piscataway, NJ, USA (1998), pp. 290–295.

33. C. Beuret, G. A. Racine, J. Gobet, R. Luthier and N. F. de Rooij, Microfabrication of 3D multidirectional inclined structures by UV

lithography and electroplating, *Proc. IEEE Micro Electro Mechanical Systems 1994*, Oiso, Japan (1994).

34. K. Suzomori, A. Koga and R. Haneda, Microfabrication of integrated FMAs using stereo lithography, *Proc. IEEE Micro Electro Mechanical Systems 1994*, Oiso, Japan (1994).

35. M. C. Carrozza, N. Croce, B. Magnani and P. Dario, Piezoelectric-drive stereolithography-fabricated micropump, *J. Micromechanics and Microengineering,* **5**(2), June (1995), pp. 177–179.

36. F. Yi, J. Wu and D. Xian, LIGA technique for microstructure fabrication, *Weixi Jiagong Jishu/Microfabrication Technology,* No 4 (December, 1993), pp. 1–7.

37. H. Lehr, W. Ehrfeld, K. P. Kaemper, F. Michel and M. Schmidt, LIGA components for the construction of milliactuators, *Proc. 1994 IEEE Symp. Emerging Tehnologies and Factory Automation,* Tokyo, Japan, (1993), pp. 43–47.

38. R. K. Kupka, F. Bouamrane, C. Cremers and S. Megtert, Micro-fabrication: LIGA-X and applications, *Appl. Surf. Sci.* **164**, 97–110 (September 2000).

39. G. Aigeldinger, P. Coane, B. Braft, J. Goettert, S. Ledger, G. L. Zhong, H. Manohara and L. Rupp, Preliminary results at the ultra deep X-ray lithography beamline at CAMD, *Proc. SPIE Vol. 4019, Design, Test, Integration, and Packaging of MEMS/MOEMS,* Paris, France (2000), pp. 429–435.

40. S. Basrour, H. Majjad, J. R. Coudevylle and M. de Labachelerie, Complex ceramic–polymer composite microparts made by microstere-olithography, *Proc. SPIE Vol. 4408, Design, Test, Integration, and Packaging of MEMS/MOEMS,* Cannes, France (2001), pp. 535–542.

41. V. K. Varadan and V. V. Varadan, Micro stereo lithography for fabrica-tion of 3D polymeric and ceramic MEMS, *Proc. SPIE Vol. 4407, MEMS Design, Fabrication, Characterization, and Packaging,* Edinburgh, UK (2001), pp. 147–157.

42. A. Cohen, U. Frodis, F.-G. Tseng, G. Zhang, F. Mansfeld and P. Will, EFAB: low-cost automated electrochemical batch fabrication of aritrary 3-D microstructures, *Proc. SPIE — The Int. Soc. Opt. Eng.* **3874**, 236–247 (1999).

43. R. T. Chen, E. R. Brown and C. A. Bang, A compact low-loss Kᴀ-band filter using 3-dimensional micromachinced integrated Coax, MicroFabric Inc, University of California Los Angeles.

44. J.R. Reid and R.T. Webster, A 55 GHz bandpass filter realized with integrated TEM transmission lines, *Int. Microwave Symp. Digest, IEEE MTT-S* (June 2006), pp. 132–135.

45. M. Burns, *Automated Fabrication: Improving Productivity in Manufacturing* (PTR Prentice Hall, 1993).

46. G. Kobe, Cubital's unknown Solider, *Automotive Industries*, August, 54–55 (1992).

47. J. L. Johnson, *Principles of Computer Automated Fabrication* (Palatino Press, 1994), Chapter 2, p. 44.

48. Cubital News Release, New Cubital-based SoliCast process offers inexpensive metal prototypes in two weeks (April, 1995).

PROBLEMS

1. Describe the process flow of the 3D System stereolithography apparatus.
2. Describe the process flow of the Objet polyjet systems.
3. Compare and contrast the laser-based stereolithography systems and the Objet polyjet systems. What are the strengths for each of the systems?
4. Describe the main differences between Perfactory and other commercial RP systems which use UV curing process.
5. Which liquid-based machine has the largest work volume? Which has the smallest?
6. Meiko Co., Ltd produces the LC and LCV series for jewelry prototyping. By comparing the machine specifications with other vendors,

discuss what you think are the important specifications that will determine their suitability for jewelry prototyping.

7. What is DMD, as found in Perfactory?

8. Which are the key materials for Bioplotting? Identify the methods of reaction for solidification.

9. Discuss the principle behind the two-laser-beams method. What are the major problems in this method?

10. As opposed to many of the liquid-based RP systems which use a photosensitive polymer, water is used in RFP. What are the pros and cons of using water?

11. How is the support of the ice part removed from the actual part?

12. Comparing the process of EFAB technology and ACCULAS, what are the advantages they individually have over each other?

Chapter 4

SOLID-BASED RAPID PROTOTYPING SYSTEMS

Solid-based rapid prototyping (RP) systems are very different from the liquid-based photo-curing systems described in Chap. 3. They are also different from one another, though some of them do use laser in the prototyping process. The basic common feature among these systems is that they all utilize solids (in one form or another) as the primary medium to create the prototype. A special group of solid-based RP systems that uses powder as the prototyping medium will be covered separately in Chap. 5.

4.1. STRATASYS' FUSED DEPOSITION MODELING (FDM)

4.1.1. Company

Stratasys Inc. was founded in 1989 in Delaware and developed the company's RP systems based on Fused Deposition Modeling (FDM) technology. The technology was first developed by Scott Cramp in 1988 and the patent was awarded in the USA in 1992. FDM uses an extrusion process to build 3D models. Stratasys introduced its first RP machine, the 3D modeler®, in early 1992. In 2002, Stratasys introduced the Dimension series that was aimed on affordable desktop 3D printer market. In 2007, Stratasys is the worldwide leader of installed RP systems and for the year shipped 44% of all RP systems in the world.[1] The company holds more than 180 granted or pending additive fabrication patents globally. Stratasys Inc. is located at 7665 Commerce Way, Eden Prairie, MN 55344-2080, USA.

4.1.2.   Products

4.1.2.1.   *FDM MC Machines*

The FDM manufacturing center (MC) machines provide customers with a comprehensive range of versatile RP systems that are also meant for direct digital manufacturing. They offer finer feature details, smooth surface, more accurate and stronger parts. These high-end systems are not only able to produce 3D models for mechanical testing, they are also able to produce functional prototypes that work as well as production parts. All FDM MC machines are supplied with Insight™ software, which imports files in the stereolithography (STL) format.[2] There are four machines in the FDM MC series, the FDM 200mc, FDM 360mc, FDM 400mc and FDM 900mc (Fig. 4.1). The 200mc and 360mc offer RP and entry level rapid manufacturing using ABS or ABS*plus* plastics, while the 400mc and the newer 900mc offer a wider range of materials, including ABS, Polycarbonate (PC), PC-ABS blend and polyphenylsulfone (PPSF) with better accuracy, finer finishing and higher repeatability, making them suitable for moderate volume direct rapid manufacturing. FDM offers two methods of support removal. WaterWorks™ is a soluble support material system and BASS™ is a manual breakaway support material system. Details of the FDM MC machines are summarized in Table 4.1.

FDM 200mc   **FDM 400mc**   **FDM 900mc**

Fig. 4.1. Stratasys' FDM MC machines (courtesy Stratasys Inc.).

Table 4.1. Specifications of Stratasys' FDM MC machines.

		Model		
	FDM 200mc	FDM 360mc	FDM 400mc	FDM 900mc
Technology		FDM		
Build size, mm (in.)	$203 \times 203 \times 305$ ($8 \times 8 \times 12$)	$355 \times 254 \times 254$ ($14 \times 10 \times 10$)	$355 \times 254 \times 254$ ($14 \times 10 \times 10$)	$914 \times 610 \times 914$ ($36 \times 24 \times 36$)
Accuracy mm (in.)		± 0.127 or ± 0.0015 mm per mm whichever is greater (± 0.005 or ± 0.0015 in. per in. whichever is greater)		
Layer thickness, mm (in.)	0.178–0.254 (0.007–0.010)	0.127–0.330 (0.013–0.005)	0.127–0.330 (0.013–0.005)	0.127–0.330 (0.013–0.005)
Support structures	Automatically generated with Insight software; WaterWorks soluble support system		Automatically generated with Insight software Soluble support for ABS-M30 and PC-ABS BASS™ breakaway for PC and PPSF	
Size, $w \times h \times d$, mm (in.)	$686 \times 1041 \times 864$ ($27 \times 41 \times 34$)	$1281 \times 895.35 \times 1962$ ($50.45 \times 35.25 \times 77.25$)	$1281 \times 1962 \times 895.35$ ($50.45 \times 35.25 \times 77.25$)	$2772 \times 1683 \times 2281$ ($109.1 \times 66.3 \times 79.8$) *With manufacturing light tower: $2772 \times 1683 \times 2281$ ($109.1 \times 66. \times 89.8$)*
Power requirements	110–120 VAC, 60 Hz, 15 A max. or 220–240 VAC, 50/60 Hz, 7 A max.	110–120 VAC, 60 Hz, 20 A max. or 220–240 VAC, 50/60 Hz, 16 A max.	110–120 VAC, 60 Hz, 20 A max. or 220–240 VAC, 50/60 Hz, 16 A max.	110–120 VAC, 60 Hz, 20 A max. or 220–240 VAC, 50/60 Hz, 16 A max.
Materials	ABS*plus* single material system	ABS-M30 single material	ABS-M30, PC, PC-ABS, PPSF multiple material system	ABS-M30, PC, PPSF multiple material system
Software		Insight™ and FDM Control Center™		

The older versions of RP systems from Stratasys, on which the FDM MC machines are based, include the Maxum, Titan, Vantage and Prodigy plus. These machines still have a large installed base worldwide and are still widely used in the industry. Prodigy Plus is the entry-level FDM machine with a build envelope of 203 × 203 × 305 mm and layer thickness varies between 0.178 and 0.33 mm. The next level are the Vantage machines that come in three variations and are based on a common platform. The Vantage I has a build envelope of 203 × 203 × 305 mm and uses ABS or PC materials. The S variant (stands for "speed") has the same envelope but builds twice as fast as the Vantage I on average. It uses ABS, PC and PC-ISO materials. The Vantage SE is both faster and has a much larger envelope of 406 × 355 × 406 mm. The Titan is essentially similar to the Vantage SE, but has more material options that include ABS, PC and PPSF and it has dual material cartridges for both build and support. The biggest machine is the Maxum. Its build platform is 600 × 500 × 600 mm.

4.1.2.2. *Dimension Series*

The Dimension series is created as an affordable office friendly RP system meant mainly for concept modeling, creating product mock-up and has some functional testing capabilities. The products are marketed under the Dimension Group as a business unit of Stratasys and its efforts are focused on the rapidly growing 3D printing marketplace by aggressively targeting new and existing industries. At the time of writing, the top product of the series, the Dimension Elite 3D printer is priced from only $32,900, while the basic 768 BST 3D printer is tagged starting at only $18,900, though both prices do not include additional options, shipping, tax, duties, etc. The Dimension Series 3D printers are easy to install and use as the machines utilize the Catalyst® EX software to create the model directly from CAD (Fig. 4.2). Based on their individual needs, users can select from five models available, namely the Dimension 768 BST, 1200 BST, 768 SST, 1200 SST and the Dimension Elite. The codes 1200 and 768 refer to their build size (in cubic inches), while BST refers to the manual Breakaway Support Technology system and SST refers to the Soluble Support Technology

Fig. 4.2. The Dimension Series 3D Printer (courtesy Stratasys Inc.).

system for automated support removal. The Dimension Elite has the same build size as that of the 768 Series but is capable of building fine features with layer thickness at a fine 0.178 mm (0.007 in.) and uses the stronger ABS*plus*, an ABS material 40% stronger than the ones used in other machines. Details of the machines can be found in Table 4.2.

4.1.3. Process

In this patented process,[3] a geometric model of a conceptual design is created on CAD software, which uses .STL or IGES formatted files. It can then be imported into the workstation where it is processed through the Insight™ software, or Catalyst® software for the Dimension series, which automatically generates the supports. Within this software, the CAD file is sliced into horizontal layers after the part is oriented for the optimum build position and any necessary support structures are automatically detected and generated. The slice thickness can be set manually anywhere between 0.178 and 0.356 mm (0.007 and 0.014 in.) depending on the needs of the models. Tool paths of the build process are then generated, which are downloaded to the FDM machine.

Table 4.2. Specifications of Dimension Series.

	Model		
	Dimension 768	Dimension 1200	Dimension Elite
Build size, mm (in.)	$203 \times 203 \times 305$ ($8 \times 8 \times 12$)	$254 \times 254 \times 305$ ($10 \times 10 \times 12$)	$203 \times 203 \times 305$ ($8 \times 8 \times 12$)
Materials	ABS or ABSplus plastic in standard natural, blue, fluorescent yellow, black, red, olive green, nectarine, or gray colors Custom colors are available (Dimension Elite uses only ABSplus)		
Support structures	BST breakaway support technology or SST soluble support removal process		SST soluble support removal
Material cartridges	One autoload cartridge with 922 cu. cm. (56.3 cu. in.) ABS or ABSplus material One autoload cartridge with 922 cu. cm. (56.3 cu. in.) support material		
Layer thickness, mm (in.)	0.245 or 0.33 (0.010 or 0.013)		0.178 or 0.245 (0.007 or 0.010)
Size, mm (in.)	$686 \times 914 \times 1041$ ($27 \times 36 \times 41$)	$838 \times 737 \times 1143$ ($33 \times 29 \times 45$)	$686 \times 914 \times 1041$ ($27 \times 36 \times 41$)
Weight, kg (lbs)	136 (300)	148 (326)	136 (300)
Software	Catalyst® EX software		
Power requirements	110–120 VAC, 60 Hz, minimum 15 A dedicated circuit or 220–240 VAC, 50/60 Hz, minimum 7 A dedicated circuit		

Modeling material is in the form of a filament, very much like a fishing line and is stored in a cartridge or in a spool. The filament is fed into an extrusion head and heated to a semi-liquid state. The semi-liquid material is extruded through the head and then deposited in ultra-thin layers from the FDM head, one layer at a time. Since the air surrounding the head is maintained at a temperature below the materials' melting point, the exiting material quickly solidifies. Moving on the X–Y plane, the head follows the tool path generated by Insight™, generating the desired layer. When the layer is completed, the head moves on to create the next layer. Two filament materials are dispensed through a dual tip mechanism in the FDM machine.

A primary modeler material is used to produce the model geometry and a secondary material, or release material, is used to produce the support structures. The software handles the location where the support deposition takes place. The release material forms a bond with the primary modeler material and with itself and can be washed away upon completion of the 3D models. A schematic diagram of the FDM process is shown in Fig. 4.3.

4.1.4. Principle

The principle of the FDM is based on the surface chemistry, thermal energy and layer manufacturing technology. The material in filament (in cartridge or spool) form is melted in a specially designed head, which extrudes it through the nozzle. As it is extruded, it is cooled and thus solidifies to form the model. The model is built layer by layer, like the other RP systems. Parameters that affect performance and functionalities of the system are material column strength, material flexural modulus, material viscosity, positioning accuracy, road widths, deposition speed, volumetric flow rate, tip diameter, envelope temperature and part geometry.

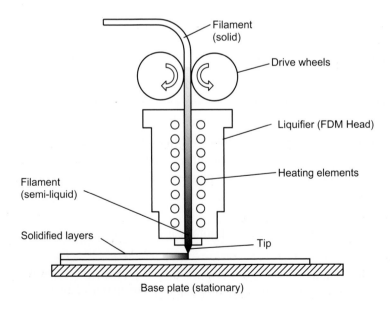

Fig. 4.3. The basic FDM process.

4.1.5. Strengths and Weaknesses

The main strengths of the FDM technology are as follows:

(1) *Fabrication of functional parts*: FDM process is able to fabricate prototypes with materials that are similar to that of the actual molded product. With ABS, it is able to fabricate fully functional parts that have 85% of the strength of the actual molded part and with ABS*plus* the strength of the parts can rival those that are injection molded. This is especially useful in developing products that require quick prototypes for functional testing.

(2) *Minimal wastage*: The FDM process builds parts directly by extruding semi-liquid melt onto the model. Thus, only those materials needed to build the part and its support are needed and material wastages are kept to a minimum. There is also little need for cleaning up the model after it has been built.

(3) *Ease of support removal*: With the use of Break Away Support System (BASS™) and WaterWorks™ Soluble Support System, or the BST and SST on dimension series, support structures generated during the FDM building process can be easily broken off or simply washed away. This makes it very convenient for users to get to their prototypes very quickly and there is very little or no post-processing necessary.

(4) *Ease of material change*: Build materials, supplied in spool or cartridge form, are easy to handle and can be changed readily when the materials in the system are running low. This keeps the operation of the machine simple and maintenance relatively easy.

(5) *Large build volume*: FDM machines, especially the FDM 900mc and the Maxum, offer larger build volume than most of the other RP systems available.

The weaknesses of the FDM technology are as follows:

(1) *Restricted accuracy*: Parts built with the FDM process usually have restricted accuracy due to the shape of the material used, i.e., the

filament form. Typically, the filament used has a diameter of 1.27 mm and this tend to set a limit on how accurate the part can be achieved. The newer FDM machines, however, have made significant improvements in easing this problem by using better machine control.

(2) *Slow process*: The building process is slow, as the whole cross-sectional area needs to be filled with building materials. Building speed is restricted by the extrusion rate or the flow rate of the build material from the extrusion head. As the build materials used are plastics and their viscosities are relatively high, the build process cannot be easily speeded up.

(3) *Unpredictable shrinkage*: As the FDM process extrudes the build material from its extrusion head and cools them rapidly on deposition, stresses induced by such rapid cooling are invariably introduced into the model. As such, shrinkages and distortions caused to the model built are common occurrence and are usually difficult to predict, though with experience, users may be able to compensate these by adjusting the process parameters of the machine.

4.1.6. Applications

FDM models can be used in the following general applications areas:

(1) *Models for conceptualization and presentation*: Models can be marked, sanded, painted and drilled and thus can be finished to be almost like the actual product.

(2) *Prototypes for design, analysis and functional testing*: The system can produce fully functional prototype in ABS or other common plastics used in the industry, like PC and PC-ABS blend. The resulting parts have 85% or higher (depending on the materials used) of the strength of the actual molded parts. Thus, actual functional testing can be carried out, especially with consumer products.

(3) *Patterns and masters for tooling*: Models produced on the FDM can be used as patterns for vacuum forming, investment casting, sand casting and molding.

4.1.7. Example

4.1.7.1. *Toyota Uses FDM for Design and Testing* [4]

Toyota, the forth-largest automobile manufacturer in the United States, produces more than 1 million vehicles per year. Its design and testing of vehicles are mainly done in the Toyota Technical Center (TTL) USA Inc.

In 1997, TTL purchased Stratasys FDM 8000 system to improve on their efficiency in design and testing. The system was able to produce prototypes with excellent physical properties and also able to build them fast. Furthermore, the system did not require any special environment to be operated on.

In the past, fabricating a prototype was costly and time-consuming in TTL. To manufacture a fully functional prototype vehicle, it required $10,000 to $100,000 to manufacture a prototype injection mold and took as long as 16 weeks. Furthermore, the number of parts required was around 20–50 pieces and thus, the conventional tooling method was unnecessarily costly.

In the Avalon 2000 project, TTL replaced its conventional tooling method with the FDM system. Although a modest 35 parts were being replaced by rapid prototypes, it was estimated that it saved Toyota more than $2 million in prototype tooling costs. Moreover, RP also helped designers to identify unforeseeable problems early in the design stage. It would have added to production costs significantly if the problems were discovered later during production stage.

The physical properties of these prototypes were not identical to those made from the conventional methods but nevertheless, as claimed by one of the staff in TTL, they were often good enough. TTL planned to increase its RP capacities by introducing additional units of the FDM systems. Its aim was to eliminate all conventional prototyping tooling and go straight to production tooling in the near future.

4.1.7.2. *Xerox Uses Dimension for Prototype Design* [5]

Xerox Limited, with its 350 engineers, develops designs for printers, copiers, multifunction devices and document centers. Before the arrival of the Dimension 3D Printer into the department, Xerox relies on the traditional machining from external agents in the prototype design process for molded components. With the Dimension 3D Printer, the engineers are able to create prototype parts in their own office at a much faster rate, resulting in significant time and cost savings. In two months with 12 h of operation everyday at 70% utilization, the Dimension 3D Printer builds 254 parts. Part files can be sent from Xerox's partner company in Toronto, Canada, to Xerox in the UK and are quickly produced by the machine.

Peter Keilty, the facility manager at Xerox headquarters in Welwyn Garden City, England, commented: *"Capitalizing on dead time gave us a significant advantage and having the ability to produce the parts overnight when everyone was asleep saved the day. Thinking bigger picture — time to market has been reduced and Dimension has played a significant part in that".*

4.1.7.3. *Hyundai Uses FDM for Design and Testing*[6]

Hyundai Mobis is a Korean company that makes original and aftermarket equipment for the automotive industry. To be one of the best in its field, it relies on good prototyping for design verification, airflow evaluation and functional testing. The company uses an FDM system for components such as instrument panels, air ducts, gear frame bodies, front-end modules and stabilizer-bar assemblies.

The quality of a vehicle is judged by the consumers on many aspects, the most important being component fit and finish. Hyundai Mobis paid a lot of attention to evaluate up to the smallest details using rapid prototypes to ensure the fit conveys a sense of quality. An example is the design verification of an instrument panel for Kia's Spectra. This instrument panel exceeded even the large build area of the company's Stratasys FDM Maxum™; hence, it was modeled in four pieces and then assembled to measure $498 \times 454 \times 1382$ mm. To ensure the quality of the model, the design team mounted it on a fixture and scrutinized it with a coordinate measuring machine (CMM) before mounting it to the cockpit assembly. The result was that over a length of 1382 mm, the greatest deviation on the model was just 0.75 mm. Mounted in the cockpit mock-up, the 27 design flaws were discovered. Although these errors were minor, collectively they would have added cost and likely delaying the project. The prototype allowed the designers to locate these design issues that were challenging to pick up in 3D CAD. When the FDM part was combined with mating components and subassemblies, the design flaws were quickly detected and repaired. As a result, Kia has garnered accolades from Car and Driver magazine for its Spectra, who wrote "... its interior fit and finish is premium".

Besides meeting Hyundai Mobis' requirements on the tight tolerance of the model, the FDM Maxum also gives them durable parts which they need for assembly and functional testing. The water-soluble support structure is very important due to the time pressures as without it, a complex component like the instrument panel would take many hours, if not days, to post-process.

Mobis' Maxum has been operating at 91% of capacity — roughly 8,000 h per year — still 60% of its prototyping work has been outsourced. Thus, the company plans to install a second FDM Maxum as they felt that the system is perfect for their design needs.

4.2. SOLIDSCAPE'S BENCHTOP SYSTEM

4.2.1. Company

Solidscape, formerly known as Sanders Prototype Inc., was established in 1994. Solidscape was a privately held company and was funded by Sanders Design Inc. (SDI), a research company, which is principally owned and managed by Solidscape's founder, Royden C. Sanders. In 2004, SDI and Sanders transferred all RP technology to Solidscape and ceased any further work in RP field. The company's current product line, the T612 and T66 Benchtop, combines thermoplastic ink-jet technology and high-precision milling. Solidscape has provided more than 2,500 systems in 20 countries. The address of the company is Solidscape, Inc., 316 Daniel Webster Highway, Merrimack, NH 03054-4115, USA.

4.2.2. Products

The T612 and T76 Benchtop were introduced to replace the T612 master modeling and T66 personal modeler system, which had earlier replaced the PatternMaster and ModelMaker system, respectively. As with other previous Solidscape systems, the Benchtop has high precision accuracy with build layer thicknesses of between 0.0127 and 0.762 mm (0.0005–0.003 in.). The T612 Benchtop has build envelope of 30.48 × 15.24 × 15.24 cm (12 × 6 × 6 in.), while the T76's is 15.24 × 15.24 × 10.16 cm (6 × 6 × 4 in.). The system is supplied with materials called InduraCast™ as the model material and InduraFill™ as the soluble support material. It also comes with the extended-build substrate called InduraBase™. The Benchtop uses ModelWorks™ software, which supports multiple slice thicknesses in a single build, as well as STL and stereolithography contour (SLC) input format. The examples of possible office layout with the T76 and T612 Benchtops can be seen in Fig. 4.4. Specifications of the two machines are listed in Table 4.3.

4.2.3.   Process

The process can be considered as a hybrid of FDM and 3D printing. The process uses two ink-jet-type print heads, one depositing the thermoplastic building material, InduraCast™ and the other depositing soluble supporting wax, InduraFill™. Like most other commercial systems, the model is built on a platform, which is lowered one layer after each layer is built. The liquefied build material cools as it is ejected from the print head and solidifies upon impact on the model. Soluble wax is deposited to provide a flat, stable surface for deposition of build material in the subsequent

(a)   (b)

Fig. 4.4. (a) The Solidscape T76 Benchtop Build Chamber and (b) the Solidscape T612 Benchtop.

Table 4.3. Specification of T76 and T612.

	Model	
	T76	T612
Build layer, in. (mm)	0.0005–0.003 (0.013–0.076)	
Surface finish (RMS)	32–63 microinches (RMS)	
Minimum feature size, in. (mm)	0.010 (0.254)	
Size of microdroplet, in. (mm)	0.003 (0.076)	
Build envelope, mm (in.)	152.4 × 152.4 × 101.6 (6 × 6 × 4)	304.8 × 152.4 × 152.4 (12 × 6 × 6)
Plotter carriage speed, cm/s (in./s)	50 (20)	
Dimension, mm (in.)	548.6 × 489.2 × 407.7 (21.6 × 19.26 × 16.05)	711.2 × 495.3 × 495.3 (28 × 19.5 × 19.5)
Software	ModelWorks™ supporting STL and SLC	
Material	InduraCast™	
Support structure	InduraFill™	

layers. After each layer is complete, a cutter removes a thin portion off the layer's top surface to provide a smooth, even surface ready for the next layer. Figure 4.5 shows the schematic of the process.

After several layers have been deposited, print heads are moved automatically to a monitoring area to check for clogging of the jet nozzles. If there is no blockage, the process continues, otherwise the jets are purged and the cutter removes any layers deposited since the last check of jets and restarts the building process from that point. This ensures the accuracy of the overall build.

4.2.4. Principle

The principle is that of building parts layer by layer, like in most other RP systems. The systems build models using a technique akin to ink-jet, applied in three dimensions. Two print heads, one applying a special thermoplastic material and the other depositing soluble supporting wax at positions where required. The twin heads shuttles back and forth much like a line printer but only in three dimensions to build one layer. When the layer is complete, the platform is distanced from the head (Z-axis) and the head begins building the next layer. This process is repeated until the

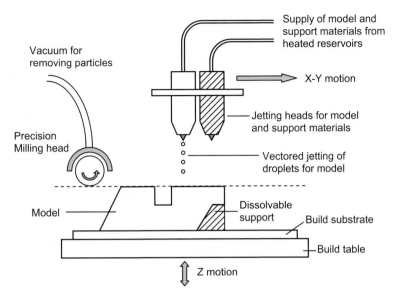

Fig. 4.5. Schematic of the Solidscape's high precision 3D modeling system.

entire concept model is complete. The main factors that influence the performance and functionalities are the thermoplastic materials, the ink-jet print head, the *X–Y* controls and the *Z*-controls.

4.2.5. Strengths and Weaknesses

The main strengths are as follows:

(1) *High precision*: The systems are able to achieve a build layer of 0.0127 mm (0.0005 in.) and produce a considerably smooth surface finish, 0.8–1.6 μm (32–63 microinches) RMS. This compares favorably with most RP machines.
(2) *Adjustable build layers*: Build layers can be adjusted, ranging from 0.013 mm (0.0005 in.) to 0.076 mm (0.0030 in.) in a single build, depending on the users' requirements and the need to speed up the build process.
(3) *Office-friendly process*: This system can be used in an office environment, without any special facilities as the process is simple, clean and

efficient. Moreover, materials used are nontoxic, which makes the process safe for office operation.

(4) *Minimal post-processing*: Prototypes fabricated using these systems require minimal post-processing and they are sufficiently strong and can be used directly for casting.

The weaknesses are as follows:

(1) *Small build volume*: The build volume is comparatively smaller than many high-end RP systems (only 30.48 × 15.24 × 15.24 mm or 12 × 6 × 6 in. for the T612). Thus, the systems are not suitable for building prototypes that are larger than these dimensions.
(2) *Limited materials*: Like most concept modelers, materials that can be processed by these systems are restricted to Solidscape's InduraCast™, InduraFill™ and InduraBase™. Thus, it puts a heavy limitation to the range of material-dependant functional capabilities of the prototypes.

4.2.6. Applications

The main application for Solidscape's systems is to produce high precision master patterns for design verification and production application. These machines are intended to function in an office environment in the immediate vicinity of the CAD workstations. The thermoplastic material provided is also suitable for investment casting.

4.2.7. Example

4.2.7.1. *Slowinski Uses Benchtop for Jeweler Design*[7]

Christopher Designs, a New-York-based jeweler design and manufacturing company, specializes in high-quality upscale platinum and gold jewelry. Christopher Slowinski, its founder and president, who holds three US patents on diamond settings and a new diamond cut, Crisscut, has discovered that these innovations and the quality of his designs are not sufficient to ensure success. To remain competitive and to come to better designs, he blends handcrafting with technology that includes milling and RP.

Slowinski's use of technology in his design and prototyping began in the early 1990s. He realized that the traditional methods could not give him the precision he required at a cost and time taken to develop the designs that would be competitive. He first adopted manual three-axis milling machines and then computer controlled (CNC) four-axis machine tools. While these achieved the quality and precision that he needed, he also found that they had limitations.

In 1997, Slowinski was introduced to the 3D modeling RP technology from Solidscape®, Inc. At first, he was concerned that the quality of the models produced would not meet his stringent requirements. Today Christopher Designs owns three Solidscape T66 systems, including T66 Benchtop. Even with the capability, Slowinski continues to use the milling machines as he often has to select the right technology for a piece for addressing the balance of speed, precision and cost.

The Solidscape T66 systems, which can build with 13 micron (0.0005 in.) layers, are used to construct master models that have complex shapes. The Solidscape master models are used to cast rubber molds for retail pieces and are used as the sacrificial model for casting of platinum and gold jewelry. According to Slowinski, they have no problem casting his jewelry from a Solidscape master model, unlike most other RP systems. Over the seven years that he has used the technology, Slowinski has seen dramatic improvements in model quality.

4.3. CUBIC TECHNOLOGIES' LAMINATED OBJECT MANUFACTURING (LOM™)

4.3.1. Company

Cubic Technologies was established in December 2000 by Michael Feygin, the inventor who developed Laminated Object Manufacturing (LOM). In 1985, Feygin set up the original company, Helisys Inc., to market the LOM RP machines. However, sales figures did not meet up to expectations[8] and the company ran into financial difficulties, Helisys Inc. subsequently ceased operation in November 2000. Currently, Cubic

Technologies, successor to Helisys Inc., is the exclusive manufacturer of the LOM RP machines. The company's address is Cubic Technologies, Inc., 2785 Pacific Coast Highway #295, Torrance, CA 90505, USA.

4.3.2. Products

Cubic Technologies offers two models of LOM RP systems, the LOM-1015Plus™ and LOM-2030H™ (see Fig. 4.6). Each of these systems uses a CO_2 laser, with the LOM-1015 Plus™ operating a 25 W laser and the LOM-2030H™ operating a 50 W laser. The optical system, which delivers a laser beam to the top surface of the work, consists of three mirrors that reflect the CO_2 laser beam and a focal lens that focuses the laser beam to about 0.25 mm (0.010 in.). The control of the laser during cutting is by means of a *XY* positioning table that is servo-based as opposed to a galvanometer mirror system. The LOM-2030H™ is a larger machine meant for building larger prototypes. The work volume of the LOM-2030H™ is 810 × 550 × 500 mm (32 × 22 × 20 in.) and that of the LOM-1015Plus™ is 380 × 250 × 350 mm (15 × 10 × 14 in.). Detailed specifications of these two machines are summarized in Table 4.4.

4.3.3. Process

The patented LOM process[9–11] is an automated fabrication method in which a 3D object is constructed from a solid CAD representation by sequentially laminating the part cross sections. The process consists of

Fig. 4.6. LOM-1015Plus™ and LOM-2030H™.

Table 4.4. Specification of LOM-1015Plus™ and LOM-2030H™.

	Model	
	LOM-1015Plus™	LOM-2030H™
Max. part envelope size, mm (in.)	381 × 254 × 356 (15 × 10 × 14)	813 × 559 × 508 (32 × 22 × 20)
Max. part weight, kg (lbs)	32 (70)	204 (405)
Laser, power and type	Sealed 25 W, CO_2 laser	Sealed 50 W, CO_2 laser
Laser beam diameter, mm (in.)	0.20–0.25 (0.008–0.010)	0.203–0.254 (0.008–0.010)
Motion control	Servo-based *X–Y* motion systems with a speed up to 457 mm/s (18 in./s)	Brushless servo-based *X–Y* motion systems with a speed up to 457 mm/s (18 in./s)
Part accuracy *XYZ*-directions, mm (in.)	±0.127 mm (± 0.005 in.)	± 0.127 mm (±0.005 in.)
Material thickness, mm (in.)	0.08–0.25 (0.003–0.008)	0.076–0.254 (0.003–0.008)
Material size	Up to 356 mm (14 in.) Roll width and roll diameter	Up to 711 mm (28 in.) roll width and roll diameter
Floor space, m (ft)	3.66 × 3.66 (12 × 12)	4.88 × 3.66 (16 × 12)
Power	Two (2) 110 VAC or Two (2) 220 VAC, 50/60 Hz, 20 A, single phase	220 VAC, 50/60 Hz, 30 A, single phase
Materials	LOMPaper® LPH series, LPS series LOMPlastics® LPX series	LOMPaper® LPH series, LPS series LOMPlastics® LPX series, LOM Composite® LGF series

three phases: pre-processing, building and post-processing. The overall LOM process is illustrated in Fig. 4.7.

4.3.3.1. *Pre-Processing*

The pre-processing phase comprises several operations. The initial steps include generating an image from a CAD-derived STL file of the part to

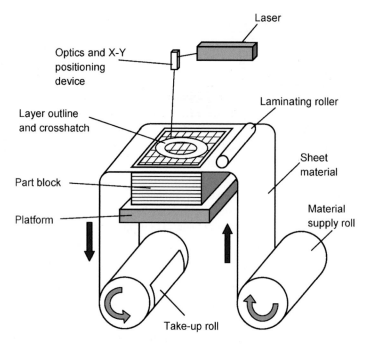

Fig. 4.7. The LOM process.

be manufactured, sorting input data and creating secondary data structures. These are fully automated by LOMSlice™, the LOM system software, which calculates and controls the slicing functions. Orienting and merging the part on the LOM system are done manually. These tasks are aided by LOMSlice™, which provides a menu-driven interface to perform transformations (e.g., translation, scaling and mirroring) as well as merges.

4.3.3.2. *Building*

In the building phase, thin layers of adhesive-coated material are sequentially bonded to each other and individually cut by a CO_2 laser beam (see Fig. 4.8). The build cycle has the following steps:

(1) LOMSlice™ creates a cross section of the 3D model measuring the exact height of the model and slicing the horizontal plane accordingly. The software then images crosshatches, which define the outer perimeter and convert these excess materials into a support structure.

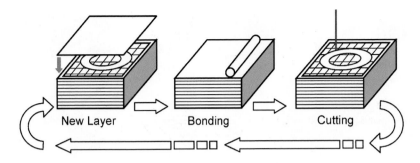

Fig. 4.8. Building process (courtesy Cubic Technologies, Inc.).

(2) The computer generates precise calculations, which guide the focused laser beam to cut the cross-sectional outline, the crosshatches and the model's perimeter. The laser beam power is designed to cut exactly the thickness of one layer of material at a time. After the perimeter is burned, everything within the model's boundary is "freed" from the remaining sheet.

(3) The platform with the stack of previously formed layers descends and a new section of material advances. The platform ascends and the heated roller laminates the material to the stack with a single reciprocal motion, thereby bonding it to the previous layer.

(4) The vertical encoder measures the height of the stack and relays the new height to LOMSlice™, which calculates the cross section for the next layer as the laser cuts the model's current layer.

This sequence continues until all the layers are built. The product emerges from the LOM machine as a completely enclosed rectangular block containing the part.

4.3.3.3. *Post-Processing*

The last phase, post-processing, includes separating the part from its support material and finishing it. The separation sequence is as follows (see Figs. 4.9(a)–4.9(d)):

(1) The metal platform, home to the newly created part, is removed from the LOM machine. A forklift may be needed to remove the larger and heavier parts from the LOM-2030H.

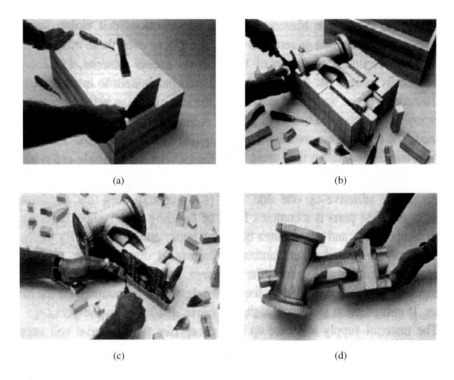

(a)   (b)

(c)   (d)

Fig. 4.9. Separation of the LOM object (courtesy Cubic Technologies, Inc.). (a) The laminated stack is removed from the machine's elevator plate. (b) The surrounding wall is lifted off the object to expose cubes of excess material. (c) Cubes are easily separated from the object's surface. (d) The object's surface can then be sanded, polished or painted, as desired.

(2) Normally a hammer and a putty knife are all that is required to separate the LOM block from the platform. However, a live thin wire may also be used to slice through the double-sided foam tape, which serves as the connecting point between the LOM stack and the platform.

(3) The surrounding wall frame is lifted off the block to expose the cross-hatched pieces of the excess material. Crosshatched pieces may then be separated from the part using wood carving tools.

After the part is extracted from surrounding crosshatches the wood-like LOM part can be finished. Traditional model-making finishing techniques, such as sanding, polishing, painting, etc., can be applied. After the part has been separated, it is recommended that it be sealed immediately with urethane, epoxy, or silicon spray to prevent moisture absorption and expansion of the part. LOM parts can also be machined using conventional machines like those for drilling, milling and turning.

4.3.3.4. *System Structure*

The LOM-1015Plus™ and LOM-2030H™ have a similar system structure which can be broken down into several subsystems: computer hardware and software, laser and optics, *X–Y* positioning device, platform and vertical elevator, laminating system, material supply and take-up (see Fig. 4.7).

The LOM software, LOMSlice™, is a true 32-bit application and is completely integrated, providing pre-processing, slicing and machine control within a single program. Z-dimension accuracy is maintained through a closed-loop real-time feedback mechanism and is calculated upon each lamination.

LOMSlice™ can also overcome STL file imperfections that violate facets of normal vector orientation or vertex-to-vertex rules,[12] or even those missing facets. In order to facilitate separation of the part from excess material, LOMSlice™ automatically assigns reduced crosshatch sizes to intricate regions.

To make it easier, faster and safer to align the laser beam, a helium–neon visible laser, which projects a red beam of light and is collinear with the live CO_2 laser beam, is used. The operator can switch on the innocuous red laser beam and the mirrors are aligned automatically.

Lamination is accomplished by applying heat and pressure by the way of rolling a heated cylinder across the sheet of material that has a thin layer of a thermoplastic adhesive on the one side. Studies[13] have indicated that the interlaminate strength of LOM parts is a function of bonding speed, sheet deformation, roller temperature and contact area between the paper and the roller. By increasing pressure of the heated roller, lamination is improved with fewer air bubbles. Increased pressure also augments the contact area and bolster interlaminate strength. Pressure is controlled by the limit switch mounted on the heated roller. Too high a compression can cause distortion in the part.

The material supply and take-up system comprises two material roll supports (supply and rewind), several idle rollers to direct the material and two rubber-coated nip rollers (driving and idle), which advance or rewind the sheet material during the pre-processing and building phases. To make material flow through the LOM systems more smoothly, mechanical nip rollers are used. The friction resulting from compressing moving material between rubber-coated rollers on both the feed and wind mechanism ensures a clean feed and avoids jamming.

4.3.3.5. *Materials*

Potentially, any sheet material with adhesive backing can be utilized in LOM. Cubic Technologies has demonstrated that plastics, metals and even ceramic tapes can be used. However, the most popular material has been Kraft paper with a polyethylene-based heat seal adhesive system because it is widely available, cost-effective and environmentally benign.[14]

In order to maintain uniform lamination across the entire working envelope, it is critical that the temperature is kept constant. A temperature control system, with closed-loop feedback, ensures that the system's temperature remains constant, regardless of its surrounding environment.

4.3.4. **Principle**

The LOM process is based on the following principles:

(1) Parts are built, layer-by-layer, by laminating each layer of paper or other sheet-form materials and the contour of part on that layer is cut by a CO_2 laser.
(2) Each layer of the building process contains the cross sections of one or many parts. The next layer is then laminated and built directly on top of the laser-cut layer.
(3) The Z-control is activated by an elevation platform, which lowers when each layer is completed and the next layer is then laminated and ready for cutting. The Z-height is then measured for the exact height so that the corresponding cross-sectional data can be calculated for that layer.

(4) No additional support structures are necessary as the "excess" material, which are cross hatched for later removal, act as the support.

4.3.5. Strengths and Weaknesses

The main strengths of using LOM technology are as follows:

(1) *Wide variety of materials*: In principle, any material in sheet form can be used in the LOM systems. These include a wide variety of organic and inorganic materials such as paper, plastics, metals, composites and ceramics. Users can select the type and thickness of materials that can meet their specific functional requirements and applications of the prototype.

(2) *Fast build time*: The laser in the LOM process does not scan the entire surface area of each cross section; rather, it only outlines its periphery. Therefore, parts with thick sections are produced just as quickly as those with thin sections, making the LOM process especially advantageous for the production of large and bulky parts.

(3) *High precision*: The feature-to-feature accuracy that can be achieved with LOM machines is usually better than 0.127 mm (0.005 in.). Through design and selection of application specific parameters, higher accuracy levels in the X–Y and Z dimensions can be achieved. If the layer does shrink horizontally during lamination, there is no actual distortion as the contours are cut post-lamination and laser cutting itself does not cause shrinkage. If the layers shrink in the transverse direction, a closed-loop feedback system gives the true cumulative part height upon each lamination to the software, which then slices the 3D model with a horizontal plane at the appropriate location.

The LOM system uses a precise X–Y positioning table to guide the laser beam; it is monitored throughout the build process by the closed-loop, real-time motion control system, resulting in an accuracy of within 0.127 mm regardless of part size. The Z-axis is also controlled using a real-time, closed-loop feedback system. It measures the cumulative part height at every layer and then slices the CAD geometry at the exact Z location. Also, as the laser cuts only the perimeter of a slice there is no need to translate vector data into raster form; therefore, the accuracy of the cutting only depends on the resolution of CAD model triangulation.

(4) *Support structure*: There is no need for additional support structure as the part is supported by its own material that is outside the periphery of the part built.

(5) *Post-curing*: The LOM process does not need to convert expensive and in some cases toxic, liquid polymers to solid plastics or plastic powders into sintered objects. Because sheet materials are not subjected to either physical or chemical phase changes, the finished LOM parts do not experience warpage, internal residual stress or other deformations.

The weaknesses of the LOM technology are as follows:

(1) *Precise power adjustment*: The power of the laser used for cutting the perimeter (and the crosshatches) of the prototype needs to be precisely controlled so that the laser cuts only the current layer of lamination and does not penetrate the previously cut layers. Poor control of the cutting laser beam can cause distortion to the entire prototype.

(2) *Fabrication of thin walls*: The LOM process is not well suited for building parts with delicate thin walls, especially in the Z-direction. This is because such walls usually are not sufficiently rigid to withstand the post-processing stresses when the cross hatched portion of the block is being removed. The operator performing post-processing tasks of separating the thin wall of the part from its support must be aware of where such delicate parts are in the model and take sufficient precaution so as not to damage these parts.

(3) *Integrity of prototypes*: The part built by the LOM process is essentially held together by the heat seal adhesives. The integrity of the part is therefore entirely dependent on the adhesive strength of the glue used and as such is limited by it. Therefore, parts built may not be able to withstand the vigorous mechanical loading functional prototypes may require.

(4) *Removal of* supports: The most labor-intensive part of the LOM process is its final phase of post-processing when the part has to be

separated from its support material within the rectangular block of laminated material. This is usually done with wood carving tools and can be tedious and time-consuming.

4.3.6. Applications

LOM can be applied across a wide spectrum of industries, including industrial equipment for aerospace or automotive industries, consumer products and medical devices ranging from instruments to prostheses. LOM parts are ideal in design applications where it is important to visualize what the final piece will look like, or to test for form, fit and function, as well as in a manufacturing environment to create prototypes, create production tooling or even small volume production.

(1) *Visualization*: LOM's ability to produce exact dimensions of a potential product lends it well for visualization purposes. LOM part's wood-like composition allows it to be painted or finished to be a good mock-up of the product. As the LOM procedure is inexpensive several models can be built, giving sales and marketing executives opportunities to utilize these prototypes for consumer testing, product introductions, packaging samples and samples for vendor quotations.

(2) *Form, fit and function*: LOM parts lend themselves well for design verification and performance evaluation. In low-stress environments LOM parts can withstand basic tests, giving manufacturers the opportunity to make changes as well as evaluate the aesthetic property of the prototype in its total environment.

(3) *Manufacturing*: LOM part's composition is such that, based on the sealant or finishing products used, it can be further tooled for use as a pattern or mold for many secondary tooling technique including investment casting, sand casting, injection molding, silicone rubber mold, vacuum forming and spray metal molding. LOM parts offer several advantages important for the secondary tooling process, namely predictable level of accuracy across the entire part; stability and resistance to shrinkage, warpage and deformity and the flexibility to create a master or a mold.

(4) *Rapid tooling*: Two-part negative tooling can be easily created with the LOM. As the material is solid and inexpensive, complicated tools can be cost effective to produce. These wood-like molds can be used for injection of wax, polyurethane, epoxy, or other low-pressure and low-temperature materials. The tooling can also be converted to aluminum or steel via the investment casting process for use in high-temperature molding processes.

4.3.7. Example

4.3.7.1. *National Aeronautical and Space Administration (NASA) and Boeing Rocketdyne Use LOM to Create Hot Gas Manifold for Space Shuttle Main Engine* [15]

One example of successful application of LOM systems in the design process would be from the Rapid Prototyping Laboratory, NASA's Marshall Space Flight Center (MSFC), Huntville, Alabama.

The laboratory was set up initially to conduct research and development in different ways to advance the technology of building parts in space by remote processing methods. However, as MSFC engineers found a lot more useful applications, i.e., production of concept models and proof-out of component designs other than remote processing when RP machines were installed, the center soon became a RP shop for other MSFC groups, as well as other NASA locations and NASA subcontractors.

The center acquired the LOM 1015™ machine in 1999 to add on to their RP systems. The machine was put through its first challenge when MSFC's contractor, Boeing Rocketdyne designed a hot gas manifold for the space shuttle's main engine. The part measured 2.40 m (8 ft) long and 0.10 m (4 in.) in diameter and was complex with many design intricacies. If conventional prototyping methods were employed, it would have required individual steel parts to be welded together to form the prototype. Such a part would have a potential of leakage at the joints and thus an alternate method was considered. The prototype was to be made from a single piece of steel but not only was such a solution expensive, the prototype built was unlikely to fit well to the main engine of the space shuttle. Eventually, engineers at Boeing decided to build the part using the LOM process at MSFC. They prepared a CAD drawing of the design and sent it over to MSFC. The design was sectioned into eight parts, each with irregular boss-and-socket built in them so as to facilitate joining of the parts together upon completion.

The whole building process took 10 days to complete, including 3 days of rework for flawed parts. It was worked on continuously. One advantage of using LOM machine is that the system can be left unattended throughout the building process and if the system runs out of paper or the paper gets jammed while building, it is able to alert the operator via wireless means. The prototype was then mounted onto the actual space shuttle for final fit check analysis. It was estimated that the company saved tens of thousands of dollars, although Boeing declined to reveal the actual cost saving. The whole process also drastically reduces the building time from two to three months to a mere 10 days.

4.4. 3D SYSTEM'S MULTI-JET MODELING SYSTEM (MJM)

4.4.1. Company

3D Systems is one of the earliest and largest RP companies in the world. Among its many products, 3D Systems also manufactures 3D printing systems meant mainly for the office environment. The multi-jet modeling (MJM) system was first launched in 1996 as a concept modeler for the office to complement the more sophisticated stereolithography apparatus (SLA) machines (see Sec. 3.1) and selective laser sintering (SLS) machines (see Sec. 5.1). The company's details are found in Sec. 3.1.1.

4.4.2. Products

The ProJet™ and InVision™ series machines are based on MJM technology with thermal material application and UV-curing. The newer ProJet™ Series machines are launched in 2008,[16] while the older InVision™ Series machines in 2005.

The ProJet™ SD3000 Printer is aimed for concept development, form and fit analysis, as well as patterns for casting or molding. The ProJet™

HD3000 (HD for high definition) is targeted for high resolution production–quality patterns for casting of jeweler and small components. This
high throughput system has resolutions and accuracies that are better than
the InVision™ HR machines. The build materials are acrylic plastics
called VisiJet HR 200 for the HD3000 and VisiJet SR 200 for the both the
HD3000 and SD3000. The material VisiJet S100, available for both
HD3000 and SD3000, is a nontoxic wax material meant for hands-free
melt-away supports. 3D System also has the ProJet™ DP set, which comprises the ProJet™ HD3000 3D Printer, a 3D Scanner and special software
customized specifically for dental lab needs. This is similar to the
InVision™ DP set in the InVision™ series. The ProJet™ machines and
their specifications are shown in Fig. 4.10 and Table 4.5.

The InVision™ Series equivalent to the HD3000, SD 3000 and DP3000
are the InVision™ HR, InVision™ XT and InVision™ DP, respectively. The
InVision™ machines and their specifications are shown in Fig. 4.11 and
Table 4.6.

The InVision™ Series also has an entry level concept modeler, the
InVision™ LD 3D modeler, which is based on a different technology of
sheet lamination. This printer is manufactured by Solidimension Ltd., an
Israel-based company and is offered by 3D Systems under its own brand
name. This will be discussed further in Sec. 4.5.

Fig. 4.10. ProJet™ HD3000 (left) and SD3000 (right) machines (courtesy 3D Systems).

Table 4.5. Specifications of 3D Systems ProJet™ series.

	Model	
	ProJet™ SD300	ProJet™ HD3000
Technology	Multi-jet modeling. Thermal material application, with UV-curing	
Mode of operation	—	High definition (HD) or ultra-high definition (UHD)
Build volume, mm (in.)	$298 \times 185 \times 203$ $(5 \times 7 \times 2)$	*HD mode*: $298 \times 185 \times 203$ $(11.75 \times 7.3 \times 8)$ *UHD mode*: $127 \times 178 \times 152$ $(5 \times 7 \times 6$ in.$)$
Maximum single model size, mm (in.)	$298 \times 185 \times 203$ $(5 \times 7 \times 2)$	*HD Mode*: $298 \times 185 \times 203$ $(11.75 \times 7.3 \times 8)$ *UHD Mode*: 6450 mm^2 $(x–y) \times 50$ mm (z); 10 in.^{2} $(x–y) \times 2$ in. (z)
Resolution, dots per inch (DPI) (xyz)	$328 \times 328 \times 606$	*HD mode*: $328 \times 328 \times 606$ *UHD mode*: $656 \times 656 \times 800$
Accuracy	0.001–0.002 in. (0.025–0.05 mm) per inch	
Electrical	100–127 VAC, 50/60 Hz, single-phase, 15 A; 200–240 VAC, 50 Hz, single-phase, 10 A	
Weight, kg (lb)	254 (560)	
Overall dimensions, mm (in.)	$737 \times 1257 \times 1504$ $(29.0 \times 49.5 \times 59.2)$	
Network	10/100 ethernet interface	
Software	ProJet™ Accelerator Software	
Input format	STL, SLC	
Build material	VisiJet® SR 200	VisiJet® SR 200 VisiJet® HR 200
Support material	VisiJet® S100	

4.4.3. Process

The ProJet™ Series uses the ProJet™ Accelerator Software to set up the "sliced" STL files for the system from the CAD software. The ProJet™ Accelerator Software is a powerful software, which allows users to verify the preloaded STL files and autofix any errors where necessary. The software

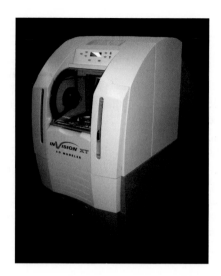

Fig. 4.11. InVision™ HR (left) and InVision™ XT (right) machines (courtesy 3D Systems).

also helps users to position the parts with its automatic part placement features so as to optimize building space and time. The Software also has automatic support generation capabilities to ease the operation of the print of the 3D model. After all details have been finalized, the data are placed in a queue, ready for the ProJet™ machine to build the model. The InVision™ Series uses InVision™ Print Client Software instead and it has similar functions of that of the ProJet™ Accelerator Software for the ProJet™ Series.

During the build process, the head is positioned above the platform. The head begins building the first layer by depositing materials as it moves in the *X*-direction. As the machine's print head contains hundreds of jets, it is able to deposit material quickly and efficiently. After a single layer pass is completed, the platform is lowered for the head to work on the next layer while the UV lamp floods the work space to cure the layer. The process is repeated until the part is finished. The schematic of the process is shown in Fig. 4.12.

4.4.4. Principle

The principle underlying MJM is the layering principle, used in most other RP systems. MJM builds models using a technique akin to ink-jet or

Table 4.6. Specifications of InVision™ series.

	Model	
	InVision™ HR	InVision™ XT
Technology	Multi-jet modeling. Thermal material application, with UV-curing	
Build volume, mm (in.)	127 × 178 × 50 (5 × 7 × 2)	298 × 185 × 203 (11.75 × 7.3 × 8)
Resolution, dots per inch (DPI) (*xyz*)	656 × 656 × 800	328 × 328 × 606
Electrical	100–127 VAC, 50/60 Hz, single-phase, 15 A; 200–240 VAC, 50 Hz, single-phase, 10 A	
Weight, kg (lb)	254 (560)	
Overall dimensions, mm (in.)	720 × 1230 × 1450 (28.3 × 48.6 × 57)	
Network	10/100 ethernet interface	
Software	InVision print client software	
Input format	STL, SLC	STL
Material	VisiJet® HR 200	VisiJet® SR 200
Support	VisiJet® S100	

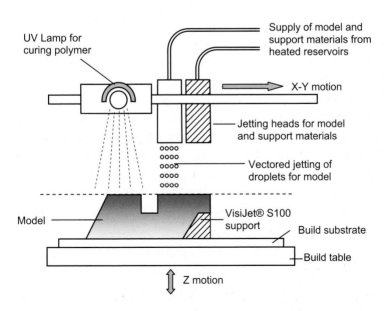

Fig. 4.12. Schematic of the multi-jet modeling system.

phase-change printing and applied in three dimensions. The print head jets are oriented in a linear array and build models in successive layers, each jet applying a special thermo-polymer material only where required. The MJM head shuttles back and forth like an ink-jet printer (*X*-axis), building a single layer of what will soon be a 3D concept model. If the part is wider than the print head, the platform will then reposition (*Y*-axis) itself to continue building the layer. The UV lamp flashes with each pass to cure the thermo-polymer deposited. When the layer is complete, the platform is distanced from the head (*Z*-axis) and the head begins building the next layer. This process is repeated until the entire concept model is complete. The main factors that influence the performance and functionalities of the MJM are the thermo-polymer materials, the UV flood lamp curing, the MJM head (number of jets and their arrangements), the *X*–*Y* controls and the *Z*-controls.

4.4.5. Strengths and Weaknesses

The strengths of the MJM technology are as follows:

(1) *Efficient and ease of use*: MJM technology is an efficient and economical way to create concept models. The large number of jets allows fast and continuous material deposition for maximum efficiency. MJM builds models directly from any STL file created with any 3D solid modeling CAD programs and no file preparation is required.
(2) *High precision*: The new MJM 3D Printer provides best-in-class part quality and accuracy with the choice of both high-definition and ultra-high-definition build modes in a single system. Its resolution of $656 \times 656 \times 800$ DPI in *xyz* orientation is one of the highest in the RP industry.
(3) *Cost-effective*: MJM uses inexpensive thermo-polymer material that provides for cost-effective modeling.
(4) *Fast build time*: As a natural consequence of MJM's raster-based design, geometry of the model being built has little effect on building time. Model work volume (envelope) is the singular determining factor for part build time.

(5) *Office-friendly process*: As the system is clean, simple and efficient, it does not require special facilities, thereby enabling it to be used directly in an office environment. Owing to its networking capabilities, several design workstations can be connected to the machine just like any other computer output peripherals.

The weaknesses of the MJM technology are as follows:

(1) *Small build volume*: The machine has a relatively small build volume as compared with most other high-end RP systems (e.g., SLA-500), thus only small prototypes can be fabricated. The HD3000, being aimed at jewelry and small components, has the smallest build volume.
(2) *Limited materials*: Material selections are restricted to 3D Systems' VisiJet® HR 200 and SR 200 thermo-polymers. This limited range of material means that many functionally based concepts that are dependent on material characteristics cannot be effectively tested with the prototypes.

4.4.6. Applications

The wide range of uses of the ProJet™ and InVision™ machines includes applications for the concept development, design validation, form and fit analysis, production of molding and casting patterns, direct investment casting of jewelry and other fine feature applications. Specifically for the ProJet™ and InVision™ DP machines, they can be used in the dental laboratory.

4.5. SOLIDIMENSION'S PLASTIC SHEET LAMINATION (PSL)/3D SYSTEM'S INVISION LD SHEET LAMINATION

4.5.1. Company

Solidimension is an Israel-based company that has been developing PVC-lamination-based 3D printer since 1999. In 2003 the company established a partnership with Japanese Graphtec Corp. and the following years, Graphtec agrees to distribute 900 units of Solidimension's XD700 3D Printer under its own brand name in Japan, Taiwan, Korea, Australia, Germany, France and the UK.[17] In 2005, 3D System became Solidimension marketing partner and commercialized its 3D printer as 3D System's InVision™ LD 3D Modeler.[18] The address of the company is Solidimension Ltd., Shraga Katz Building, Be'erot Itzhak, 60905, Israel.

4.5.2. Products

The main product of Solidimension is the SD300, a low-cost entry-level desktop 3D printer aimed at the concept modeling process as well as form and fit analysis in an office environment. The corresponding system sold by 3D Systems is known as InVision™ LD 3D Modeler (see Sec. 4.4). The SD 300 uses the SolidVC®, whereas the InVision™ LD uses the VisiJet® LD 100, both of these are thermoplastic (polyvinylchloride, PVC-based) sheets in the form of sheet rolls, together with adhesive and release agent or masking fluid. 3D Systems improved the support removal system by introducing an easy "peel-away" process called EZPeel®, which allows the user to remove the support simply by peeling away the zigzag support. The SDview® software supplied enables the user to manipulate, position and edit parts, as well as to split large models into multiple components to be glued manually as part of the post-processing. It supports STL input files. The Solidimension SD300 and the 3D Systems InVision™ LD are shown in Fig. 4.13 and their specifications summarized in Table 4.7.

4.5.3. Process

The concept modeling process begins when the STL model data are loaded into SDview®software. Within the software, the user is able to edit and orientate the model, as well as perform the splitting of the model if the model is found to be larger than the maximum build size. The software then slices the model and generates the required cutting and gluing action for the model and the support EZPeel®. The processed data are then sent to the machine through a USB connection. The machine then

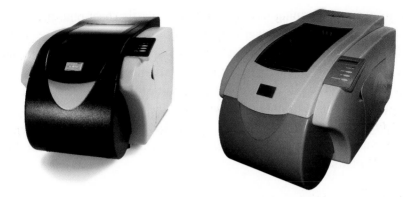

Fig. 4.13. Solidimension SD300 (left) and 3D Systems InVision™ LD (right) (courtesy Solidimensions Ltd and 3D Systems).

Table 4.7. Specification of SD300 and Invision™ LD.

	Model	
	SD300	InVision™ LD
Technology	Sheet lamination	
Accuracy	±0.1 mm	±0.2 mm
Build volume, mm (in.)	160 × 210 × 135	160 × 210 × 135
	(6.29 × 8.26 × 5.31)	(6.29 × 8.26 × 5.31)
Layer thickness, mm (in.)	0.168	0.15 (0.0059)
Electrical	100–120 VAC, 50/60 Hz; 200–240 VAC, 50/60 Hz	
Weight, kg (lb)	36 (79.3)	
Overall dimensions, mm (in.)	465 × 770 × 420 (18.3 × 30.3 × 16.5)	
Input format	STL, 3DS	
Material	SolidVC® PVC Material	VisiJet LD 100

builds the model layer by layer from the bottom up with the layer build up cycle.

The layer build up cycle consists of the following steps:

(1) *Preparation*: The machine first prepares a new PVC plastic sheet in the buffer.
(2) *Application of glue*: Glue is then applied over the entire area of the new plastic sheet.

(3) *Ironing*: The ironing unit then irons the plastic sheet onto the model block, joining a new layer to the model.
(4) *Cutting*: The new plastic layer is then cut by the cutting knife unit according to the pattern of the layer. Only the single layer is cut.
(5) *Trimming*: The plastic sheet is then trimmed by the trim knife and the supply is rolled back in preparation for the next layer.
(6) *Masking*: The final step of the layer build up cycle is the application of masking fluid by the anti-glue pens near the model to create the EZPeel® support.

This cycle is repeated until the model is complete. The components and parts of the SD300 used in the layer build up cycle are illustrated in Fig. 4.14. The material near the model is laminated and cut in such a way that it forms a continuous spiral-like form from the top-most layer to the bottom-most layer. During post-processing, larger blocks of waste material on the

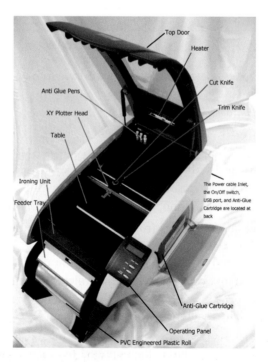

Fig. 4.14. Operating components of the SD300 3D Printer (courtesy Solidimensions Ltd.)

outer part can be separated directly because of the straight cut at the edges. The support material near the model can be peeled off by just pulling the upper-most layer. The spiral made from the layers in a zigzag fashion is like a thread on a spool being pulled to the axis of the spool cylinder. This forms the EZPeel® support removal system and it simplifies greatly the removal of the support material.

4.5.4. Principle

The main principle of the PSL system is similar to that of the LOM, which is building the object layer by layer by laminating sheets of plastic into a model block. Each layer is glued to the previous layer and the contour of the part is cut for every layer. The unique feature of the PSL that differentiates it from other solid-based systems is the laminating and cutting of the material around the model to form the EZPeel® support removal that makes post-processing very easy and simple.

4.5.5. Strengths and Weaknesses

The strengths of the Solidimension's PSL system are as follows:

(1) *Support structures*: There is effectively no need for additional support structure as the part is supported by its own material at the periphery of the part built. The removal of the part is also made easy and simple by the use of EZPeel® support peel away during post-processing.
(2) *Cost-effective*: The thermoplastic material used is inexpensive and that provides for cost-effective modeling. The machine has also one of the lowest costs in the 3D printer market.
(3) *Office-friendly process*: The system is clean, efficient and safe without the use of toxic chemicals. Post-processing is relatively clean and the PVC waste can be disposed of safely. This makes the system suitable for an office environment.
(4) *Fast build time*: As the cutting of the patterns of each layer is at the part peripheral and not the entire surface area, thus each layer can be built very quickly. This greatly shortens the overall process time.

The weaknesses of the Solidimension's PSL system are as follows:

(1) *Small build size*: As a compact desktop machine, a small build volume is expected. The ability to split big parts inside SDview® software is a helpful addition, although the durability of the resulting combined parts depends largely on the glue used and the gluing technique employed by the user.

(2) *Internal voids*: Internal voids cannot be produced on a single build since it is impossible to remove the material in the internal voids in this process. The model may need to be split into multiple parts to produce the void.

(3) *Inability to vary layer thickness*: The thickness of the layer is dependent on the thickness of the sheet material supplied, 0.168 for the SD300 and 0.15 for the InVision™ LD.

4.5.6. Applications

The Solidimension's PSL system is mainly for conceptual modeling and visualization during the design stage that do not require high accuracy or large build size. The system produces tough plastic material that can also be used for testing fit and form.

4.6. KIRA'S PAPER LAMINATION TECHNOLOGY (PLT)

4.6.1. Company

Kira Corporation Ltd was established in February 1944 and has developed a wide range of industrial products since then. Its line of products includes CNC machines, automatic drilling/tapping machines, recycle units and systems and folding bicycles. Kira joined in the RP industry by developing two RP machines, the KSC-50N (PLT-A3) and the PLT-A4. These two systems use a technology known as Paper Lamination Technology (PLT), formerly known as Selective Adhesive and Hot Press (SAHP).[19] The speed and reliability of both machines have ranked the systems as one of the better systems. The current PLT technology line continues with the

Katana Rapid Mockup System. The address of the company is Kira Corporation Ltd, Tomiyoshishinden, Kira-Cho, Hazu-gun, Aichi 444-0592, Japan.

4.6.2. Products

Kira Corporation, at the time of writing, only produces Katana PLT-20 Rapid Mockup System, the name reflecting its speed in producing prototypes (Fig. 4.15). A simple digital camera mock-up can be produced in just over an hour. Users may select the paper thicknesses of choice, although for a single build, the paper thickness needs to be the same. Different from its predecessor, the Katana uses paper supply in the form of a paper roll. The specifications of the machine are summarized in Table 4.8.

4.6.3. Process

The Katana PLT-20 Rapid Mockup System consists of the following hardware: a PC, a photocopy printer, a paper alignment mechanism, a hot press and a mechanical cutter plotter.[20] The Katana PLT-20 Rapid

Fig. 4.15. Katana PLT-20 Rapid Mockup system (courtesy of Kira Corporation).

Table 4.8. Specifications of Katana PLT-20 (courtesy Kira Corporation).

Model	PLT-20
Technology	Paper Lamination Technology (PLT)
Materials	KATANA paper
Max. build volume, mm	$180 \times 280 \times 150$
Resolution, mm	0.025 (X,Y), 0.1 (Z) or 0.16 (Z)
Precision, mm	± 0.5 (X,Y), $\pm 3.0\%$ (Z)
Power requirement	AC100/110 V $\pm 10\%$ single phase 15 A
System size, mm	$860 \times 660 \times 1330$
Weight, kg	320 kg
Data control unit	PC/AT compatible PC

Mockup System makes a plain paper layered solid model. The process is somewhat similar to that of the LOM process, except that no laser is used and a flat hot-plate press is used instead of rollers. Called Paper Lamination Technology (PLT) and formerly known as Selective Adhesive and Hot Press (SAHP), the process includes six steps: generating a model and printing resin powder, hot pressing, cutting the contour, completing block, removing excess material and post-processing (see Fig. 4.16).

First, the 3D data (STL files) of the model to be built is loaded onto the PC. The model is then oriented within the system with the help of the software for the best orientation for the build. Once achieved, the system software will proceed to slice the model and generate the printing data based on the section data of the model. The plain paper roll is fed onto the machine and trimmed. The resin powder or toner is applied on the paper from the paper roll using a typical laser stream printer and is referred to as the Xerography process (i.e., photocopying). The printed area is the common area of two consecutive sections of the model.

A paper alignment mechanism then adjusts the printed sheet of paper onto the previous layer on the model. The table then extends to a hot press over model block with the printed sheet pressed to a hot plate at high pressure. The temperature-controlled hot press melts the toner (resin powder), which adheres the sheets together. The hot press also flattens the top surface and prevents the formation of air bubbles between sheets. The PC

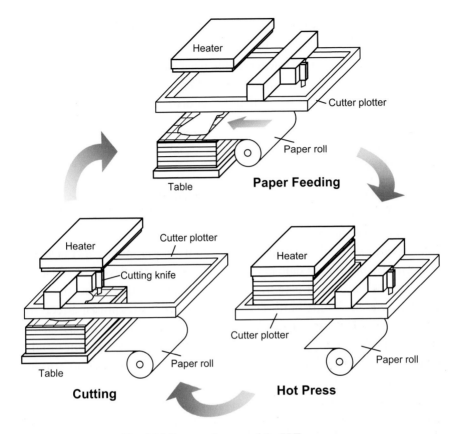

Fig. 4.16. Process diagram of the PLT process.

measures the amount of the movement up to the hot plate to compensate for any deviation of the sheet thickness.

The PC then generates plotting data based on the section data of the model. A precision cutter cuts the top layer of the block along the contour of the section as well as parting lines from which excess papers are removed later. These steps are repeated until the entire model is built. When printing, hot pressing and cutting are completed the model block is removed from the machine and unnecessary portions of the paper are disconnected quickly and easily sheet by sheet. When the model is complete, its surface may then be finished by normal wood-working or mechanical means.

The tensile strength and bending strength of the material made by the earlier SAHP process have been shown to be approximately one-half those of wood models. However, enhancement to the PLT process has enabled model hardness to improve significantly and it has been reported that hardness up to 25% better than the equivalent wood model has been achieved.

4.6.4. Principle

The principle of the process is based partly on the photocopy principle, conventional mechanical layering and precision cutting techniques. A typical laser stream printer is used for printing and a resin powder instead of print toner is used as a toner, which is applied to the paper in the exact position, indicated by the section data to adhere the two adjacent layers of paper.

Three factors, cutter plotter, temperature and humidity, affect the accuracy of the model being built. The accuracy of the cutter plotter affects the accuracy of the model in the X- and Y-directions. The shrinkage of the model occurs when the model is cooled down in the hot press unit. Expansion of the model occurs when the model is exposed to varying humidity conditions in the hot press unit.

4.6.5. Strengths and Weaknesses

The key strengths in using PLT, as described by Kira Corporation[21] are as follows:

(1) *Flatness*: The PLT process uses a flat plate and high pressure to bond the layers together. Each layer is pressed with a flat hot plate and the model remains flat during the entire build process. Since the block is released after cooling, there is minimal internal strain and therefore there is little or no curling in the final model.

(2) *Surface smoothness*: The PLT process uses a computerized knife to cut sheets of paper, which results in a smooth surface for the model built. Better surface finishing can be further attained by simple sanding tools but are seldom necessary. Prototypes can be sanded, cut or coated according to the user's needs.

(3) *Hardness*: High lamination pressure used in the PLT process has resulted in products that are 25% harder than the equivalent wood model and this is often strong enough for most prototyping applications.

(4) *Support structures*: Additional support structures are not necessary in the PLT process as the part is supported by its own material that is outside the perimeter of the cut-path. These are not removed during the hot press process and thus act as natural supports for delicate or overhang features.

(5) *Office-friendly process*: Kira's machines can be installed in any environment where electricity is available. There is no need for special facilities or utilities for running the Katana. Moreover, the process is safe as no high-powered laser or hazardous materials are used.

The weaknesses of using PLT are as follows:

(1) *Inability to vary layer thickness*: Fabrication time is slow in the Z-direction as the process builds prototypes layer by layer and the height of each layer is fixed by the thickness of the paper used. Thus, the speed of build cannot be easily increased as the thickness of the paper used cannot be varied.

(2) *Fabrication of thin walls*: Like the LOM process, the PLT process is also not well suited for building parts with delicate thin walls, especially walls that are extended in the Z-direction. This is due to the fact that such walls are joined transversely and may not have sufficient strength to withstand the post-processing. However, the flat hot plate used in the process may be able to pack the heat seal adhesive resin a little better to limit the problem.

(3) *Internal voids*: Models with internal voids cannot be fabricated within a single build as it is impossible to remove the unwanted support materials from within the "void".

(4) *Removal of supports*: Similar to that of the LOM process, the most labor-intensive part of the build comes at the end — separating the part from the support material. However, as the support material is not adhesively sealed, the removal process is simpler, though wood working

tools are sometimes necessary. The person working to separate the part needs to be cautious and aware of the presence of any delicate parts on the prototype so as not to damage the prototype during post-processing.

4.6.6. Applications

The PLT has been used in many areas, such as the automobile, electric machine and component, camera and office automation machine industries. Its main application area is in conceptual modeling and visualization. In Japan, Toyota Motor Corp., NEC Corp. and Mitsubishi Electric Corp. have reportedly used the Kira Corporation's Katana to model their products.

4.7. CAM-LEM'S CL 100

4.7.1. Company

CAM-LEM (Computer-Aided Manufacturing of Laminated Engineering Materials) Inc., a privately owned company in the USA, commercializes a RP technology akin to Cubic Technology's LOM. The process uses the "form-then-bond" laminating principle where the contour of the cross section is first cut before laminating it to the previous layers. This technology was developed by a team of researchers from Case Western Reserve University. The company's strategy is not to sell the RP machine developed but rather, CAM-LEM intends to act as a service provider to interested customers. The company's address is CAM-LEM, Inc., 1768 E. 25th St., Cleveland, OH 44114, USA.

4.7.2. Products

The RP systems developed by CAM-LEM Inc. is called CL-100 (see Fig. 4.17) and it is able to produce parts up to $150 \times 150 \times 150$ mm in size. One distinct feature of CL-100 is that the system is able to build metal and ceramic parts with excellent mechanical properties after the sintering process. The layer thickness ranges from 100 microns to 600 microns and

Fig. 4.17. CAM-LEM's CL100 RP system (courtesy CAM-LEM Inc.).

higher and up to five different materials can be incorporated into a single automated build cycle.

4.7.3. Process

The CAM-LEM approach (see Fig. 4.18), like other RP methods, originates from a CAD model decomposed into boundary contours of thin slices.[22] In the CAM-LEM process, these individual slices are laser cut from sheet stock of engineering material (such as "green" ceramic tape) as per the computed contours. The resulting part-slice regions are extracted from the sheet stock and stacked to assemble a physical 3D realization of the original CAD description. The assembly operation includes a tacking procedure that fixes the position of each sheet relative to the pre-existing stack. After assembly, the layers are laminated by warm isostatic pressing (or other suitable method) to achieve intimate interlayer contact, promoting high-integrity bonding in the subsequent sintering operation. The laminated "green" object is then fired (with an optimized heating schedule) to densify the object and fuse the layers and particles within the layers into a monolithic structure. The result is a 3D part, which exhibits not only correct geometric form but also functional structural behavior.

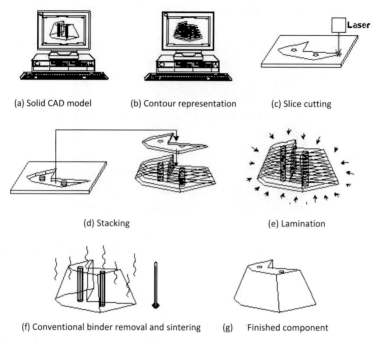

(a) Solid CAD model   (b) Contour representation   (c) Slice cutting

(d) Stacking   (e) Lamination

(f) Conventional binder removal and sintering   (g)   Finished component

Fig. 4.18. Schematic of CAM-LEM process (courtesy CAM-LEM, Inc.).

4.7.4. Principle

The CAM-LEM process is based on the same principle applied to most solid-based RP systems. Objects are built layer by layer and the laminated objects can be fabricated from a wide variety of engineering materials. The sections that form the layers are cut separately from sheet stock with a CO_2 laser and are then selectively extracted and stacked precisely. Multiple material types can be used within a single build. This process allows for the formation of interior voids and channels without manual waste removal, thus overcoming the problem of entrapped volume that plagues most other RP systems. The distinct characteristic of this process is the separation of the geometric formation process from the material process, thus providing users with more flexibility. The crucial factors affecting the quality of the model built are the laser cutting, the indexing and tacking, the alignment of the stacking process, the binding process and the sintering process.

4.7.5. Strengths and Weaknesses

The key strengths of using CAM-LEM technology are as follows:

(1) *Allow for the formation of interior voids and channels*: The CAM-LEM process separates the laser-cutting process from the stacking and lamination process, thus allowing for the formation of interior voids and channels, thereby eliminating the problem of entrapped volumes that troubles many other RP systems.

(2) *Laser power adjustment*: The cutting laser power of CAM-LEM technology does not need to be precisely adjusted because the process uses the "form-then-bond" laminating principle, where the contour of the cross section is first cut before laminating it to the previous layers. Thus, it eliminates the problem of laser burning into the previous layers.

(3) *High-quality prototypes*: As the technology uses the "form-then-bond" principle, it ensures that the layers are free from fine grains of unwanted materials before bonding them to the previous layer. This is highly desirable as the unwanted materials trapped in between layers would affect the mechanical property of the final product and such a situation should be eliminated so as to fabricate high quality prototypes.

(4) *Adjustable build layers*: The CAM-LEM process allows prototypes to be built using different material thickness, which could effectively speed up the process. Regions with large volume are built with thicker sheets of paper and surfaces that require a smooth surface finish are built with thinner sheets of paper.

The weaknesses of using CAM-LEM technology are as follows:

(1) *Significant shrinkage*: The main disadvantage in using CAM-LEM's technology is that the prototype will shrink in size by around 12–18%, which makes the dimensional and geometric control of the final prototype difficult.

(2) *Precise alignment*: The process requires high accuracy from the system to align the new bonding layer to the previous layer before bonding it.

Any slight deviation in alignment from previous layers will not only affect the accuracy of the model but also its overall shape.

(3) *Lacks natural supports*: While the process eliminates the problem of entrapped volume, it does require users to identify the locations of supports for the prototype, especially for overhanging features. As this process only transfers the desired layers to bond with the previous layers, all unwanted materials, which could have been used as supports, are thus left behind.

4.7.6. Applications

The CAM-LEM process has been used mainly to create rapid tools for manufacturing. Functional prototypes and even production of ceramic and metal components have been built.

4.8. OTHER NOTABLE SOLID-BASED RP SYSTEMS

4.8.1. EnnexCorporation's Offset Fabbers

4.8.1.1. *Introduction*

Ennex Corporation started as Ennex Fabrication Technologies in 1991 by a successful physicist and computer entrepreneur, Marshall Burns. Burns was then joined by a team of experienced and enthusiastic professionals to develop a range of state-of-the-art RP machines, which focused mainly on digital manufacturing technology. The company named these machines as "digital fabricator" or simply "fabber". A "fabber", as defined by the company, is a *"factory in a box" that makes things automatically from digital data.*[23] Besides developing this technology, the company also provides valuable consulting advice to related companies that deal with "fabbers". Ennex Corporation is located at 549 Landfair Ave., Los Angeles, CA 90024-2172, USA.

4.8.1.2. *The Genie® Studio Fabber*

One of the "fabbers", which the company has developed, is the Genie® Studio Fabber, based on a technology known as "Offset™ Fabbing".

The technology is somewhat similar to that of Kira Corporation's Paper Laminating Technology (PLT). The main difference between the two technologies is that "Offset™ Fabbing" uses the "form-then-bond" fabrication[24] method. The working principle of this fabrication method is simple. It uses a mechanical knife to cut the outline of the layer and after the sheet of material is cut to the required shape, it is then transferred and laminated onto the previous layer. The process of cutting and laminating of the materials is repeated while the model is being built layer by layer. Although Ennex Corp first announced "Offset™ Fabbing" technology in 1996, the company has still yet to commercialize its first Genie® Studio Fabber. However, a working prototype has been built in the company's development laboratory. Ennex has claimed that its output will be fast, up to ten times or more than the leading systems.

4.8.1.3. *Process*

The process of "Offset™ Fabbing" can be described as follows (see Fig. 4.19):

(1) Thin fabrication material (any thin film that can be cut and bonded to itself by means of adhesive) is rested on a carrier.
(2) A horizontal plotting knife is used to cut the outlines of cross sections of the desired object into the fabrication material without cutting through the carrier.
(3) The plotter can also cut parting lines and outlines for support structures.

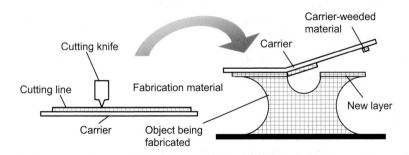

Fig. 4.19. Schematic of Ennex "Offset™ Fabbing" process.[23]

(4) The film is then "weeded" to remove some or all of the "negative" material.

(5) The film is inverted so that the carrier is facing up and the cut pattern is brought into contact with the top of the growing object and bonded to it.

(6) The carrier is then peeled off to reveal the new layer just added and a fresh surface is ready to bond to the next layer.

4.8.1.4. *Strengths and Weaknesses*

The strengths of "Offset™ Fabbing" technology are as follows:

(1) *Wide variety of materials*: The process is not restricted to a few materials. Basically, any thin film that can be cut and bonded itself can be used for the process. Paper and thermoplastic sheets are typical examples.

(2) *Minimal shrinkage*: As the bonding layers do not experience any significant change in temperature during the fabrication process, the shrinkage of the parts built is minimal.

(3) *Office-friendly process*: The process is clean and easy to use in ordinary office environment. Furthermore, the process is safe as it does not involve the use of laser or hazardous materials.

The weaknesses of Offset™ Fabbing technology used on the Genie Studio Fabber are as follows:

(1) *Need for precise cutting force*: The cutting force of the cutting knife must be precise, as a deep cut will penetrate into the carrier while a shallow cut will not give a clean cut to the layer resulting in tearing and sticking.

(2) *Need for precise alignment*: The process requires high accuracy from the system to align the new bonding layer to the previous layer before bonding it. Precise position is critical as any slight deviation in alignment from previous layers will affect not only the accuracy of the model but also the overall shape of the model.

(3) *Wastage of materials*: The weeded (unwanted) materials cannot be reused after the layers have been bonded with the previous layer and

hence a significant amount of materials are wasted during the process.

4.8.1.5. *Applications*

The "Offset™ Fabbing" process can be used mainly to create concept models for visualization and proofing during the early design process and are intended to function in an office environment near the CAD workstations. Functional prototypes and tooling patterns intended for rapid tooling can also be built.

4.8.2. Shape Deposition Manufacturing Process

4.8.2.1. *Introduction*

While most RP processes based on the discretized layer-by-layer process are able to build almost any complex shape and form, they suffer from the very process of discretization in terms of surface finish as well as geometric accuracies. The Shape Deposition Manufacturing (SDM) process, first pioneered by Prof. Fritz Prinz and his group at Carnegie Mellon University and later Stanford University, is a RP process that overcomes these difficulties by combining the flexibility of the additive layer manufacturing process with the precision and accuracy attained with the subtractive CNC machining process. Shot-peening, microcasting and a weld-based material deposition process can be further combined within a CAD–CAM environment using robotic automation to enhance the capability of the process.[25] The SDM process, though well researched and has many capabilities, is yet to be commercialized.

4.8.2.2. **Process**

The SDM process is a RP process that systematically combines the advantages of layer-by-layer manufacturing with the advantages of precision material removal process.[26] The process is illustrated by Fig. 4.20.

Materials for the individual segments of the part are first deposited at the deposition station to form the layer of part (see Fig. 4.20(a)). One of

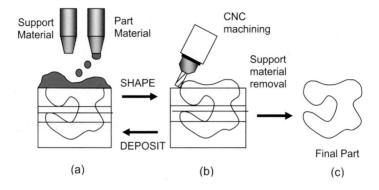

Fig. 4.20. Shape deposition manufacturing building process.

the several deposition processes is a weld-based deposition process called microcasting,[25] and the result is a near-net shape deposition of the part for that layer. The part is then transferred to the shaping station, usually a 5-axis CNC machining center, where material is removed to form the desired or net shape of the part (Fig. 4.20(b)). After the shaping station, the part is transferred to a stress relief station, such as shot-peening, to control and relieve residual stress build-up due to the thermal process during deposition and machining. The part is then transferred back to the deposition station where complementary shaped, sacrificial support material is deposited to support the part. The sequence of depositing part or support material is dependent on the geometry of the part as explained in the following paragraph. The process is repeated until the part is complete, after which the sacrificial support material is removed and the final part is revealed (Fig. 4.20(c)).

One major difference that SDM has is that unlike most RP processes, the CAD model of the part is decomposed into slices or segments that maintain the full 3D geometry of the outer surface. The major advantage of this strategy of decomposing shapes into segments, or "compacts" as called by the inventors, is that it eliminates the need to machine undercut features. Instead, such features are formed by depositing support or part material as appropriate onto previously deposited and shaped segments (see Fig. 4.20(a)). For example, if the part (like in a "U"-shape) over-hangs the support, then the support material for that layer will be deposited first. Conversely, if the part is clear of the support in the upward direction (as in Fig. 4.20(b)), then the part material for the layer will be deposited first. As such, this flexibility of sequencing part-support material deposition totally eliminates the need to machine difficult undercut features. The total layer thickness is not set arbitrarily but is dependent on the local geometry of the part and the deposition process constraints. Because of the alternating deposition-shaping sequences, discretization steps that plague many other RP systems are also eliminated.

Microcasting is a nontransferred metal inert gas (MIG) welding process, which deposits discrete, super-heated molten metal droplets to form dense metallurgical-bonded structures. Apart from the microcasting deposition process, several other alternative processes are also available to deposit a variety of materials in SDM. These processes and their processing materials are summarized in Table 4.9.[27]

4.8.2.3. *Strengths and Weaknesses*

The strengths of SDM are as follows:

(1) *Wide variety of materials*: The process is capable of handling a wide variety of materials, including stainless steel, steel alloys, metals, thermoplastics, photo-curable plastics, waxes, ceramics, etc. There is very little limitation or constraint on the type of materials that this process can handle.
(2) *Ability to build heterogeneous structures*: In addition to the RP of complex shapes, the SDM process is also able to fabricate multi-material structures and it also permits pre-fabricated components to be embedded within the built shapes. These provide the process with the unique capability of fabricating heterogeneous structures, which will enable the manufacture of novel product designs.
(3) *Variable layer thickness*: As the CAD model of the part is decomposed into slices or segments that maintain the full 3D geometry of the outer surface, the layer thickness is varied. The actual layer thickness will depend on the local geometry of the part and the deposition process constraints.

Table 4.9. Deposition process on the SDM.[27]

Deposition process	Description	Part materials	Support materials
Microcasting	An arc is established between a tungsten electrode of an MIG system and a feedstock wire that is fed from a charged contact tip. The wire melts in the arc, forming a molten droplet, which, having accumulated sufficient molten material, falls from the wire to the part. The droplets remain super-heated and have enough energy to remelt locally the part to form metallurgical bonding on solidification. A laminar curtain of shielding gas prevents oxidation	Stainless steel	Copper
Extrusion	Materials are deposited with an extruder with a single screw drive	Thermoplastics, ceramics	Water-soluble thermoplastics
Two-part resin system	Polyurethanes and epoxy resins are deposited as two-part resin-activator systems	Polyurethanes, epoxy resins	Wax
Hot wax dispensation	Waxes are deposited with a hot-melt extrusion system. Waxes can be used as either part or support material	Wax	Wax, water-soluble photo-curable resins
Photo-curable resin dispensation	Photo-curable resins are deposited with a simple syringe pumping system. This is usually used as support material for wax parts	Photoresins	Water-soluble photo-curable resins
MIG welding	MIG welding is used to deposit directly onto the part for fast deposition	Steel alloys	Copper
Thermal spraying	Thermal spraying is used to spray thin layers of "high-performance" materials. Plasma sprayers are typically used	Metals, plastics, and ceramics	

(4) *Direct creation of functional metal shapes*: The process is one of only few that are able to directly create or fabricate functional metal shapes without the secondary processes needed in other RP systems.

(5) *Ease of creating undercut features*: The decomposition strategy used in the SDM process overcomes the many difficulties in building undercuts that constraints many other RP systems. As such, highly complex shapes and forms can be built with little problem.

The weaknesses of the SDM are as follows:

(1) *Need for precise control of automated robotic system*: The process requires moving the part from station to station (deposition, shaping, shot-peening, etc.). Such a process demands precise control in placing the part accurately for each layer built. Invariably, because of the movement, errors can accumulate over several thicknesses built and thus affect the overall dimensional accuracy of the part. Also the control of the deposition process has to be precise to prevent creation of voids and causing excess remelting.

(2) *Thermal stresses due to temperature gradient*: The microcasting process of melting metals and precisely depositing them on the part to solidify with the previous layer results in a steep temperature gradient. This introduces thermal stresses into the part and can result in distortion. Controlled shot-peening can alleviate the problem but cannot entirely eliminate it. A careful balance of the temperature gradient, the associated internal stress build up, the thermal energy accumulation and their relief has to be achieved in order for SDM to produce quality precision parts.

(3) *Need for a controlled environment*: As several deposition and shaping processes can be incorporated into the SDM, the environment within which the system functions has to be controlled. For example, MIG welding and other hot work can result in waste gases discharge that have to be dealt with.

(4) *A large area is required for the system*: Owing to the multiple capability of the SDM process and the several stations (deposition, shaping, etc.) needed to accomplish the building of the parts, a relatively large area

is necessary for the entire system. This would mean that the system is not suitable for any office environment.

4.8.2.4. *Applications*

With its capability to handle multiple materials and deposition processes, the SDM can be applied to many areas in many industries, especially building finished parts and heterogeneous products. One example of finished parts is the direct building of ceramic silicone nitrite component for aircraft engines, which is tested and remains unscathed on a jet engine, rig at up to 1250°C. The SDM process has also successfully fabricated heterogeneous products like an electronic device by building a nonconductive housing package and simultaneously embedding interconnecting electronic components within the housing. This is especially useful in fabricating purpose-build devices for special applications like the wearable computer. Many interesting heterogeneous products with multiple materials embedded within the product have also been tested.[27]

REFERENCES

1. T. Wohlers, Wohlers Report 2008: State of the Industry, Annual Worldwide Progress Report, Wohlers Association, Inc. (2008).

2. Rapid Prototyping Report, Stratasys's New Insight Software, **11**(3), CAD/CAM Publishing, Inc., March 2001, p. 7.

3. S. Crump, The extrusion of fused deposition modeling, *Proc. 3rd Int. Conf. Rapid Prototyping* (1992), pp. 91–100.

4. Rapid Prototyping Report, Toyota Uses Fused Deposition Modeling to Bypass Prototype Tooling, **9**(8), CAD/CAM Publishing, Inc. (September 1999), pp. 1–3.

5. Stratasys Inc., Transatlantic Copying from Xerox [On-line], available at http://www.dimensionprinting.com/successstories/xerox.aspx (June 2008).

6. Stratasys Inc., Stratasys FDM Case Studies (Automotive) — Hyundai Mobis [On-line], Available at http://intl.stratasys.com/media.aspx?id=416 (June 2008).

7. Solidscape Inc., Jewellery Market Case Studies — Christopher Designs, [On-line], available at http://www.solid-scape.com/christopher_designs_case_ study.pdf (June 2008).

8. T. Wohlers, Wohlers Report 2000: Rapid Prototyping and Tooling, State of the Industry, Wohlers Association, Inc. (2000).

9. M. Feygin, Apparatus and Method for Forming an Integral Object from Laminations, U.S. Patent No. 4,752,352 (21 June 1988).

10. M. Feygin, Apparatus and Method for Forming an Integral Object from Laminations, European Patent No. 0,272,305 (2 March 1994).

11. M. Feygin, Apparatus and Method for Forming an Integral Object from Laminations, U.S. Patent No. 5,354414 (11 October 1994).

12. Industrial Laser Solutions, Application Report, NASA Builds Large Parts Layer by Layer [On-line], Available at http://www.industrial-lasers.com/ archive/1999/05/ 0599fea3.html (September 2001).

13. T. Wohlers, Wohlers Report 2001, Industrial Growth, Rapid Prototyping and Tooling State of the Industry, Wohlers Association, Inc. (2001).

14. Rapid Prototyping Report, Stratasys's New Insight Software, **11**(3), CAD/CAM Publishing, Inc. (March 2001), p. 7.

15. Rapid Prototyping Report, Stratasys Introduces Genisys Concept Modeller, **6**(3), CAD/CAM Publishing, Inc. (March 1996), pp. 1–4.

16. Reuters, On-line News, 3D Systems to Launch New, Professional, High-Definition 3-D Printer [On-line], Available at http://www.reuters.com/article/pressRelease/idUS200564+07-Jan-2008+BW200 80107 (June 2007).

17. T. Wohlers, Wohlers Talk, Solidimension Finally on a Roll [On-line], Available at http://wohlersassociates.com/blog/2004/08/solidimension-finally-on-a-roll/, August 2004.

18. BNET Business Network, 3D Systems Unveils its Plan to Launch the InVision LD 3-D Printer — A New Affordable Desk-Top 3-D Printer [On-line], Available at http://findarticles.com/p/articles/ mi_m0EIN/is_2005_ April_6/ai_n13559041 (September 2005).

19. E. Inui, S. Morita, K. Sugiyama and N. Kawaguchi. SHAP — A plain 3-D printer/plotter process, *Proc. Fifth Int. Conf. Rapid Prototyping*, Dayton, Ohio, USA (1994), pp. 17–26.

20. Kira Corporation, The Katana Process and Feature [On-line], available at http://www.rapidmockup.com/eg/menu2_5_e.htm (June 2007).

21. Kira Solid Center, Characteristics: Why PLT Is So Good [On-line], available at http://www.kiracorp.co.jp/EG/pro/rp/RP_characteristic.html (August 2001).

22. CAM-LEM, Inc., The CAM-LEM process [On-line], available at http:// www.camlem.com/camlemprocess.html (November 2005).

23. Ennex Corporation, All About Fabbers [On-line], available at http://www. ennex.com/fabbers/index.sht (August 2001).

24. M. Burns, K. J. Hayworth and C. L. Thomas, Offset® Fabbing, *Solid Freeform Fabrication Symposium*, University of Texas at Austin, USA (1996).

25. R. Merz, F. B. Prinz, K. Ramaswami, M. Terk and L. E. Weiss, Shape Deposition Manufacturing, *Solid Freeform Fabrication Symposium*, University of Texas at Austin, USA (1994).

26. The Shape Deposition Manufacturing Process: Methodology [On-line], available at http://www-2.cs.cmu.edu/~sdm/methodology.htm (September 2001).

27. The Shape Deposition Manufacturing Process [On-line], available at http://www-2.cs.cmu.edu/~sdm (September 2001).

PROBLEMS

1. Describe the process flow of Cubic's Laminated Object Manufacturing.
2. Describe the process flow of Stratasys' Fused Deposition Modeling.

3. Compare and contrast the laser-based LOM™ process and the FDM systems. What are the advantages (and disadvantages) for each of the systems?

4. Describe the critical factors that will influence the performance and functions of
    (i)   Cubic's LOM™,
    (ii) Stratasys FDM,
    (iii) Kira Corporation's PLT,
    (iv) 3D Systems' MJM,
    (v)   Solidscape's Benchtop System,
    (vi) Solidimensions' PSL.

5. Compare and contrast 3D Systems'MJM System with Solidscape's Benchtop System. What are the advantages (and disadvantages) for each of the systems?

6. Compare and contrast 3D Systems'MJM machines with its own liquid-based SLA® machines. What are the advantages (and disadvantages) for each of the systems?

7. Compare and contrast Cubic's LOM™ process with Kira Corporation's PLT and Ennex Corporation's Offset™ Fabber. What are the advantages (and disadvantages) for each of the systems?

8. What are the advantages and disadvantages of solid-based systems compared with liquid-based systems?

9. In the LOM™ systems, what do you think are the factors that limit the work volume of the systems?

Chapter 5
POWDER-BASED RAPID PROTOTYPING SYSTEMS

This chapter describes the special group of solid-based rapid prototyping (RP) systems, which primarily use powder as the basic medium for prototyping. Some of the systems in this group, such as Selective Laser Sintering (SLS), bear similarities with liquid-based RP systems described in Chap. 3, i.e., they generally have a laser to "draw" the part layer by layer, but the medium used for building the model is a powder instead of photo-curable resin. Others, such as three-dimensional printing (3DP) and multi-phase jet solidification (MJS), have similarities with the solid-based RP systems described in Chap. 4. The common feature among these systems described in this chapter is that the material used for building the part or prototype is invariably powder-based.

5.1. 3D SYSTEMS' SELECTIVE LASER SINTERING (SLS)

5.1.1. Company

3D Systems was founded by Charles W. Hull and Raymond S. Freed in 1986 commercializing the SLA systems. 3D Systems acquired DTM Corporation, the original company that first introduced the SLS® technology, in August 2001. DTM was first established in 1987. With financial support from BFGoodrich Company and based on the technology that was developed and patented at the University of Texas, Austin, USA, DTM shipped its first commercial machine in 1992. It had worldwide exclusive license to commercialize the SLS® technology till its acquisition by 3D Systems. The address of 3D Systems' head office is 333 Three D Systems Circle Rock Hill, SC 29730, USA.

5.1.2. Products

3D Systems has introduced several generations of the SLS system. The current generation consists of the Sinterstation® HiQ™ Series SLS® System and Sinterstation® Pro SLS® System. The Sinterstation® HiQ™ Series is the upgrade of the Sinterstation® 2500plus and Vanguard™ SLS systems. The Sinterstation® HiQ™ Series core technology is the ability to accurately monitor build temperatures throughout the build volume, automatically calibrating on a layer-by-layer basis, resulting in good part quality and highly consistent mechanical properties throughout the part geometry. Part of this series is the Sinterstation® HiQ™ + HS Series SLS® system. This system is capable of fabricating very high quality (HiQ) parts in a shorter time frame. The standard HiQ™ Series uses a 30 W CO_2 laser with a Beam Delivery System (BDS). However, the HiQ™ + HS Series employs a 50 W CO_2 laser with high-speed Celerity BDS thus explaining the high-speed (HS) capability.

The other system available is the Sinterstation® Pro SLS® System. Two models, the Sinterstation Pro 230 SLS system and the Sinterstation Pro 140 SLS system, are available. The Sinterstation Pro 230 is designed as a high-capacity, heavy-duty automated rapid manufacturing system, which can accommodate build volumes of up to 230 L. The Sinterstation® Pro SLS® System, on the other hand, are simple-to-use, integrated with a smart accessory package consisting of a material cartridge, removable build chamber, finished part retrieval station and automated material recycling module, which provides for a clean, efficient, cost effective and automated round-the-clock "lights out" operating environment. The SLS® is the only RP process with the capability to directly produce a variety of engineering thermoplastic, metallic, ceramic and composite materials all from one system. Figure 5.1 shows two of the Sinterstations from 3D Systems. Table 5.1 summarizes the main specifications of the Sinterstations.

5.1.3. Process

5.1.3.1. *The SLS® Process*

The SLS® process creates 3D objects, layer by layer, from CAD-data using powdered materials with heat generated by a CO_2 laser within

Fig. 5.1. 3D Systems' Sinterstation® Pro SLS® Systems and Sinterstation® HiQ™ Series SLS® System (courtesy 3D Systems).

Table 5.1. Summary specifications of 3D Systems' Sinterstation® HiQ™ Series SLS® and Sinterstation® Pro SLS® systems.

	Model			
	Sinterstation® HiQ™ Series SLS®		Sinterstation® Pro SLS®	
	HiQ™ system	HiQ™ + HS system	Pro 140	Pro 230
Laser type	CO$_2$			
Laser power (W)	30	50	70	
Maximum scan speed (m/s)	5 (Standard Beam Delivery System)	10 (High-Speed Celerity™ BDS)	10	
Build volume, *XYZ*, mm (liters), in.	381 × 330 × 457 (57) [15 × 13 × 18]		550 × 550 × 460 (140) [~22 × 22 × 18]	550 × 550 × 750 (250) [~22 × 22 × 30]
CAD interface	STL			
Power supply	240 VAC, 12.5 kVA, 50/60 Hz, 3-phase or 380 VAC, 12.5 kVA, 50/60 Hz, 3-phase		208 VAC, 638 A, 3-phase	

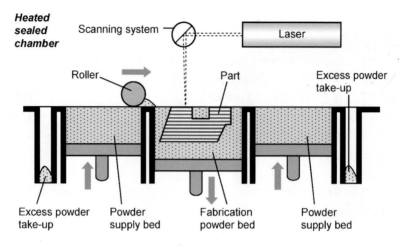

Fig. 5.2. Schematic of the selective laser sintering (SLS) process.

the Sinterstation®. CAD data files in the .STL file format are first transferred to the Sinterstation® systems where they are sliced. From this point, the SLS® process (see Fig. 5.2) begins and operates as follows:

(1) A thin layer of heat-fusible powder is deposited onto the part-building chamber.
(2) The bottom-most cross-sectional slice of the CAD part to be fabricated is selectively "drawn" (or scanned) on the layer of powder by a CO_2 laser. The interaction of the laser beam with the powder elevates the temperature to the point of melting, fusing the powder particles to form a solid mass. The intensity of the laser beam is modulated to melt the powder only in areas defined by the part's geometry. Surrounding powder remains a loose compact and serves as natural supports.
(3) When the cross section is completely "drawn", an additional layer of powder is deposited via a roller mechanism on top of the previously scanned layer. This prepares the next layer for scanning.
(4) Steps 2 and 3 are repeated, with each layer fusing to the layer below it. Successive layers of powder are deposited and the process is repeated until the part is complete.

As SLS® materials are in powdered form, the powder not melted or fused during processing serves as a customized, inherent built-in support structure. Thus, there is no need to create additional support structures within the CAD design and therefore no post-build removal of these supports is required. After the SLS® process, the part is removed from the build chamber and the loose powder simply falls away. SLS® parts may then require some post-processing or secondary finishing, such as sanding, lacquering and painting, depending upon the application of the prototype built.

The Sinterstation® Pro SLS® system contains these hardware components[1]:

(1) *Sinterstation Pro SLS System*: Manufactures part(s) from 3D CAD data.
(2) *Rapid change module (RCM)*: Build module mounted on wheels for quick and easy transfer between the Sinterstation, the offline thermal station (OTS) and the Break-Out Station (BOS).
(3) *Nitrogen generator*: Delivers a continuous supply of nitrogen to the SLS system to keep the fabrication inert and prevents oxidation.
(4) *Offline thermal station (OTS)*: Preheats the RCM before it is loaded into the system and controls the RCM cool-down process after a build has been completed.
(5) *Break-out station (BOS)*: The built parts are extracted from the powder cake here. The nonsintered powder automatically gets sifted and transferred to the integrated recycling station (IRS).
(6) *Integrated recycling station (IRS)*: The IRS automatically blends recycled and new powder. The mixed powder is automatically transferred to the SLS system.
(7) *Intelligent powder cartridge (IPC)*: New powder is loaded into the IRS from a returnable powder cartridge. When the IPC is connected to the IRS, electronic material information is automatically transferred to the SLS system.

The software and system controller for Sinterstation® HiQ™ Series SLS® System includes the proprietary SLS system software working on

Microsoft's Windows XP operating system. The software that comes with Sinterstation® Pro SLS® system includes the following:

- Build Setup and Sinter (included).
- Sinterscan™ (optional) software provides more uniform properties in X- and Y-directions and improved surface finish.
- RealMonitor™ (optional) software provides advanced monitoring and tracking capabilities.

5.1.3.2. *Materials*

The material used in SLS® system can be broadly classified into three groups: DuraForm® materials, Laserform™ materials and CastForm™ materials.[2]

The DuraForm® group consists of five types of materials: DuraForm® GF plastic, DuraForm® PA plastic, DuraForm® EX plastic, DuraForm® Flex plastic and DuraForm® AF plastic. These are discussed as follows:

(1) *DuraForm® GF plastic*: These are glass-filled polyamide (nylon) material for tough real-world physical testings and functional applications. The features of the material are as follows: excellent mechanical stiffness, elevated temperature resistance, dimensional stability, easy-to-process and relatively good surface finish. Applications for the material include housings and enclosures, consumer sporting goods, low-to-medium batch size manufacturing, functional prototypes, parts requiring stiffness and thermally stressed parts.

(2) *DuraForm® PA plastic*: These are durable polyamide (nylon) material for general physical testing and functional applications. The features of the material are as follows: excellent surface resolution and feature detail, easy-to-process, compliant with USP Class VI testing, compatible with autoclave sterilization and good chemical resistance and low moisture absorption. Applications for the material include producing complex, thin-wall ductwork, e.g., motorsports, aerospace, impellers and connectors, consumer sporting goods, vehicle dashboards and grilles, snap-fit designs, functional prototypes that approach end-use

performance properties, medical applications requiring USP Class VI compliance, parts requiring machining or joining with adhesives.

(3) *DuraForm® EX plastic*: These are impact-resistant plastic offering the toughness of injection-molded thermoplastics and are suitable for rapid manufacturing. They are available in either natural (white) or black colors. Features of the material are as follows: it offers the toughness and impact resistance of injection-molded ABS and polypropylene. Applications for the material include complex, thin-walled ductwork, motorsports, aerospace and unmanned air vehicles (UAVs), snap-fit designs, hinges, vehicle dashboards, grilles and bumpers.

(4) *DuraForm® Flex plastic*: This is a thermoplastic elastomer material with rubber-like flexibility and functionality. Features of the material are as follows: flexible, durable with good tear resistance, variability of Shore A hardness using the same material, good powder recycle characteristics, good surface finish and feature detail. The applications for the material include athletic footwear and equipment, gaskets, hoses and seals, simulated thermoplastic elastomer, cast urethane, silicone and rubber parts.

(5) *DuraForm® AF plastic*: These are polyamide (nylon) material with metallic appearance for real-world physical testing and functional use. Features of the material are as follows: metallic appearance with nice surface finish, good powder recycle characteristics, excellent mechanical stiffness, easy-to-process and dimensional stability. Applications for the material include housings and enclosures, consumer products, thermally stressed parts and plastic parts requiring a metallic appearance.

The Laserform™ group consists of three types of materials, LaserForm™ A6 (steel) material, LaserForm™ ST-200 material and LaserForm™ ST-100 material:

(1) *LaserForm™ A6 metal*: These provide the creation of complex metal parts suitable for rapid tooling and rapid manufacturing. Tooling can be quickly produced with conformal cooling channels and other complex metal parts. Features of the material include good surface finish, compatibility with machining, EDM processing and polishing, high surface hardness, excellent thermal conductivity and good "green" part

strength. Applications for the material include complex tooling inserts for injection molding and die-casting, conformal cooling or heating channels integrated into tool designs, smaller and complex geometry metal parts and low-volume metal part manufacturing.

(2) *LaserForm*™ *ST-200 material*: This is a special stainless steel composite developed to produce durable, fully dense metal parts and tooling inserts for injection-molding and die-casting applications. Features of the material are as follows: durable and functional and high-quality inserts with complex geometries. Applications for the material include direct metal parts fabrication, tooling inserts for injection molding and die casting, ideal for complex geometries and features.

(3) *LaserForm*™ *ST-100 material*: This is also a powdered stainless steel. Features of the material are as follows: part durability, suitable for metal parts and mold inserts directly from CAD files without casting or machining, able to produce up to 100,000 parts per tool as a part of a two-step process. Applications for the material include making durable metal mold inserts, bridge tooling, short run of metal parts and prototypes and complex geometries and features.

Finally, the CastForm™ PS material: This directly produces complex investment casting patterns without tooling. Features of the material includes functions such as foundry wax and "foundry friendly", low-residual ash content (less than 0.02%), short burnout cycle, easy-to-process plastic and good plastic powder recycle characteristics. Applications of the material include creating complex investment casting patterns, indirectly producing reactive metals such as titanium and magnesium, near net-shaped components, low-melt-temperature metals such as aluminum, magnesium and zinc, ferrous and nonferrous metals, smaller parts can be joined to create very large patterns and sacrificial, expendable patterns.

5.1.4.   Principle

The SLS® process is based on the following two principles:

(1) Parts are built by sintering when a CO_2 laser beam hit a thin layer of powdered material. The interaction of the laser beam with the powder

raises the temperature of the powder to the point of melting, resulting in particle bonding, fusing the particles to themselves and the previous layer to form a solid. This is the basic principle of sinter bonding.

(2) The building of the part is done layer by layer. Each layer of the building process contains the cross sections of one or many parts. The next layer is then built directly on top of the sintered layer after an additional layer of powder is deposited via a roller mechanism.

The packing density of particles during sintering affects the part density. In studies of particle packing with uniform-sized particles[3] and particles used in commercial sinter bonding,[4] packing densities are found to range typically from 50% to 62%. Generally, the higher the packing density, the better will be the expected mechanical properties. However, it must be noted that scan pattern and exposure parameters are also major factors in determining the part's mechanical properties.

5.1.4.1. *Sinter Bonding*

In the process of sinter bonding, particles in each successive layer are fused to each other and to the previous layer by raising their temperature with the laser beam above the glass-transition temperature. The glass-transition temperature is the temperature at which the material begins to soften from a solid state to a jelly-like condition. This often occurs just prior to the melting temperature at which the material will be in a molten or liquid state. As a result, the particles begin to soften and deform owing to its weight and cause the surfaces in contact with other particles or solid to deform and fuse together at these contact surfaces. One major advantage of sintering over melting and fusing is that it joins powder particles into a solid part without going into the liquid phase, thus avoiding the distortions caused by the flow of molten material during fusing. After cooling, the powder particles are connected in a matrix that has approximately the density of the particle material.

As the sintering process requires the machine to bring the temperature of the particles to the glass-transition temperature, the energy required is considerable. The energy required to sinter bond a similar layer thickness of material is approximately between 300 and 500 times higher than that required for photo-polymerization.[5,6] This high power requirement can be reduced by using auxiliary heaters to raise the powder temperature to just below the sintering temperature during the sintering process. However, an inert gas environment is needed to prevent oxidation or explosion of the fine powder particles. Cooling is also necessary for the chamber gas.

The parameters that affect the performance and functionalities are the properties of the powdered materials and its mechanical properties after sintering, the accuracy of the laser beam, the scanning pattern, the exposure parameters and the resolution of the machine.

5.1.5. Strengths and Weaknesses

The strengths of the SLS® process include:

(1) *Good part stability*: Parts are created within a precise controlled environment. The process and materials provide for functional parts to be built directly.

(2) *Wide range of processing materials*: In general, any material in powder form can be sintered on the SLS. A wide range of materials including nylon, polycarbonates, metals and ceramics are available directly from 3D Systems, thus providing flexibility and a wide scope of functional applications.

(3) *No part supports required*: The system does not require CAD-developed support structures. This saves the time required for support structure building and removal.

(4) *Little post-processing required*: The finish of the part is reasonably fine and requires only minimal post-processing such as particle blasting and sanding.

(5) *No post-curing required*: The completed laser sintered part is generally solid enough and does not require further curing.

(6) *Advanced software support*: The New Version 2.0 software uses a Windows NT-style graphical user interface (GUI). Apart from the basic features, it allows for streamlined parts scaling, advanced nonlinear parts scaling, in-progress part changes, build report utilities and is available in foreign languages.[7]

The weaknesses of the SLS® process include:

(1) *Large physical size of the unit*: The system requires a relatively large space to house it. Apart from this, additional storage space is required to house the inert gas tanks that are required for each build.
(2) *High power consumption*: The system requires high power consumption due to the high wattage of the laser required to sinter the powder particles together.
(3) *Poor surface finish*: The as-produced parts tend to have poorer surface finish due to the relatively large particle sizes of the powders used.

5.1.6. Applications

The SLS® system can produce a wide range of parts in a broad variety of applications, including the following:

(1) *Concept models*: Physical representations of designs used to review design ideas, form and style.
(2) *Functional models and working prototypes*: Parts that can withstand limited functional testing, or fit and operate within an assembly.
(3) *Polycarbonate (RapidCasting™) patterns*: Patterns produced using polycarbonate and then cast in the metal of choice through the standard investment casting process. These build faster than wax patterns and are ideally suited for designs with thin walls and fine features. These patterns are also durable and heat resistant.
(4) *Metal Tools (RapidTool™)*: Direct rapid prototype of tools of molds for small or short production runs.

5.1.7. Examples

5.1.7.1. *Los-Angeles-Based TESTA Architecture/Design Utilizes SLS for Large-Scale Models of Carbon Tower Prototype*

In the "Extreme Textiles: Designing for High Performance" exhibition, the design team of TESTA Architecture/ Design utilized SLS® systems to build the prototype of the Carbon Tower Prototype.

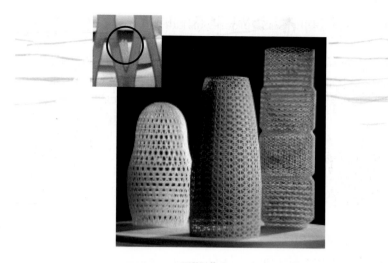

Fig. 5.3. Carbon Tower Prototype with mirco-detailed parts and features. Inset shows the human figures built into the model to indicate the scale of model (courtesy 3D Systems).

The Carbon Tower Prototype demonstrates a new construction system for an all-carbon and glass-fiber high-rise office building that is woven, knitted and braided. This 40-story prototype, engineered with ARUP software, is several times lighter and stronger than conventional steel and concrete structures. Given the intricate nature of the tower design, it was not possible to construct architectural models with traditional modeling materials and techniques. Using 3D Systems' SLS® technology, the very large model measuring 152.4 cm (60 in.) by 35.6 cm (14 in.) in diameter was fabricated in only five interlocking pieces from digital files prepared by TESTA.[8]

It has been proven that using SLS to fabricate complex and intricate structures rapidly is possible. Figure 5.3 shows such a part prototyped of the carbon tower built with mirco-detailed parts.

5.1.7.2.   *Frontloading — The Crossblade Outside Mirror*

General contractors Bertrandt developed the Smart Crossblade for Smart GmbH in just six months and built an exclusive series of 2000 with Binz, their production partners.[9] The Smart Crossblade was designed based on

Fig. 5.4. The Smart Crossblade. The inserts show the model of the door pillar with SLS parts (lighter middle portion) as the mounting of the mirror base and the new mirror system of the Smart Crossblade (courtesy 3D Systems).

the concept of a Smart convertible platform, which resulted in a final car design that had no roof, no convertible top, no screens and no doors (Fig. 5.4). The new concept raised a number of issues, one of which regarded the mirror system. The designer decided to develop a new mirror system to match this new concept. The SLS® system and DuraForm™ PA material were used in the building the model for mounting the upper and lower shells (holding the rear-view mirrors) and attaching them to the door pillars. This helped shorten the lead time required for the injection molding tools.

As the development of the entire car was tight, it was necessary to perform functional testing, even before the injection molding tool was made. A pendulum impact test ECE R 46 was required to investigate where the mirrors would fold when a collision occurs with a pedestrian in order to minimize injuries. A laser sintered part was built and positioned on the test rig. After a few minor modifications, the prototype passed the tests for the driver's side. Thus, actual injection molded parts were fabricated for the final ECE R 46 testing.

5.1.7.3. *4D Concepts for Hekatron*

Hekatron, a German-based company in Sulzburg, is a maker of safety systems and OEM supplier for a string of renowned firms. The company developed a 220-V socket for a smoke alarms and needed a several thousand units for initial sampling and market surveys. The real challenge turned out to be the plastic casing (90 mm diameter) and its intricate interiors with many openings, contra-rotating cores, thin-walled ribs — in short, a component with a highly complex geometry that needed to be made within the shortest possible time.

4D Concepts in Gross-Gerau was appointed by Hekatron to do the task. 4D Concepts used the Vanguard HS SLS system with LaserForm steel material, allowing them to make complicated and difficult tool geometry within a few weeks and at competitive terms. Just three weeks after receiving the 3D CAD data (IGES) from Hekatron's development department, 4D Concepts delivered the first injection molded components. Using SolidWorks software, the provider first delivered the mold design drawn from the original data and then used their 3D Systems SLS system to make both mold halves in the dimensions $120 \times 120 \times 120$ mm (Fig. 5.5). The actual build time was less than 24 h.

The Vanguard HS SLS system was used to build the required geometry layer by layer using a CO_2 laser with the LaserForm ST-200 steel powder. The sintered inserts were then infiltrated with bronze to obtain a dense mold. When compared directly with a conventionally made solid steel mold, the cost was lowered by a staggering 40%. The benefits of this method were particularly evident when it came to geometries requiring a great deal of erosion.[10]

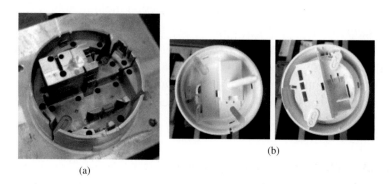

(a)

(b)

Fig. 5.5. (a) Metal injection mold produced by SLS® and (b) plastic parts produced by the metal mold (courtesy 3D Systems).

5.1.8. Research and Development

Primary research continues to focus on new and advanced materials while further improving and refining SLS® process, software and system. A new software module, trademarked OptiScan™, was introduced to reduce the build time of parts up to 34%. OptiScan™ works solely with the Laser Sintering Software and optimizes the scanning process by combining cross sections on each layer to reduce the time taken to build a part.

5.2. Z CORPORATION'S THREE-DIMENSIONAL PRINTING (3DP)

5.2.1. Company

Originally invented, patented and developed at the Massachusetts Institute of Technology (MIT) in 1993, 3DP™ technology forms the basis of Z Corporation's licensed prototyping process. Z Corp. pioneered the commercial use of 3DP technology, developing 3D printers that leading manufacturers used to produce early concept models and product prototypes for a broad range of applications.

Z Corporation was incorporated in 1994 by Hatsopoulos, Walter Bornhost, Tim Anderson and Jim Brett. It commercialized its first 3D printer, the Z™402 System, based on the 3DP technology in 1997. In 2000 Z Corp. launched its first color 3D printer and subsequently introduced high definition 3D printing (HD3DP) in 2005. Z Corp.'s distributors are found in Australia, Benelux area, France, Germany, Hong Kong, Italy, Japan, Korea, Malaysia, Russia, Singapore, South Africa, Spain, Taiwan and the United Kingdom. Its worldwide headquarters is at 32 Second Avenue Burlington, MA 01803 USA.

5.2.2. Products

Z Corporation's current products are the ZPrinter® 310 Plus, ZPrinter® 450 and Spectrum Z™510 systems (see Fig. 5.6, specifications are summarized in Table 5.2). The ZPrinter® 310 Plus is the enhancement of the entry level ZPrinter 310, which incorporates many key features of its premium

Fig. 5.6. From the left, the ZPrinter® 310 Plus, ZPrinter® 450 and Spectrum Z™510 systems (courtesy Z Corporation).

Spectrum Z™510 printer — including HD3DP, a heated build chamber and advanced firmware. The addition of a heated build chamber allows the shortening of print time by up to 25%. With the incorporation of firmware the finish of the printed parts is smoothened and the build resolution is enhanced by up to 50%.[11]

The ZPrinter® 450 simplifies the printing process with complete automated operations. New features include automatic setup, powder loading and self-monitoring of materials and print status. This 3D printer also automatically removes and recycles loose powder. Its automated operation and new quick-change material cartridges reduce touch time by 40%. The ZPrinter® 450 also produces color in a new way with a single tri-color print head instead of multiple print heads, reducing cost and enabling 66% faster changes. It is quiet, uses safe build materials, employs "negative pressure" and a closed-loop system to contain airborne particles and produces zero liquid waste. New "no-touch" powder and binder cartridges enable clean loading of build materials. An integrated fine-powder removal chamber reduces the footprint of the system. Users can control the ZPrinter® 450 from either the desktop or 3D printer itself. New ZPrint™ software lets users monitor powder, binder and ink levels from their desktops and remotely read the machine's LCD display. Similar to its predecessors, the ZPrinter® 450 produces models and prototypes 5 to 10 times faster than competing printers at half the cost.

Table 5.2. Specifications of Z Corporation's 3D Printers (source from Z Corp.)

	Model		
	ZPrinter® 310 Plus	ZPrinter® 450	Spectrum Z™510
Build speed		2–4 layers/min.	
Build size, mm	203 × 254 × 203	203 × 254 × 204	254 × 356 × 203
Material options	High performance composite, elastomeric, direct casting, investment casting	High performance composite	High performance composite, elastomeric, direct casting
Layer thickness (mm):	0.089–0.203	0.089–0.102	0.089–0.203
Resolution	300 × 450 dpi	300 × 450 dpi	600 × 540 dpi
Number of printheads, jets	1 print-head, 304 jets	2 print-heads, 604 jets	4 print-heads, 1216 jets total
System software	Z Corp.'s proprietary system software accepts solid models in STL, VRML and PLY file formats as input. ZPrint software features3D viewing, text labeling and scaling functionality		
Equipment dimensions (mm × mm × mm):	740 × 860 × 1090	1220 × 790 × 1400	1070 × 790 × 1270
Equipment weight (kg)	115	193	204
Power requirements	115 V, 4.3 A or 230 V, 2.4 A	100 V, 14.4 A or 115 V, 14.0 A or 230 V, 6.2 A	100 V, 7.8 A or 115 V, 6.8 A or 230 V, 3.4 A
Network connectivity	TC/ IP 100/10 base T		
Workstation compatibility	Windows® 2000 Professional and Windows® XP Professional		
Regulatory compliance	CE, CSA		

The Spectrum Z510 System introduces HD3DP™ for RP to designers, engineers and product developers. In addition to providing high-fidelity, 24-bit color, the system supports 600 × 540 dpi print-head resolution, large build sizes (254 × 356 × 203 mm or 10 × 14 × 8 in.), improved surface finishes, smaller feature resolution and lower operating costs. Color capabilities allow for better representation of assemblies, enhanced texture mapping and vibrant product coloring.[12] The system's software also enables labeling for improved communication of designs. Figure 5.6 shows a photograph of Z Corporation's ZPrinter® 310 Plus, ZPrinter® 450 and Spectrum Z™510 systems.

Z Corporation also has an accessory, the ZW4 Automated Waxer, which allows printed parts to be infiltrated with paraffin wax to enhance strength, provide uniform part finish and color, or to create patterns suitable for investment casting.

One of the important advantages of Z Corp.'s 3D printing systems is the array of material properties available. Users can infiltrate resins into the part allowing the part to take on the physical properties of a cured resin. This capability provides users with greater versatility without having to change the primary materials in the 3D printer. For concept and visualization models, users can infiltrate parts with wax or fast-curing, one-part resins. Additionally, infiltrants can be used to significantly improve the durability, humidity resistance and high-temperature properties of Z Corp. parts during airflow, vibration and other rigorous environmental tests. The infiltration can create parts with the qualities of rubber too.[13] Users can also infiltrate parts produced as tooling or fixtures with high-strength epoxy, creating very hard, rigid parts in a fraction of the time that it takes to have them machined. The 3D printed molds can also be used for prototype blow molding and thermoforming, saving time and prototype tooling costs.

Z Corporation recently launched a new high-performance composite material that introduces the world's easiest, safest and greenest post-processing option for finishing monochrome 3D printed models by a quick mist with tap water. Monochrome parts made with zp®140 require no further treatment or equipment to reach their finished strength. The process is quick, perfectly safe and virtually cost-free. In addition to the water curing capability, zp®140 delivers the brightest whites (180% whiter than before), fastest production and lowest cost per finished part. It hardens in less than half the time of previous materials, enabling users to print detailed parts faster and show them off sooner.

5.2.3. Process[14]

The 3DP technology creates 3D physical prototypes by solidifying layers of deposited powder using a liquid binder. The 3DP process is shown in Fig. 5.7.

(1) The machine spreads a layer of powder from the feed box to cover the surface of the build piston. The printer then prints binder solution onto the loose powder, forming the first cross-section. For multi-colored parts, each of the four print heads deposits a different color

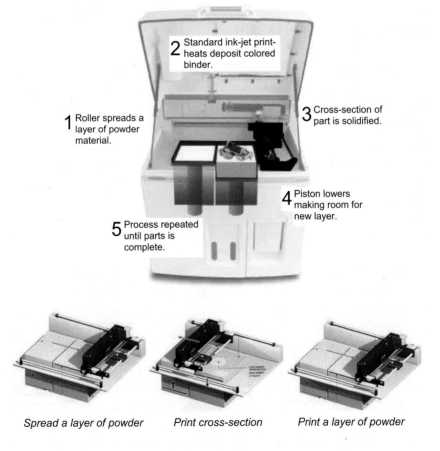

2 Standard ink-jet print-heats deposit colored binder.

1 Roller spreads a layer of powder material.

3 Cross-section of part is solidified.

4 Piston lowers making room for new layer.

5 Process repeated until parts is complete.

Spread a layer of powder   *Print cross-section*   *Print a layer of powder*

Fig. 5.7. Illustration of 3D printing (courtesy Z Corporation).

binder, mixing the four color binders to produce a spectrum of colors that can be applied to different regions of a part.

(2) The powder is glued together by the binder at where it is printed. The remaining powder remains loose and supports the following layers that are spread and printed above it.

(3) When the cross section is complete, the build piston is lowered, a new layer of powder is spread over its surface and the process is repeated. The part grows layer by layer in the build piston until the part is complete, completely surrounded and covered by loose powder. Finally, the build piston is raised and the loose powder is vacuumed away, revealing the complete part.

(4) Once a build is complete, the excess powder is vacuumed away and the parts are lifted from the bed. Once removed, parts can be finished in a variety of ways to suit your needs. For a quick design review, parts can be left raw or "green". To quickly produce a more robust model, parts can be dipped in wax. For a robust model that can be sanded, finished and painted, the part can be infiltrated with a resin or urethane.

5.2.3.1. *Materials* [13]

(1) *High performance composite material*: This can be used to make strong, high-definition parts and is the material of choice for printing color parts. It consists of a heavily engineered plaster material with numerous additives that maximize surface finish, feature resolution and part strength. This material is ideal for high strength requirements, delicate or thin-walled parts, color printing and accurate representation of design details.

(2) *Snap-fit material*: This has been optimized for infiltration with the Z-Snap™ epoxy to create parts with plastic-like flexural properties, which are ideal for snap-fit applications. It is a plaster-based system that produces parts with a more porous matrix, allowing them to absorb a greater quantity of the Z-Snap infiltrant.

(3) *Elastomeric material*: This has been optimized for infiltration with an elastomer to create parts with rubber-like properties. The material system consists of a mix of cellulose, specialty fibers and other additives

that combine to provide an accurate part capable of absorbing the elastomer.

(4) *Investment casting material*: This can be used to quickly fabricate parts that can be dipped in wax to produce investment casting patterns. The material consists of a mix of cellulose, specialty fibers and other additives that combine to provide an accurate part while maximizing the absorption of wax and minimizing residue during the burnout process.

(5) *Direct casting material*: This can be used to create sand casting molds for nonferrous metals. This material is a blend of foundry sand, plaster and other additives that have been combined to provide strong molds with good surface finish. It is designed to withstand the heat required to cast nonferrous metals.

5.2.4. Principle

3DP creates parts by a layered printing process and adhesive bonding, based on sliced cross-sectional data. A layer is created by adding another layer of powder. The powder layer is selectively joined where the part is to be formed by "ink-jet" printing of a binder material. The process is repeated layer by layer until the part is complete.

As described in Sec. 5.1.4, the packing density of the powder particle has a profound impact on the results of the adhesive bonding, which in turn affect the mechanical properties of the model. Similar to the powders used on the SLS, packing densities are from 50% to 62%.[4] When the ink droplet impinges on the powder layer, it forms a spherical aggregate of binder and powder particles. Capillary forces will cause adjacent aggregates, including that of the previous layer, to merge. This will form the solid network, which will result in the solid model. The binding energy for forming the solid comes from the liquid adhesive droplets. This energy is composed of two components, one is its surface energy and the other is its kinetic energy. As this binding energy is low, it is about 10^4 times more efficient than sinter binding in converting powder to a solid object.[3]

Parameters that influence the performance and functionalities of the process are the properties of the powder, the binder material and the accuracy of the XY table and Z-axis control.

5.2.5.   Strengths and Weaknesses

The key strengths of the 3DP systems are as follows:

(1) *High speed*: 3DP are HS printers. Each layer is printed in seconds, reducing the prototyping time of a hand-held part to 1–2 h.
(2) *Versatile*: Parts are currently used for automotive, packaging, education, footwear, medical, aerospace and telecommunications industries. Parts are used in every step of the design process for communication, design review and limited functional testing. Parts can be infiltrated if necessary, offering the opportunity to produce parts with a variety of material properties to serve a range of modeling requirements.
(3) *Simple to operate*: The office compatible Z-Corp. system is straight-forward to operate and does not require a designated technician to build a part. The system is based on the standard, off the shelf components developed for the ink-jet printer industry, resulting in a reliable and dependable 3D printer.
(4) *Minimal wastage of materials*: Powder that is not printed during the cycle can be reused.
(5) *Color*: It enables complex color schemes for RP parts from a full 24-bit palette of colors to be made possible.

The limitations of the 3DP systems include the following:

(1) *Limited functional parts*: Relative to the SLS, parts built are much weaker, thereby limiting the functional testing capabilities.
(2) *Poor surface finish*: Parts built by 3D printing have relatively poorer surface finish and post-processing is frequently required.

5.2.6.   Applications

The 3DP process can be used in the following areas:

(1) *Concept and functional models*: Creating physical representations of designs used to review design ideas, form and style. With the infiltration of appropriate materials, it can also create parts that are used for functional testing, fit and performance evaluation.

(2) *CAD-casting metal parts*: CAD-casting is a term used to connote a casting process where the mold is fabricated directly from a computer model with no intermediate steps. In this method, a ceramic shell with integral cores may be fabricated directly from a computer model. This results in tremendous streamlining of the casting process.

(3) *Direct metal parts*: Metal parts in a range of material including stainless steel, tungsten and tungsten carbide can be created from metal powder with 3DP process. Printed parts are post-processed using techniques borrowed from metal injection molding.

(4) *Structural ceramics*: 3DP can be used to prepare dense alumina parts by spreading submicron alumina powder and printing a latex binder. The green parts are then isostatically pressed and sintered to densify the component. The polymeric binder is then removed by thermal decomposition.

(5) *Functionally gradient materials*: 3DP can create composite materials as well. For example, ceramic mold can be 3D printed, filled with particulate matter and then pressure infiltrated with a molten material. Silicon carbide reinforced aluminum alloys can be produced directly by 3D printing a complex SiC substrate and infiltrating it with aluminum, allowing localized control of toughness.

5.2.7. Examples

5.2.7.1. *Sports Shoe Industry*

The 3D printer has been used by designers, marketers, manufacturers and managers in the footwear industry. Leading athletic shoe companies, such as Adidas, New Balance and Wolverine, have used this RP system to radically reduce prototype development time and communicate in new ways. With the introduction of full color printing, appearance of prototypes can be close to the actual product, so that more comprehensive communication on design can be done (Fig. 5.8).[15] Shoe industries are faced with constantly changing consumer preferences and have to react quickly to stay ahead of the business. With the 3D printer, lead times are drastically reduced, beating the competition to the shelves with the latest design trends while avoiding an excess inventory of unwanted designs.

Fig. 5.8. Actual production shoe with full-color prototype created with the Z Corporation system (courtesy of Z Corporation).

5.2.7.2. *Automotive Industry*

Leading automotive companies, such as Ford, Benteler, F1 Racing and Porsche, have been utilizing Z Corporation systems to enhance internal communication on product concepts. Functional testing can be done on the parts printed with Z Corporation's printer with the use of infiltration resins. Infiltrated parts can be machined, drilled and tapped. Additionally, infiltrants can be used to significantly improve the durability, humidity resistance and high-temperature properties of Z Corp. parts during airflow, vibration and other rigorous testing. Z Corp.'s ZCast™ Direct Metal Casting process allows engineers to pour metal directly into a mold printed on a Z Corp. 3D Printer, eliminating the need for a pattern (Fig. 5.9). In addition, Z Corp. parts can be used as patterns in the traditional investment casting and sand casting processes for the production of metal prototype parts.[15]

5.3. EOS'S EOSINT SYSTEMS

5.3.1. Company

EOS was founded in 1989 and is today a world leader in laser sintering systems. Laser sintering is the key technology for manufacturing — the

Fig. 5.9. Brembo Brake prototype printed on Spectrum Z™510 (courtesy of Z Corporation).

fast, flexible and cost-effective production of products, patterns, or tools. The technology accelerates product development and optimizes production processes. EOS gained market leadership by assisting numerous industries in Europe, North America and Asia in making use of laser sintering in their product development and manufacturing. Its address is at EOS GmbH Electro Optical Systems, Robert-Stiring-Ring 1, D-82152 Krailling, Germany.

5.3.2. Products

EOSINT systems share identical software but offer a different spectrum of building materials. The systems EOSINT P and FORMIGA P are used specifically for rapid fabrication of plastic parts. EOSINT M 270 focuses specifically for direct sintering of metal powder. EOSINT S 750 is used to build sand cores and molds for metal casting. Table 5.3 provides the specifications of the EOS RP systems.

The FORMIGA P 100 represents laser sintering in the compact class. With a build envelope of $200 \times 250 \times 330$ mm, the FORMIGA P 100 produces plastic products from polyamide or polystyrene within a few hours and

Table 5.3. Specifications of EOSINT machines (courtesy EOS GmbH).

	Model				
	FORMIGA P 100	EOSINT P 390	EOSINT P 730	EOSINT M 270	EOSINT S 750
Effective build volume, mm	200 × 250 × 330	340 × 340 × 620	700 × 380 × 580	250 × 250 × 215	720 × 380 × 380
Building speed (material dependent)	Up to 24 mm height/h	Up to 35 mm height/h	Up to 35 mm height/h	2–20 mm³/s	Up to 2,500 cm³/h
Layer thickness (material dependent)	Typically 0.1 mm	Typically 0.1–0.15 mm	Typically 0.12 mm	20–100 μm	0.2 mm
Laser type	CO_2, 30 W	CO_2, 50 W	CO_2, 2 × 50 W	Yb-fiber laser, 200 W	CO_2, 2 × 100 W
Precision optics	F-theta lens	F-theta lens	F-theta lens	F-theta lens, high-speed scanner	2 × F-theta lens, 2 × high-speed scanner
Scan speed, m/s	Up to 5	Up to 6	Up to 2 × 6	Up to 7.0	Up to 3.0
Power supply	16 A	32 A	32 A	32 A	32 A
Power consumption	2 kW (nominal)	2 kW (nominal)	3.5 kW (nominal)	5.5 kW (max)	6 kW (avg.), 12 kW (max.)
Nitrogen generator	Integrated (optional)	Integrated (optional)	Integrated	Standard	
Compressed air supply	Min 6,000 hPa; 0.2 m³/h	Min 5,000 hPa; 6 m³/h	Min 6,000 hPa; 20 m³/h	7,000 hPa; 20 m³/h	
Approximate installation space, mm	3,200 × 3,500 × 3,000	4,300 × 3,900 × 3,000	4,800 × 4,800 × 3,000	3,500 × 3,600 × 2,500	4,500 × 4,600 × 2,700
PC	Current Windows operating system				
Software	EOS RP Tools; Magics RP (Materialise)				
CAD interface	STL. Optional: converter to all common formats				
Network	Ethernet				
Certification	CE	CE, NFPA	CE, NFPA	CE, NFPA	CE, NFPA

directly from CAD data. The machine is ideally suited for the economic production of small series and individualized products with complex geometry — requirements that apply among others to the medical device industry as well as for high-value consumer goods.[16] At the same time, it provides capacity for the quick and flexible production of fully functional prototypes and patterns for plaster, investment and vacuum casting. With turnover times of less than 24 h, the FORMIGA P 100 integrates itself perfectly in a production environment that requires the highest level of flexibility. The system distinguishes itself also by having comparatively lower investment costs.

The EOSINT P 390 offers economical solutions for a broad range of applications. It is a highly productive system for processing thermoplastics. The machine manufactures plastic products of any complexity from polyamide or polystyrene materials. Especially for individualized products or for products with complex geometries, the machine unleashes its full potential. IntelliScan 20, a digital scanner of the latest generation, exposes the respective layers in a so-far unknown speed and stands for highest precision. The build volume of the machine is 340 × 340 × 620 mm. Thus, the system also creates larger plastic components in a single process. The volume also allows the efficient production of a broad range of plastic goods from medical device, automotive to aerospace industries.

EOSINT P 730 is an advancement of the EOSINT P 700, the worldwide first double-laser system for laser sintering of plastics. It is one of the largest plastic laser sintering systems available on the market. EOSINT P 730 is up to 40% more productive compared with the EOSINT P 700 and combines increased productivity with highest part quality. The system is ideally suited for the economic production of small series and individualized products, especially with complex geometries. At the same time, it provides the capacity for fast and flexible creation of prototypes or patterns for investment and vacuum casting. Within a very short time, the machine also produces large and complex plastic products or castings. Figures 5.10 and 5.11 show the FORMIGA P 100, EOSINT P 390 and EOSINT P 730 machines.

EOSINT M 270 (see Fig. 5.12) builds metal parts using Direct Metal Laser Sintering (DMLS). The technology fuses metal powder into a solid

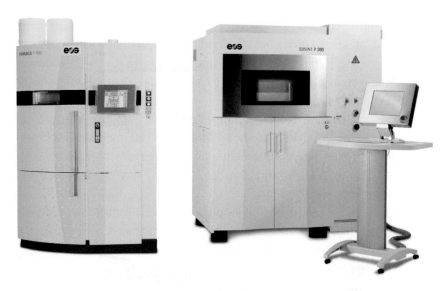

Fig. 5.10. The FORMIGA P 100 machine (left) and the EOSINT P 390 machine (right) (courtesy EOS GmbH).

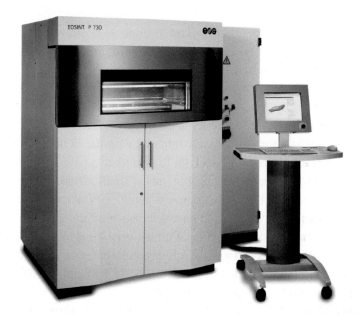

Fig. 5.11. The EOSINT P 730 machine (courtesy EOS GmbH).

Fig. 5.12. The EOSINT M 270 machine (courtesy EOS GmbH).

part by melting it locally using a focused laser beam. Even highly complex geometries are created directly from 3D CAD data automatically, in just a few hours and without any tooling. It is a net-shape process, producing parts with high accuracy and detail resolution, good surface quality and excellent mechanical properties. A wide variety of materials can be processed by the EOSINT M 270, ranging from light alloys via steels to superalloys and composites. EOS has developed novel alloys especially for the DMLS process and has also optimized and qualified standard industrial materials such as stainless steels for this machine. The material currently used is a special alloy mixture comprising mainly bronze and nickel, developed by Electrolux Rapid Prototyping and licensed exclusively to EOS. This metal can be sintered at without pre-heating and exhibits negligible net shrinkage during the sintering process. The DirectMetal™ 50-V2 is a fine-grained steel-based metal powder with a maximum particle size of 50 microns used for DirectTool™ applications

Fig. 5.13. EOSINT S 750 machine (courtesy EOS GmbH).

on the EOSINT M laser sintering system for very high precision, good detail resolution and smooth finishing.[17] Further materials are continually being developed and qualified.

EOSINT S 750 (see Fig. 5.13) is the only double-laser sintering system worldwide for the processing of Croning molding material. Using the DirectCast method, the system builds cores and molds for sand casting at a build speed of up to 2,500 cm³/h without any additional tooling. Sand parts of any complexity are built layer by layer, with high accuracy, detail resolution and surface quality. The maximum part size adds up to 720 × 380 × 380 mm. The resulting cores or core packages are realized with significant savings in time and costs compared with conventional technologies. Usually they also consist of fewer parts, which are thus assembled faster and more precisely.

DirectCast with EOSINT S 750 enables the production of castings in batch sizes that would be extremely laborious, economically unviable or even impossible to manufacture with conventional techniques. It enables the production of high-quality castings for the engine development, for pumps or hydraulic applications. These castings can be used as fast, cost-effective prototypes or as final products. The technology allows foundries to cater for new trends such as spare parts on demand. EOSINT S 750 uses different Croning sands, which are commonly used in foundries. These sands have been optimized by EOS for the DirectCast application. Laser sintering of foundry sand achieves excellent results for light-weight constructions using aluminum or magnesium. The technology also opens up new applications for cast iron and steel.

5.3.3. Process

The EOSINT process applies the following steps to creating parts: processing data, preparing new layers, scanning and removal of unsintered powder[18]:

(1) Firstly, the part is created in a CAD system on a workstation. Next, the CAD data are processed by EOS's software EOSOFT and converted to the cross-sectional format that EOSINT machines use to control the sintering process.
(2) At the build stage, a new layer of powder covers the platform. The laser scans the new powder layer and sinters the powder together according to the cross-sectional data. Simultaneously, the new layer is joined to the previous layer.
(3) When the sintering of the cross section is complete, the elevator lowers and another new layer is prepared for the next step. The processes are repeated till the part is completed.
(4) The powder around the part is then removed.

The EOSINT system typically contains a Silicon Graphics workstation and an EOSINT machine including a working platform, a laser and an optical scanner calibration system.

5.3.3.1. *Materials*

(1) *PA 2200*: Fine polyamide. This is a polyamide fine powder. This is the typical material for fully functional prototypes with high-end

finishes from the process. They can easily withstand high mechanical and thermal load.

(2) *PrimePart*: Fine polyamide. This is suitable for the rapid and cost-effective manufacture of models, functional prototypes, end products as well as spare parts that have a balanced relationship between mechanical strength and elasticity over a wide temperature range.

(3) *PA 2210 FR*: Fine polyamide. This is for the manufacture of flame resistant parts with high mechanical properties. PA 2210 FR contains a chemical flame retardant. In the case of fire, a carbonating coating develops at the surface of the part, isolating the plastic below. PA 2210 FR is free of halogens.

(4) *PA 3200 GF*: Glass-filled fine polyamide. Typical applications of the material are strong housings and thermally stressed parts.

(5) *Alumide®*: Aluminum filled fine polyamide. A typical application for Alumide® is the manufacture of stiff parts of metallic appearance for applications in automotive manufacture (e.g., wind tunnel tests or parts that are not safety relevant), for tool inserts for injecting and molding small production runs, for illustrative models (metallic appearance), etc. Surfaces of parts made of Alumide® can be finished by grinding, polishing, or coating. An additional advantage is that low-tool-wear machining is possible.

(6) *CarbonMide*: Carbon fiber filled polyamide. CarbonMide® has outstanding mechanical properties characterized by extreme stiffness and strength. Typical applications of the material are fully functional prototypes with high end finish for wind tunnel tests or other aerodynamic applications.

(7) *PrimeCast 101*: Polystyrene. The typical application for the material is the production of lost patterns for the plaster casting process. Generally, PrimeCast 101 is also suitable for ceramic shell casting. However, special measures against shell cracking are necessary. Another application of PrimeCast 101 is the production of master patterns for vacuum casting.

(8) *DirectMetal 20*: This is a very fine-grained, bronze-based, multi-component metal powder. The resulting parts offer good mechanical

properties combined with excellent detail resolution and surface quality. The surfaces can be easily post-processed by shot-peening and can be easily polished. The special powder mixture contains different components, which expand during the laser sintering process, partially compensating for the natural solidification shrinkage and thereby enabling a high part accuracy to be achieved. It is ideal for most prototype injection molding tools (DirectTool) and for many functional metal prototype applications (DirectPart). It is particularly suitable for larger tools and parts and also offers a broad window of usable process parameters including a wide range of achievable mechanical properties and build speeds. Standard parameters use 20 μm layer thicknesses for the skin and 60 μm layers for the core, but for faster building the entire part can be built using 40 μm layers for the skin and 80 μm layers for the core. Areas built with core parameters have a porous structure, but the combination of skin and core produces a strong total part. Parts built from DirectMetal 20 are also corrosion resistant.

(9) *DirectSteel H20*: This is a very fine-grained steel-based, multi-component metal powder, which offers high strength, hardness, wear resistance and surface density. The resulting parts have properties similar to conventional tool steels and can be polished to an excellent, pore-free surface finish. This material is particularly suitable for DirectTool applications such as injection molds, pressure die-casting tooling and other applications where high strength, wear resistance and surface quality are important. Standard parameters use 20 μm layer thickness for the outer skin and 60 μm layers for the core. To achieve the high density and hardness, the skin area is completely melted, which results in slower build speed than for DirectMetal 20. Mechanical properties are generally higher in the building plane (*XY*) than perpendicular to the building plane (*Z*). Typical applications include heavy duty injection molds and inserts, with achievable tool life of up to millions of parts; die casting molds for small series of up to several thousand parts in light alloys; metal stamping and other heavy duty tooling applications; direct manufacturing of heavily loaded functional prototypes.

(10) *EOS StainlessSteel 17-4*: This is a pre-alloyed stainless steel in fine powder form. Its composition corresponds to US classification 17-4 PH and European 1.4542 and fulfills the requirements of AMS 5643 for Mn, Mo, Ni, Si, C, Cr and Cu. The resulting part has very good corrosion resistance and mechanical properties, especially excellent ductility in laser processed state and is widely used in a variety of engineering applications. It is ideal for many part-building applications (DirectPart) such as functional metal prototypes, small series products, individualized products, or spare parts. Standard processing parameters use full melting of the entire geometry with 20 μm layer thickness, but it is also possible to use skin and core building style to increase the build speed. Using standard parameters, the mechanical properties are fairly uniform in all directions. The resulting part can be welded, machined, microshot-peened, polished and coated if required. Unexposed powder can be reused without restriction or refreshing. Typical applications include parts requiring high corrosion resistance, sterilizability and particularly high toughness and ductility.

(11) *EOS CobaltChrome MP1*: This is a fine powder mixture for laser sintering on EOSINT M 270 systems, which produces parts in a cobalt–chrome–molybdenum-based super alloy. This material has excellent mechanical properties, is both corrosion resistance and temperature resistance. Such alloys are commonly used in biomedical applications such as dental and medical implants. They are also used for high-temperature engineering applications such as in aero engines. The chemistry of EOS CobaltChrome MP1 conforms to the composition UNS R31538 of high carbon CoCrMo alloy. It is nickel-free (<0.1% nickel content), sterilizable and suitable for biomedical applications. The laser-sintered parts are characterized by a fine, uniform crystal grain structure. They fully meet the requirements of ISO 5832-4 and ASTM F75 for cast CoCrMo implant alloys, as well as the requirements of ISO 5832-12 and ASTM F1537 for wrought CoCrMo implants alloys except remaining elongation. The remaining elongation can be increased by hot isostatic pressing (HIP) to meet this standard. Typical applications include prototype or one-off biomedical implants, e.g., spinal and

dental; parts requiring high mechanical properties at elevated temperatures (500–1000°C) and with good corrosion resistance, e.g., turbines and parts for engines and parts having very small features such as thin walls and pins, which require particularly high strength or stiffness.

(12) *EOS CobaltChrome SP1*: This is a fine powder mixture, which produces parts in a cobalt–chrome–molybdenum-based superalloy. In addition to excellent mechanical properties, corrosion resistance and temperature resistance, it has been especially developed to fulfill the requirements of dental restorations which have to be veneered with dental ceramic material. Typical applications include dental restorations (crowns, bridges etc.)

(13) *EOS Titanium Ti64/Ti64ELI. Titanium Ti64*: This is a pre-alloyed Ti6AlV4 alloy in fine powder form. This well-known light alloy has excellent mechanical properties and corrosion resistance combined with low specific weight and biocompatibility. The ELI (extra-low interstitial) version has particularly low levels of impurities. Typical applications include parts requiring a combination of high mechanical properties and low specific weight, e.g., structural and engine components for aerospace and motor racing applications; biomedical implants.

(14) *EOS MaragingSteel MS1*: This is a maraging steel in fine powder form. Its composition corresponds to US classification 18 Maraging 300, European 1.2709 and German X3NiCoMoTi 18-9-5. This kind of steel is characterized by having very high strength combined with high toughness. It is easily machinable after the building process and can be easily post-hardened up to approximately 55 HRC by a simple thermal age-hardening process. This kind of steel is conventionally used for complex tooling as well as for high-performance industrial parts, e.g., in aerospace applications. Typical applications include heavy duty injection molds and inserts for molding, with achievable tool life of up to millions of parts; die casting molds for small series of up to several thousand parts in light alloys; direct manufacture of heavily loaded functional metal prototypes.

(15) *Foundry Sand*: Ceramics 5.2 is a phenolic resin-coated aluminum silicate sand (synthetic mullite). The material is suited for generative

fabrication of complex sand cores and sand molds for all casting applications. Owing to its high heat capacity and low-temperature expansion, this ceramic sand can be used for high-temperature casting. The material is suited for generative fabrication of complex sand cores and sand molds for all casting applications.

5.3.4. Principle

The principle of the EOSINT systems is based on the laser sintering principle and layer manufacture principle similar to that of the SLS® described in Sec. 5.1. The parameters that influence the performances and functionalities of the EOSINT systems are properties of the powder materials, the laser and the optical scanning system, the precision of the working platform and working temperature.

5.3.5. Strengths and Weaknesses

The key strengths of the EOSINT machines are as follows:

(1) *Good part stability*: Parts are created in a precise controlled environment. Functional parts can be built directly.
(2) *Wide range of processing materials*: A wide range of materials including polyamide, glass-filled polyamide composite, polystyrene, metals and foundry sands are available for a wide range of applications.
(3) *Support structures not required*: The systems do not require support structures or use only simplified support structures as in the case of the Direct Croning Process®. This simplifies the post-processing.
(4) *Little post-processing required*: The surface finish for the as-produced part is very good, thus requiring only minimal post-processing.
(5) *High accuracy*: For the EOSINT P system, the polystyrene it uses can be laser-sintered at a relatively low temperature, thereby resulting in low shrinkage and high inherent building accuracy.
(6) *Large parts can be built*: The large build volume allows for relatively larger and taller parts to be built. Large single parts can be built at one go rather than building smaller parts to be joined together later.

The limitations of the EOSINT systems are as follows:

(1) *Dedicated systems*: Only dedicated systems for plastic, metal and sand are available, respectively.
(2) *High power consumption*: The EOSINT systems require relatively high laser power in order to directly sinter the powder, especially metallic ones.
(3) *Large physical size*: A relatively large space is required for the system.

5.3.6. Applications

(1) *Concept models*: Physical representations are used to visualize design ideas, forms and style.
(2) *Functional models and working prototypes*: Parts that can be used for fit and limited functional testing. The system is suitable for the automotive, aerospace, machine tools and other industries for consumer products.
(3) *Wax and styrene cast patterns*: Patterns produced in wax and then cast in metal of choice using the investment casting process. Styrene patterns, on the other hand, can be evaporated pattern cast.
(4) *Metal tools*: The main application of the EOSINT system is in rapid tooling.[19] It is used primarily for creating tools for investment casting, injection molding and other similar manufacturing processes.

5.3.7. Examples

5.3.7.1. *Volkswagen's Use of EOSINT M 270 to Build Shifter Knob for New Concept Car*

Volkswagen's concept vehicle GX3 is a cross between a sports car and a motorcycle. The design goal is the stainless steel gear lever or shifter knob (see Fig. 5.14) that Volkswagen calls its "center jewel". The design team wanted something extraordinary; hence, it set out to put a modern twist on the VW Golf GTI's golf ball-shaped shifter knob. The design team used

Fig.5.14. Shifter knob produced on the EOSINT M 270 (courtesy EOS GmbH).

the EOSINT M 270 to fabricate a 17-4 stainless steel shifter knob CNC machining was not able to produce. Derek Jenkins, chief designer of Volkswagen North America, was amazed at how well the shifter knob was machined.[20]

5.3.8.   Research and Development

EOS is working on more thermoplastic materials and post-processing and secondary processes for new applications and even wider choice of materials for its EOSINT P machines. The same can be said for EOSINT M in that more laser sintering materials are to be developed for expansion of its applications and choice of materials. EOS is also researching actively into the environmental friendliness of the materials and processes they use.[21]

5.4.   OPTOMEC'S LASER ENGINEERED NET SHAPING (LENS)

5.4.1.   Company

Optomec is a privately held company founded in 1982, with corporate head-quarters located in Albuquerque, New Mexico and a recently expanded applications facility in Saint Paul, Minnesota. Since 1997, Optomec has focused on commercializing a direct fabrication process, the Laser Engineered Net Shaping (LENS®), originally developed by Sandia National Laboratories. Optomec delivered its first commercial system to Ohio State University in 1998. In 2004, Optomec released their first commercial M³D system. The address of Optomec Inc. is 3911 Singer Boulevard, N.E., Albuquerque, NM 87109, USA.

5.4.2. Products

Optomec's two commercial systems, LENS® and M³D® systems, use additive manufacturing technologies to deliver cost-effective development, production and repair of a wide range of end-products. The LENS® systems, based on Optomec's proprietary technology (see Sec. 5.4.3), are used to fabricate and repair high-value metal components from aircraft engine parts to medical implants. At the nano- and microscales, M³D® system, based on a different process (see Sec. 5.4.7), is used to produce high-density circuitry, embedded components and biomedical devices.

With the release of the LENS® Model 850-R and related software and hardware modules, the LENS® systems are now in their third generation. These systems feature the Laser LENS® process, a technology that builds or repairs parts using metal powders to form fully dense objects to give excellent material properties. This technique can be used with a wide variety of metals including titanium, tool steels, stainless steels, copper and aluminum. The LENS® Model 850-R system contains the following hardware components: (1) Argon Recirculation Unit, (2) Laser Power Supply, (3) Hermetic Chamber, (4) Workstation and (5) Glove Box. Table 5.4 shows a summary of the models and specifications of the LENS® systems.

5.4.3. Process

The LENS® process builds components in an additive manner from powdered metals using an Nd:YAG laser to fuse powder to a solid as shown in Fig. 5.15. It is a free-form metal fabrication process in which a fully

Table 5.4. Summary specifications of Optomec's LENS® systems (courtesy Optomec Inc.).

Model	LENS® Model 850-R
Process	LENS
Build volume, *XYZ*, mm	$900 \times 1500 \times 900$
Laser type	fiber laser
Laser power	500 W up to 2 kW
Materials	LENS® 316 Stainless Steel, 316 SS Anneal bar, LENS® Inconel® 625, IN 625 Annealed bar, LENS® Ti-6Al-4V, Ti-6Al-4V Annealed Bar

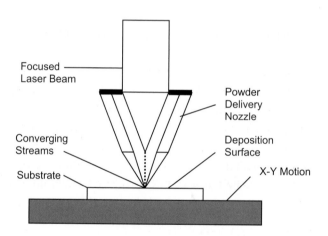

Fig. 5.15. Optomec's LENS process.

dense metal component is formed. The LENS process comprises the following steps[22,23]:

(1) A deposition head supplies metal powder to the focus of a high powered Nd:YAG laser beam to be melted. This laser is typically directed by fiber optics or precision angled mirrors.
(2) The laser is focused on a particular spot by a series of lenses and a motion system underneath the platform moves horizontally and

laterally as the laser beam traces the cross section of the part being produced. The fabrication process takes place in a low-pressure argon chamber for oxygen-free operation in the melting zone, ensuring good adhesion.

(3) When a layer is completed, the deposition head moves up and continues with the next layer. The process is repeated layer by layer until the part is completed. The entire process is usually enclosed to isolate the process from the atmosphere. Generally, the prototypes need additional finishing but are fully dense products with good grain formation.

5.4.4. Principle

The LENS® process is based on the following two principles:

(1) A high powered Nd:YAG laser focused onto a metal substrate creates a molten puddle on the substrate surface. Powder is then injected into the molten puddle to increase material volume.

(2) A "printing" motion system moves a platform horizontally and laterally as the laser beam traces the cross section of the part being produced. After formation of a layer of the part, the machine's powder delivery nozzle moves upward prior to building next layer.

5.4.5. Strengths and Weaknesses

The key strengths of the LENS® systems are as follows:

(1) *Superior material properties*: The LENS® process is capable of producing fully dense metal parts.[24] Metal parts produced can also include embedded structures and superior material properties. Microstructure produced is also relatively good.

(2) *Complex parts*: Functional metal parts with complex features can be produced on the LENS® system.

(3) *Reduced post-processing requirements*: Post-processing is minimized, thus reducing cycle time.

The limitations of the LENS® system are as follows:

(1) *Limited materials*: The process is currently narrowly focused to produce only metal parts.
(2) *Large physical unit size*: A relatively large area is required for the unit.
(3) *High power consumption*: The power consumed for using the laser is high.

5.4.6. Applications

The LENS® technology can be used in the following areas:

(1) building mold and die inserts;
(2) producing titanium parts in the race car industry;
(3) fabricating titanium components for biological implants;
(4) producing functionally gradient structures;
(5) component repair;
(6) producing titanium for the aerospace industry;
(7) manufacturing cermets and carbon nanotube-reinforced nickel.[25,26]

Figure 5.16 shows the illustrations of metallic automobile and medical parts produced using LENS®. The LENS® system can also be integrated with conventional processes to create unique hybrid manufacturing solutions.

Fig. 5.16. Automobile (left) and medical (right) LENS produced metal parts (courtesy Optomec Inc.).

For example, LENS® can be used for feature enhancement to an existing component by adding layers of wear-resistant material or other surface treatments. LENS® can also produce value-added features on existing parts, such as adding a boss or flange to a large casting component.

LENS® offers unique flexibility in the geometries and material range it supports, enabling the use of innovative design concepts such as hollow and internal structures and functional material gradients that incorporate mechanical property transitions within a single part.

5.4.7. Maskless Mesoscale Material Deposition (M³D)

M³D® is an additive manufacturing solution that reduces the overall size of electronic systems by using nanomaterials to produce fine feature circuitry and embedded components without the use of masks or patterns. The resulting functional electronics can have line widths and pattern features as small as 10 μm and as large as 100 μm or more — successfully bridging the gap between existing screen-printing and thin-film lithography capabilities.[27]

Optomec's M³D® systems are used in the development and fabrication of the next generation of microelectronic devices. The systems can also be used to repair production defects and legacy electronics and further have the flexibility to be used for life science and biomedical applications. Optomec's M³D® system comes in three configurations: (1) Standalone System, (2) Benchtop System and (3) Deposition System:

(1) *M³D® Standalone System*: This system includes full M³D Deposition System embodied within a ruggedized base and enclosure. The system is primarily suitable for low-volume fabrication of electronic circuitry and components. It is also used for product development, prototyping and life sciences research.

(2) *M³D® Benchtop System*: This system includes the M³D Deposition System together with motion control and software. It is primarily configured as a cost-saving laboratory-based unit for product development, research and development purposes and materials research applications. It is also suitable for pre-production process development.

(3) *M³D® Deposition System*: This system includes the Deposition Subsystem, PCM and Ultrasonic/Pneumatic Atomizer. It allows for

custom integration of M³D® capabilities within the user's specific production setting.

5.4.7.1. *M³D® Concept and Process*

The M³D® process generates a mist that atomizes a source material.[28] The resulting aerosol stream contains particles that are refined in a virtual impactor and further treated on the fly to provide optimum process flexibility. The material stream is then focused using a flow guidance deposition head, which creates an annular flow of sheath gas to collimate the aerosol. The coaxial flow serves to focus the material stream. Patterning is carried out by precised translation of the flow guidance head while the substrate position remains fixed. After deposition, the materials may undergo thermal or chemical post-treatments to attain final desired electrical and mechanical properties and adhesion to the substrate. M³D® can locally process the deposition using a laser treatment process that permits the use of substrate materials with very low-temperature tolerances, such as polymers. The end result is a high-quality thin film (as fine as 10 nm) with excellent edge definition and near-bulk properties.

This system allows "high-tech" applications such as printed circuit boards with embedded passives and components, hybrid manufacturing for electronic devices, semiconductor packaging, life sciences manufacturing, biocompatible electronics, diagnostics, drug delivery and tissue engineering.

5.4.8. Research and Development

The LENS®Technology Group's future plans include continued research into embedded structures, thermally conductive materials, single crystal applications, gradient materials, metal matrix composites, mold repair and modification and ways to increase deposition rate. Research also include direct-write vapor sensors on polymeric substrates and magnetic nanomaterial deposits.[29,30] Optomec is also developing software for a five-axis head, allowing the process to handle more difficult geometries. The new head will permit the deposition of metal in areas that have walls that are 90° to each other.

5.5. ARCAM'S ELECTRON BEAM MELTING (EBM)

5.5.1. Company

Arcam AB, a Swedish technology development company, was founded in 1997. The company's main activity is concentrated in the development of the Electron Beam Melting (EBM) technique for the production of solid metal parts directly from metal powder based on a 3D CAD model. The fundamental development work for Arcam's technology began in 1995 in collaboration with Chalmers University of Technology in Gotherburg, Sweden. The Arcam EBM technology was commercialized in 2001. The company's address is Krokslatts Fabriker 30, SE-431 37 Molndal, Sweden.

5.5.2. Products

In April 2007, Arcam introduced a new larger EBM system. The new Arcam A2 features a choice of two builds tanks, enabling the production of a 75% larger build than the present Arcam EBM S 12. The two build tanks allow the user to choose between wide and high builds. Depending on the build tank selected, the build envelope is either $200 \times 200 \times 350$ mm or a diameter of 300 mm with a height of 200 mm. The Arcam A2 features a completely new power supply, enabling improved beam control and a more advanced heat model to increase build speed and precision. A new software was also introduced, with new features such as automatic calibration, for enhanced accuracy. With the new Arcam A2, they targeted the aerospace market, racing industry and general industry with a machine fulfilling these industries' need for the production of larger components.

The EBM process is used to produce metal parts directly from a CAD model. The Arcam EBM S12 includes the following hardware components:

- electron beam gun with sweeping system;
- vacuum chamber with fabrication tank and powder holder/setter;
- vacuum pumps;
- monitor;

Table 5.5. Specifications of the Arcam A2.

Model	ARCAM A2
Dimension, mm	$1850 \times 900 \times 2200$ ($W \times D \times H$)
Weight, kg	1420
Build volume, mm	$250 \times 250 \times 400$ or $350 \times 350 \times 250$
Layer thickness, mm	0.05–0.2
EB scan speed, m/s	> 1000 m/s
Electron position accuracy, nm	± 0.025 mm
Part accuracy, mm	± 0.3 mm
Calibration	Automatic
Cooling	Automatic start
Power supply	3×400 V, 32 A, 7 kW
Process computer	PC, XP Professional
CD interface	Standard: STL
Network	Ethernet 10/100
Certification	CE

- linear device;
- high voltage unit;
- electronic control system;
- control unit.

Table 5.5 shows the specifications for the Arcam A2. Figure 5.17 shows a photograph of the Arcam A2 system. Currently, four metal powders are available, namely Ti6Al4V (ASTM 136), Ti6Al4V ELI (ASTM F136), Titanium Grade 2 (ASTM F67) and CoCr Alloy (ASTM F75).

5.5.3. Process

The Arcam ECM process consists of the following steps:

(1) The part to be produced is first designed in a 3D CAD program. The model is then sliced into thin layers, approximately a tenth of a millimeter thick.
(2) An equally thin layer of powder is scraped onto a vertically adjustable surface. The first layer's geometry is then created through the layer of

Fig. 5.17. Arcam A2 system (courtesy Arcam AB, Sweden).

powder melting together at those points directed from CAD file with a computer-controlled electron beam.

(3) Thereafter, the building surface is lowered and the next layer of powder is placed on top of the previous layer. The procedure is then repeated so that the object from the CAD model is shaped layer by layer until a finished metal part is completed.

5.5.4. Principle

The EBM process is based on the following two principles[31]:

(1) Parts are built up when an electron beam is fired at metal powder. The computer-controlled electron beam in vacuum melts the layer of powder precisely as indicated by CAD model with the gain in electron kinetic energy.

(2) The building of the part is accomplished layer by layer. A layer is added once the previous layer has melted. In this way, the solid details are built up as thin metal slices melted together.

The basis for the Arcam Technology is essentially EBM. During the EBM process, the electron beam melts metal powder in a layer-by-layer process to build the physical part in a vacuum chamber. The Arcam EBM machines use a powder bed configuration and are capable of producing multiple parts in the same build. The vacuum environment in the EBM machine maintains the chemical composition of the material and provides an excellent environment for building parts with reactive materials such as titanium alloys. The electron beam's high power ensures a high rate of deposition and an even temperature distribution within the part, which gives a fully melted metal with excellent mechanical and physical properties.

5.5.5.   Strengths and Weaknesses

The key strengths of the Arcam EBM systems are as follows:

(1) *Superior material properties*: The EMB process produces fully dense metal parts that are void free and have excellent strength and material properties.
(2) *Excellent accuracy*: The vacuum provides a good thermal environment that results in good form stability and controlled thermal balance in the part, greatly reducing shrinkage and thermal stresses. The vacuum environment also eliminates impurities such as oxides and nitrides.
(3) *Excellent finishing*: The high-energy density melting process results in parts that have excellent surface finishing.
(4) *Good build speed*: High-energy density melting with a deflecting electron beam also results in an HS build and good power efficiency.

The limitations of the Arcam EBM system are as follows:

(1) *Need to maintain the vacuum chamber*: The process requires a vacuum chamber that has to be maintained as it has direct impact on the quality of the part built.
(2) *High power consumption*: The power consumed for using the electron beam is relatively high.

(3) *Gamma rays*: The electron beam used in the process can produce gamma rays. The vacuum chamber acts as a shield to the gamma rays; thus, it is imperative that the vacuum chamber has to be properly maintained.

5.5.6. Applications

(1) *Rapid manufacturing*: Rapid manufacturing includes the fast fabrication of the tools required for mass production, such as specially shaped molds, dies and jigs.[32]
(2) *Medical implants*: The technology offers production of small lots as well as customization of designs, adding important capabilities to the implant industry. A unique feature is the possibility of building parts with designed porosity and scaffolds, which enables the building of implants with a solid core and a porous surface to facilitate bone ingrowths. The ability to directly build complex geometries makes this technology viable for the manufacture of fully functional implants. The process uses standard biocompatible materials such as Ti6Al4V ELI, Ti Grade 2 and Cobalt–Chrome.
(3) *Aerospace*: This RP system offers a significant potential cost savings for the aerospace industry and enables designers to create completely new and innovative systems, applications and vehicles.

5.5.7. Examples

5.5.7.1. *An Implant Manufactured with Arcam EBM was Used to Reconstruct a Young Girl's Skull*

A young girl suffered a tragic car accident and needed to remove a large part of the skull bone to relieve the pressure on the brain. A titanium alloy implant was fabricated within 12 h using the Arcam EBM S12 machine (see Fig. 5.18). The mechanical properties were excellent with 100% density of the material. The implant produced with EBM featured a chemical composition within stipulated standards, fully solid material with fine microstructure, high ductility and good fatigue characteristics. The entire surgical operation took only 2 h as the implant fitted perfectly without

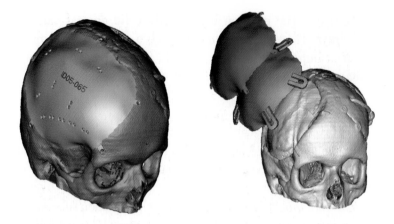

Fig. 5.18. CAD image of the young girls' skull and implant (courtesy Arcam AB, Sweden).

complications. From this case, it was shown that EBM implants can be more rapidly and cost effectively manufactured compared with CNC machined ones. The design aspect is smoother and less complicated with an EBM manufactured part and the surface is almost perfect for the implant.

5.6.   CONCEPT LASER GmbH's LaserCUSING

5.6.1.   Company

Concept Laser GmbH was founded in Lichtenfels in 2000 with a vision to optimize the SLS process. Based on this vision, Frank Herzog, the firm's Managing Director and his team of engineers developed a completely new concept strategy to perform laser processing. With the encouragement and support from the The Hofmann Group, a name standing for competence and 40 years of experience in the field of rapid tooling, the M3 linear machine was invented. The firm owns several patents related to this technology. Concept Laser GmbH is located at An der Zeil 8, 96215 Lichtenfels, Germany.

5.6.2.   Product

Figures 5.19 and 5.20 show Concept Laser's modular laser processing systems. The machine consists of the laser station containing the laser and

Fig. 5.19. Concept Laser's M3 linear system (courtesy Concept Laser GmbH).

Fig. 5.20. Concept Laser's M1 and M2 LaserCUSING® systems (courtesy Concept Laser GmbH).

the axes that are powered by high-powered dynamic linear motors. Three technology modules allow effective time and cost-saving work in addition to the flexibility of choosing and exchanging of technologies. The system specifications are presented in Table 5.6.

With the Concept Laser technology, the tool-less manufacture of components with very complex geometries for one-off metal products or

Table 5.6. Concept Laser's LaserCUSING system specifications.

	M1 Cusing	M2 Cusing	M3 Linear
Number of axles	—	—	5 mec.
Motor precision	15 μm	15 μm	15 μm
Build envelope, mm (x, y, z)	250 × 250 × 250	250 × 250 × 280	250 × 250 × 250 up to 300 × 250 × 350 × 300 mm
Layer thickness	20–80 μm	20–50 μm	20–80 μm
Production speed	2–10 cm³/h	2–20 cm³/h	2–20 cm³/h
Laser system	Fiber laser 200 W (cw)	Fiber laser 200 W (cw)	Diode-pumped solid state laser 100 W (cw + pulsed); Fiber laser 200 W (cw) (optional)
Max. scan speed	7 m/s	7 m/s	7 m/s
Focus diameter	70–200 μm	70–200 μm	70–200 μm
Power/compressed air supply	7.4 kW; 3/N/PE AC 400 V, 32 A 5 bar	7.0 kW; 3/N/PE AC 400 V, 32 A 5 bar	7.4 kW; 3/N/PE AC 400 V, 32 A 7–8 bar
Inert gas supply	Nitrogen; external generator (optl.)	Nitrogen; external generator (optl.)	Nitrogen; external generator (optl.)
Inert gas consumption	< 1 m³/h	approx. 2.5 m³/h	approx. 2.5 m³/h
Dimensions, mm $(W \times D \times H)$	2362 × 1535 × 2308 mm	2440 × 1630 × 1992 mm	2670 × 1990 × 2180 mm
Weight	1500 kg	1500 kg	2300 kg
Operating conditions	15–35°C	15–35°C	15–35°C
LaserCUSING® materials	Stainless steel, hot-work steel, nickel-based alloy (others on request)	Stainless steel, hot-work steel, aluminum, titanium alloy, nickel-based alloy	Stainless steel, hot-work steel

small batches is now a real alternative to conventional machining or casting processes.[33] Components manufactured by LaserCUSING® systems have the same properties as cast or machined products.

The M1 LaserCUSING® is a low-cost version relative to the larger laser processing machine such as the M3 linear system.

The M2 LaserCUSING® is based on the Concept Laser technology and in addition meets the special requirements of processing the reactive powder materials out of aluminum and titanium alloys. The system technology in M2 LaserCUSING®is fitted with corresponding sensors and measuring systems that fulfill the latest explosion and fire-protection regulations. The rest-oxygen and inert gas atmosphere is constantly monitored and controlled in the entire machine. The specially developed fiber laser gives very high-resolution details and mechanical properties in combination with the LaserCUSING® process at room temperature.

Powder materials of titanium alloys are characterized by their high oxygen affinity and thus have to be stored and processed without atmospheric oxygen, i.e., in an inert atmosphere. Oxygen would otherwise be bonded into the powder and lead to much worse mechanical parameters for the finished products or even component failures. This makes their processing, though also storage and handling in a gas atmosphere indispensable. With the M2 LaserCUSING® handling station, up to two materials can be stored. In the M2 LaserCUSING® handling station in an inert atmosphere, this station also has an integrated lock system and an inert powder extraction unit with integrated screening process. The powder is thus extracted from the build chamber and returned to the storage chamber within a matter of minutes automatically, removing the risk of the machine operator coming into contact with the powder.

The M3 linear is probably the largest metal machine available in the market and is supplied as a modular concept with a choice of two insert modules. This machine realizes three laser technologies: LaserCUSING®, 3D laser material erosion and laser marking. The M3 linear produces large-volume mold inserts in build envelopes of $250 \times 250 \times 250$ mm up to $300 \times 350 \times 300$ mm. The M3 linear uses a beam deflection system that is controlled via a system of linear direct drives and mirrors. This ensures accuracy in the process, which remains constant over a large build envelope and avoids elliptical beam cross sections when the deflection is to the edge area. The patented exposure strategy makes it possible to generate solid and large-volume components with a low degree of internal stresses.

5.6.3.   Process

The LaserCUSING® process allows the layer-by-layer construction of components from almost all metallic materials. Metallic powder is fully fused layer-by-layer to produce a 100% component density with a high-energy laser. A specially developed exposure strategy allows the generation of solid and large-volume components without any deformation.

A patented surface post-treatment process directly after the construction process ensures the highest surface quality and hardness. A number of technical refinements such as automatic powder evacuation, a separately developed coating system and a circulating inert gas atmosphere enhance the performances of the process.

The software developed by Concept Laser is unique in that it allows not only deep engraving but also a true 3D engraving (laser erosion) on free form surfaces. The software calculates the volume of material to be removed and this is then evaporated layer by layer using a laser. This piece of technology from Lichtenfels can be used as a supplement to milling — there is no more need for complicated programming or the production of electrodes. A laser measuring sensor integrated in the software permits automatic depth control in the process.

5.6.3.1.   *Materials*

Concept Laser materials are produced in a powder form and have been developed specifically for the LaserCUSING® process. The compositions of the materials as well as the nature of the powder and the distribution of the powder fraction have been optimized for this method and the process control. All Concept Laser powder materials are 100% compatible for reuse in subsequent construction processes. No fresh material has to be added. Typical layer thicknesses for all materials are 20–50 μm. Some of the materials used are introduced in more detail as follows:

(1) *CL 60DG — Hot-work steel*: This is a material suitable for the production of tool inserts, in pressure die cast molds for light metal alloys. This material is a hot work steel with tensile strengths up to 1.800 N/mm^2 and hardness up to 52 HRC.

(2) *CL 50WS — Hot-work steel*: This material is a very tough with excellent mechanical properties. Its main field of use is series injection molding of plastic parts. This material can be used to produce large components that are only limited by the dimensions of the machine's built envelope. Components made with this material have tensile strengths of approximately 1,100 N/mm^2 and a hardness of 40–42 HRC directly after the process. Subsequent annealing can increase these parameters to >1,600 N/mm^2 and up to 54 HRC.

(3) *CL 20ES — Stainless steel*: This is a powder material to produce acid and rust-resistant assemblies or tool components for pre-production tools. The tensile strength is approx. 650 N/mm^2 and the hardness approx. 220 HB 30.

(4) *CL 90RW and 91 RW — Stainless Hot-work steel*: These two materials have the characteristics of a hard stainless steel with high chrome content. The powder materials are for the production of tool components for the serial injection molding of packaging and medical products. Components made with these materials have tensile strengths of approx. 850 N/mm^2 and a hardness of 35–40 HRC directly after the process. Subsequent annealing can increase these parameters to >1,100 N/mm^2 an up to 45–48 HRC for CL 90RW and to >1,700 N/mm^2 an up to 48–50 HRC for CL 91RW.

(5) *CL 30AL, CL 31AL (AlSi10Mg) — Aluminum-based alloy*: These are aluminum powder materials that can be employed for high mechanical and dynamic load. The powder material can be used optimally for the production of technical prototypes or small series.

(6) *CL 40TI (TiAl6V4) — Titanium-based alloy*: CL 40TI is a titanium powder material for the production of technical lightweight components and medical implants.

(7) *CL 100NB (Inconel 718) — Nickel-based alloy*: This is a powder material made out of a nickel-based alloy for the production of heat-resistant components in the automotive and aerospace industry.

5.6.4. Principles

Based on new laser technologies and a completely new process, it became possible to overcome the weaknesses of laser sintering. Using the patented exposure strategy and original materials, solid and large-volume components, such as mold inserts, can be produced rapidly.[33] The material properties are identical to those of the original material and allow these components to be employed under production conditions. The term CUSING was coined from the words CONCEPT and FUSING. LaserCUSING® is based on the fusion of single-component metallic powder materials using a laser. This "generative" method makes it possible to assemble components layer by layer from virtually all weldable materials (e.g., stainless steel and hot-work steel), by completely fusing the metal powder layer by layer. It has shown that it can overcome the internal stress and deformation problems almost entirely and to achieve 100% component density. The typical layer thickness is between 20 and 50 µm. The LaserCUSING® method is considered an excellent link in the process chain between rapid tooling and traditional tool and mold making. The patented exposure strategy also allows the low-deformation generation of solid and large-volume components.

Generating cores and inserts with cooling ducts has been built in practice. This ability allows for the production of highly complicated 3D molds with such cooling ducted cores/inserts. In other words, the cooling ducts, which conventionally could only be introduced to a limited extent or at great expense, are now adapted exactly to the contours of the mold insert during the process. Thus, considerably shorter cycle times are achievable. The compactness of the components ensures that no cooling water can escape. The deformations on the injection-molded part are minimized due to optimized mold cooling.

5.6.5. Strengths and Weaknesses

The strengths of the LaserCUSING® systems are as follows:

(1) *Cost effective production of metal parts*: These new technologies are ideal in tool and mold making as well as for prototyping metal parts.
(2) *Flexible technology*: The technology modules allow a fast and simple change from one technology to the next. Set-up costs are cut by 50%.
(3) *High accuracy and laser beam quality*: This is due to the high laser beam quality through the small scanning range, with the scanning head being positioned directly over the component by linear motors for precision.
(4) *Excellent material properties*: LaserCUSING® can achieve 100% density of single-component metallic powder materials. This is made possible by the unique patented exposure strategy that allows for complete fusion of the metal powder. Unused metal powders can also be 100% reused.
(5) *3D erosion on free form surfaces*: New software specially developed for 3D erosion permits the realization of innovative product ideas. A laser measuring sensor to check the erosion depth is integrated in the "intelligent" software. Calculations and feedback on the erosion depth ensure "truly" 3D material erosion — on any shaped surface.
(6) *Automatic powder exchange*: A powder extraction apparatus provides both better accessibility to the machine and automatic powder exchange. The operator of the M1 Cusing has no direct contact with the metal powder.
(7) *Safe processing of reactive materials*: The M2 Cusing provides safe processing of reactive materials by powder handling in an inert atmosphere. The safety is regulated to CE and explosion protection standards.

The weaknesses of the LaserCUSING® systems are as follows:

(1) *Large physical size of the unit*: The system requires a large space to house.

(2) *High power consumption*: The system consumes high power due to the high wattage of the laser required to perform direct metal powder sintering.

5.6.6. Applications

(1) *Functional models and working prototypes*: Parts that can be used for fit and full functional testing. The system is suitable for the automotive, aerospace, machine tools and other industries for industrial products.
(2) *Metal tools and inserts*: The main application of the LaserCUSING® system is in rapid tooling. It is used primarily for creating superior tools and inserts with or without optimized cooling channels and ducts for injection molding and other similar tooling and manufacturing processes.

5.6.7. Examples

5.6.7.1. *Innovative Rowenta Iron*

The stimulating partner in terms of know-how and competence to achieve these goals was found with the full-service-engineering provider Hofmann Werkzeugbau. The successful result of this cooperation led to the introduction of the Rowenta Perfect DX 9100 that takes a leading position among steam irons (Fig. 5.21). The mode of operation of the device is based on "Intra Steam," a new method that enables ironing with pulsed steam. The steam penetrates and moisturizes the textile fibers evenly due to the short, intermittent pulses. For this to be achieved, highly complex geometric cores with cooling ducts had to be integrated within the injection mold.

5.6.7.2. *Intensive Mold Cooling with LaserCUSING®*

The LaserCUSING by Concept Laser GmbH was able to minimize the deformations on the injection-molded part due to the introduction of optimum mold cooling ducts in the mold (Fig. 5.22). There was also a significant reduction in the amount of reworking required on the mold contours. The accuracy achieved prior to secondary treatment was $\pm 50 \, \mu m$. The majority of surfaces only required rework with fine finish to obtain a higher degree of accuracy. The roughing-down process and pre-finishing

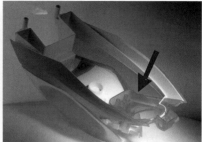

Fig. 5.21. Rowenta Perfect DX 9100 (left) and the internal geometry of the core insert with cooling ducts (right and arrowed) (courtesy Concept Laser GmbH and Hofmann Innovation Group).

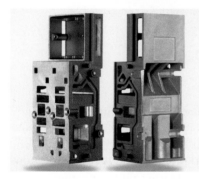

Fig. 5.22. Valve block (left) and mold insert (right) for series injection mold (courtesy Laser Concept GmbH).

process were omitted completely. As the distortion of the part was substantially reduced, the path from initial prototyping to creating a part ready for mass production was thus shorter. All these resulted in a considerable saving in time and costs.

5.7. MCP-HEK TOOLING GMBH'S REALIZER II (SELECTIVE LASER MELTING)

5.7.1. Company

MCP-HEK has had more than 50 years experience in designing and supplying process technologies, equipment and proprietary materials foremost to

Fig. 5.23. MCP-HEK's Realizer system (courtesy MCP-HEK Tooling GmbH).

the aircraft and automotive industry; work holding and material solutions for machining turbine blades and optical lenses; prototype sheet metal forming systems; metal spray mold processes for plastics tooling. One of the most popular processes was the MCP Vacuum Casting, which the company introduced as the very first supplier in 1987. The company introduced the MCP-Realizer SLM (see Fig. 5.23) based on the Selective Laser Melting (SLM) Technology to build tools and single individual engineering and implant parts in pure, dense steel cobalt–chrome and titanium from CAD data files. The MCP-HEK Tooling GmbH head office address is Kaninchenborn 24-28, 23560 Lübeck, Germany.

5.7.2. Products

The main product from MCP-HEK Tooling GmbH is the Realizer II system. Based on the SLM technology, this rapid tooling machine makes it possible to produce 100% dense metal parts from customary metal

Table 5.7. Specifications for the MCP-HEK Realizer II SLM.

Model	MCP Realizer II SLM
Build speed	7.000 cm^3/h (average)
Min. build thickness	> 20 to 50 µm (depends on powder size)
Max. volume powder	25 dm^3, i.e., 120 kg steel
Build envelope, mm	250 × 250 × 240
Laser	Ø 20–300 µm, 30–200 W/M^2 <1,1
External dimensions, mm	Approx. 1900 × 2600 × 2500
Weight, kg	400
Power supply	400 V/3-Phase/50/60 Hz/16 A

powder. The parts or specialty tools are built layer by layer (30 µm thickness). Metal powder (e.g., stainless steel 1.4404) is locally melted by an intensive infrared laser beam that traces the layer geometry. This makes it possible to build the fine details like thin vertical walls of less than 100 µm thickness. Directly after the production process the manufactured parts or tools show a surface roughness of approximately 10–30 µm for Rz. Other machine specifications are listed in Table 5.7.

The system is ergonomically designed, rigid and with a maintenance friendly construction. It gives good accessibility to the build chamber for multiple applications. Actual operating data are shown on a closed-circuit video system. Alternatively, the build chamber and build process are visible from the outside. Software is provided to control the STL data files off-line. The software is used for visualization, dimensioning, support building, fixing, editing, etc. The FuSCo system software is used to read STL- or slice-data and for running the Realizer II SLM. Build times can be reduced by utilizing the Dense Shell™ software to produce strong parts and tool inserts with hollow channels.

5.7.3. Process

The SLM process follows many of the conventions of the layer-based manufacturing processes. What differs is that the system can be used on a wide range of powdered metal materials. A component is split into layers and each of those layers is built on top of each other and fused together, layer after layer, until the part's form is built.

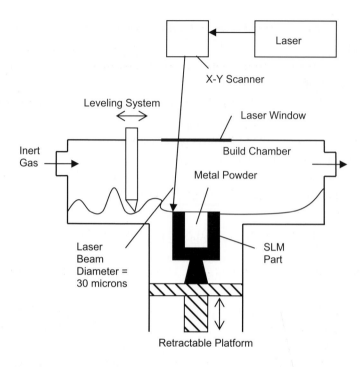

Fig. 5.24. Schematic diagram of the selective laser melting (SLM) process.

The SLM process uses a laser beam, controlled using optic lenses to pass a laser spot across the surface of a layer of powdered metal to build each layer. The metal powder is melted rather than just simply sintered together, thus giving parts that are 100% dense (or solid). The resulting part has much greater strength and dimensional accuracy than parts that are build by laser sintering.[34,35] The schematic diagram in Fig. 5.24 illustrates the SLM process.

5.7.4.   Strengths and Weaknesses

The main strengths of the MCP-HET SLM are as follows:

(1) *High-quality metal parts*: Builds high-quality parts and tooling inserts from metal powders. Allows homogeneous build up of components and tool cavities up to 100% density depending on requirements.

(2) *Large range of metal materials*: The system can be used to build almost any type of metal: stainless steel tool steel, titanium, aluminum, cobalt–chrome, various nonferrous metals. The system can also be used to shape gold.

(3) *Fast and low cost*: This is because no post-processing such as heat treatment or infiltration required.

(4) *High accuracy*: High resolution process, dimensionally accurate, low-heat generation no distortion of parts.

(5) *Complex geometries*: Produces tools and inserts with internal under-cuts and channels for conformal cooling.

The major weaknesses of the SLM process are as follows:

(1) *Large physical size of the unit*: The system requires a large space to house.

(2) *High power consumption*: The system consumes high power due to the high wattage of the laser required to perform direct metal powder sintering.

(3) *Relatively slow process*: Even with a build rate up to 20 cm^2 per hour, it is still a lengthy process compared with high-speed machining.[34]

5.7.5. Applications

Industrial applications include it being part of the process chain for building sheet metal forming and stamping tools. The SLM process can also be used for rapid tooling in building inserts, core and cavities for plastic injection molding applications. Other applications include building functional proto-types and parts with specialized or dedicated metal, e.g., aluminum alloy, for the aerospace and biomedical industries.

5.7.6. Examples

5.7.6.1. *Planning Osseointegrated Implants*

Osseointegrated implants such as titanium screws[36] are used to drive into bones for dentures and facial prostheses. When planning where to place such implants, accurate RP models allow the depth of bone to be assessed, improving the selection of drilling sites before surgery. Although in many cases this process has improved the accuracy and reduced the theater time, the drawback is that it incurs significant time and cost to produce the anatomical model, which is often damaged by test drilling.

To address this issue, the research team from PDR and Morriston Hospital decided to plan the implant sites entirely on computer and only use RP to make templates to guide the surgeon in theater. This allowed the clinicians to explore many different options without damaging the RP model of the skull. To be successful, the RP process has to produce strong, durable, dimensionally accurate parts that utilize materials that are safe to use in theater while withstanding the demands of the surgical environment and sterilization procedures.

Initial trials utilizing the SLA and a medical standard resin proved that this approach can be successful. PDR then identified a new generative SLM process that can produce accurate parts directly from metal powders, in this case 316L stainless steel, a commonly used material for medical devices. As SLM produces fully dense stainless steel parts it can provide accurate, strong and durable surgical guides that can withstand contact with aggressive surgical instruments such as drills and oscillating saws. The material can also be sterilized by a number of commonly used processes like high-temperature autoclaving. These advantages make SLM ideal for this approach and will enable PDR and its clinical partners to further explore and develop the application of computer generated surgical guides and templates (Fig. 5.25).

5.8. PHENIX SYSTEMS' PM SERIES (LS)

5.8.1. Company

Phenix Systems is a French company that specializes in the design, production and sales of rapid manufacturing systems using the laser sintering of metal and ceramic powders. The laser sintering process in solid phase was developed in the early 1990s by a study group specializing in heterogeneous materials at Ecole Nationale Supérieure de Céramique Industrielle (ENSCI) in Limoges. This patented technology has gone

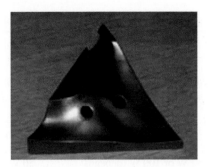

Fig. 5.25. A stainless steel drilling surgical template produced by Selective Laser Melting RP Technology (courtesy MCP-HEK Tooling GmbH).

Fig. 5.26. Phenix Systems' PM250 (left) and PM100 (right) rapid prototyping systems (courtesy Phenix Systems).

through much research and development work by Phenix Systems concerning materials and software in order to reach a wide variety of industrial activities. The Phenix Systems head office address is at Parc Européen d'Entreprises, Rue Richard Wagner, 63200 Riom, France.

5.8.2. Products

The company has two models: the PM250 and the PM 100 systems (see Fig. 5.26). The PM250 system has a larger working volume as compared

Table 5.8. Specifications for the Phenix System PM250 and PM100 systems.

	Model	
	PM250	PM100
Accuracy of parts produced	± 50 μm per 120 mm (metal and ceramic parts)	
Level of detail	150 μm (metal parts) 300 μm (ceramic parts)	
Laser	50 W or 100 W (depends on customer specifications)	50 W or 100 W (depends on customer specifications)
Build envelope	Ø250 mm, 300 mm height	Ø100 mm, 100 mm height
Production speed	1–10 mm³/s	1–10 mm³/s
Materials	*Metals*: stainless steel, tool steels, nickel, etc.	*Metals*: stainless steel, tool steels, nickel, etc.
	Ceramics: alumina, mullite, zirconia, etc.	*Ceramics*: alumina, mullite, zirconia, etc.

with the PM100. Both systems can process either metallic or ceramic powders. Table 5.8 shows the specifications for the Phenix System PM250 and PM100 systems.

The PM100 system is the first laser sintering equipment adapted mainly for the manufacture of small parts. It is also a tool, which is highly geared toward research and development and also training, as it provides all the resources of the process at the least possible cost.

5.8.3. Process

This generative manufacturing process implements the combined effect of a fiber optic-based laser and a furnace on metallic or ceramic powders thereby making provision for nuances to conform to the materials openly available in the industrial market sectors. Figure 5.27 shows the schematic diagram of the Phenix Systems' process. The main stages of the procedure are as follows:

(1) Processing from a 3D file then generating the manufacturing files of each scanned layer.
(2) Initialization of the equipment depending on the materials being used.

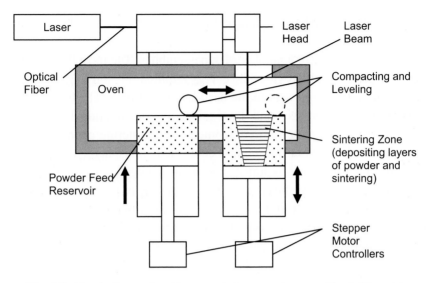

Fig. 5.27. Phenix Systems' rapid prototyping process (courtesy Phenix Systems).

(3) Production of a powder layer with a chosen thickness.
(4) Sintering of the scanned section of the part.
(5) Automatic repetition of the two previous stages until the complete production of the part.

5.8.4.   Principles

When certain metal and ceramic powders are subjected to local high temperatures caused by the laser beam, they tend to undergo sintering and strengthening. On the Phenix Systems machines, these are achieved as the laser beam follows the trajectory that matches the target trajectory. The sintering operation is continuous, with the depositing of successive layers of powder. The process is extremely accurate as it is possible to achieve details of 150 μm for metal parts and 300 μm for ceramic parts. The useful cylindrical production volume depends on the machine: for the PM 250 (the larger model), this is 250 mm diameter with a height of 300 mm. The smaller machine has a work volume of 100 mm for both the diameter and height. The manufacturing procedure is fully automated. The operator works from the CAD file for the part and sets the manufacturing parameters, depending on the type of powder and the desired outcome.

The machines have a number of subassemblies. The most important is the laser head, which provides the energy required for sintering. The laser beam (50 or 100 W) is sent to the head by optical fiber. The head moves horizontally over the oven and has a galvanometric directional system for the beam to produce the desired geometry. The sintering operation takes place inside the oven at 900°C, in a controlled atmosphere. The powder, held in a feed reservoir, is transferred by a mobile scrapper system and deposited in successive layers in the reservoir where the sintering is performed. The powder is uniformly spread in the reservoir; only the area subjected to the laser beam is sintered. The bottom of each of the reservoir is equipped with vertical movement pistons: by controlling these pistons it is possible to ensure the powder is flush on the surface. As there are no speed and torque requirements and only positioning is to be achieved, stepping motors are used, together with incremental encoders in order to guarantee the precision (with accuracy for the parts of $\pm$ 50 μm).

5.8.5. Strengths and Weaknesses

The key strengths of the Phenix Systems are as follows:

(1) *Small and accurate parts*: The Phenix System focuses on building small and accurate metal and ceramic parts.
(2) *Use of standard powders*: One of the main advantages is that the system is able to use standard powders rather than only limited to supplied powders as with that of the competition.
(3) *Low laser power*: The system uses relatively low laser power to sinter metal and ceramic powders because of the application of high working temperatures (e.g., 900°C for tool steels). Also, the use of fiber optics focuses the beam and minimizes losses due to beam diffusion.

The main weaknesses of the Phenix Systems are as follows:

(1) *High temperature work envelopes required*: Care has to be taken as the general temperature of the work area will be extremely hot.

(2) *Post-processing required*: For metals, further polishing is required while for ceramics, post-sintering in the furnace is still required.

5.9. SINTERMASK TECHNOLOGIES AB's SELECTIVE MASK SINTERING (SMS)

5.9.1. Company

Sintermask Technologies AB produces systems based on a patented technology called Selective Mask Sintering (SMS). The machine produces RP parts noticeably faster and more cost efficient compared with similar techniques. The technology is invented by Ralf Larson, one half of the research company Larson Brothers Co. that developed products such as Toner Jet® in the subsidiary Array Printers AB and Hot Plot® in Sparx AB. Hot Plot® was an early RP machine that was introduced commercially in 1989 that produced simple models in thin sheets of Styrofoam. In 1993, Ralf Larson filed for a patent for a method that welded metal powder in thin layers with an electron beam and with high degree of efficiency into metal parts and injection molding tools. The work was conducted in cooperation with Chalmers University of Technology in Gothenburg, Sweden. The patent was approved in 1997 and the company Arcam AB was founded (see Sec. 5.5). In 1996, Ralf developed the SMS based on infrared heating lamps instead of lasers. Speedpart was founded in 2000 to commercialize the technology and the intellectual proprietary rights to the technique were transferred to Sintermask Technologies in conjunction with the first investment from Venturos. Since May 2000, Sintermask has worked to fully develop and commercialize the technique. Sintermask Technologies AB is located at Krokslätts Fabriker 30, 2, 431 37 Mölndal, Sweden.

5.9.2. Product

The machine from Sintermask Technologies is the Pollux 32 (see Fig. 5.28), which builds 3D parts of any shape and size within a volume of $210 \times 300 \times 500$ mm directly from 3D CAD files. The parts and tools are built at a resolution of 0.1 mm and speeds of between 20 and 35 mm per hour.

Fig. 5.28. Sintermask Technologies AB's Pollux 32 (courtesy Sintermask Technologies AB).

The machine is ideal for rapid manufacturing of large solid parts and tools or mass production of smaller parts. The machine does not contain lasers, which eliminates the need for expensive replacement lasers and need for inert gases around the build area. This also significantly lowers the operational cost compared with other laser-based RP systems.

Similar to most powder-based systems, the parts produced in the build tank are supported by unused build material; there is no need to generate supports for building the parts. The parts can be removed from the machine as soon as the building process is finished. Sintermask Pollux 32 and the supplied material powder are ideally suited for rapidly producing impact resistant and durable tools for vacuum forming of plastic sheets. The tools can be used for vacuum forming of both prototypes and smaller production series. Since the material is porous, there is no need to drill vacuum conducts in the tool. Table 5.9 shows the specifications for the machine.

Table 5.9. Specifications for the Sintermask Pollux 32 machine.

Model	Sintermask Pollux 32
Build speed	10–20 s per layer
Build envelope, mm	$210 \times 300 \times 500$ ($W \times D \times H$)
Resolution	50–120 μm
Build height per hour	20–35 mm
Dimensions	125 cm $\times$ 210 cm $\times$ 170 cm ($W \times D \times H$)
Power supply	3×230 V (400V phase to phase, 230 V phase to neutral)
Power consumption	Average 3 kW; peak 11 kW
Materials	According to requirements of the specific application
Software	Windows XP
Connection	Ethernet TCP/IP

5.9.3. Process

Sintermask Technologies' systems are based on the SMS process.[37] It is a method, a machine and a powder material for rapid production of plastic parts from 3D-CAD drawings. The parts are made layer by layer at a thickness of between 50 and 120 μm and speed of 10–20 s per layer, resulting in a build rate of 20–35 mm per hour. Unlike most other RP systems, the build speed is independent of the build area, making it a very important solution for the production of solid details and mass production of parts.

The SMS process is illustrated schematically in Fig. 5.29. The glass plate first moves over the toner supply to collect the toner image and carrying the toner image toward the build cylinder (see Fig. 5.29(a)). The build cylinder has already been coated with a layer of fresh supply of powder by the powder spreading mechanism (powder wiper and spreader). Once the glass plate gets into position (see Fig. 5.29(b)), an infrared radiation source is turned on to expose the entire layer at once over the glass plate. The powder in the masked area remains unsintered while those in the exposed areas are. On completion, the glass plate

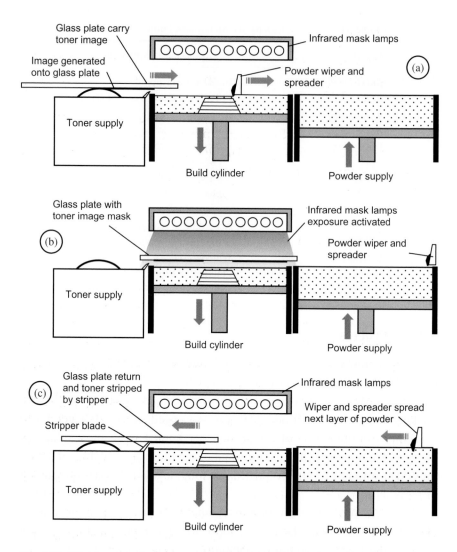

Fig. 5.29. Machine operation: (a) creating the mask and spreading powder, (b) exposure for sintering and (c) cleaning toner off glass plate.

returns and is then stripped of the toner and makes ready to print the next layer (see Fig. 5.29(c)).

Sintermask believes their technology is about four times as fast as competing laser-based powder systems and would be about one third the costs.

5.9.4. Principles

The SMS process fuses entire layers of parts in a single operation. The machine utilizes components similar to both SLS and solid ground curing (SGC, see Sec. 3.12.2). Building material powders are handled in much the same way as in SLS. However, each layer is imaged in a process much like SGC. Each slice is electrostatically printed as a negative image using toner deposited on a glass plate. The imaging method is similar to that used in photocopiers. The toner image on the glass plate is then used as an exposure mask for the layer.

5.9.5. Strengths and Weaknesses

The key strengths of the Sintermask SMS are as follows:

(1) *Fast and efficient build times*: The SMS process build entire layer in a single exposure, thus drastically reducing cycle time and part costs. Even with increased number of parts placed with the build envelope, it does not increase the cycle time, thus making the process an efficient one.
(2) *No support structures required*: One of the main advantages is that the system is able to use standard powders rather than only limited to supplied powders as with that of the competition.
(3) *No laser and nitrogen environment necessary*: The system uses relatively low-power infrared lamps to sinter metal and ceramic powders without the use of high-power lasers. The machine also operates safely without the need for a nitrogen supply.
(4) *Easy to make thick-walled components and compact structures*: As the entire layer is sintered at the same time, it is easy to build thick-walled components and compact structures.

The main weaknesses of the Sintermask SMS are as follows:

(1) *Limited range of materials*: The materials that can be processed by the SMS process are currently limited to a PA supplied by Sintermask. However, Sintermask has suggested many materials can

be included, e.g., from polyethylene terephthalate (PET) to polyetheretherketone (PEEK).

(2) *Poor surface finish*: For metals, further polishing is required while for ceramics, post-sintering in the furnace is still required.

5.9.6. Applications

Sintermask produces systems for rapid productions of tools, prototypes and manufactured parts. Tools and parts can be produced in matter of hours directly from 3D CAD as opposed to weeks using conventional methods. Sintermask's machine can also be used for small series tool-less production and spare parts on demand.

One example of application in tooling is for vacuum forming, a widely used production method for plastic part. A plastic sheet is heated and placed above a tool created by Sintermask. When vacuum is applied under the tool, the plastic sheet is formed around the tool creating a negative pattern of the tool. The method is ideal for production of working prototypes as well as serial production of parts. Since the material is porous, there is no need to drill vacuum conducts in the tool made by Sintermask. Figure 5.30 shows a picture of a tool produced by Sintermask and the corresponding vacuum formed plastic parts.

With the Sintermask, working prototypes can be built in matter of hours. Since the speed of the machine is independent of build area, it is ideally suited for building large parts or many parts simultaneously.

Vacuum formed plastic parts

Vacuum form tool made by Sintermask

Fig. 5.30. Vacuum formed tool (left) produced by Sintermask and corresponding vacuum formed plastic parts (right) (courtesy Sintermask Technologies AB).

5.9.7. Research and Development

Sintermask Technologies is developing more materials based on glass-filled nylon for building prototypes. The company is aiming at tool-less production applications where its speed offers a strong advantage. At 10 s per layer, the machine should be able build at a considerable rate over heights in the Z-direction. Since many smaller parts can be built at once, the approach is envisioned to be a very rapid and economical means for producing small volumes of parts.

5.10. 3D-MICROMAC AG's MicroSINTERING

5.10.1. Company

3D-Micromac AG corporation, a supplier of customized laser micro-machining facilities, has gained an established position in the international market over the past years. The company is subdivided into the four divisions High End Laser Systems, High End Laser Applications, High End Laser Tools and High End Laser Services. The primary focus of 3D-Micromac's system is in microsintering. The address of the company's headquarters is Annaberger Straße 240, D-09125 Chemnitz, Germany.

5.10.2. Product

The product of 3D-Mircomac is the MicroFORM that makes use of the MicroSINTERING technology. This is a processing technology for SLS under vacuum conditions specially developed for the production of prototypes as well as small and medium lots. This novel technology is capable of generating freeform microparts made from metals and alloys with unequalled precision, shape diversity and flexibility. It is done by overcoming the problems of oxidation and humidity by running the SLS process in a vacuum environment. The density of the micropart can be controlled and density gradients and compositional gradients can be breeded by varying the sintering parameters. The Micro-SINTERING procedure and related systems are covered by a number of patents. Table 5.10 shows the specifications for the MicroFORMING system.

Table 5.10. Specifications of 3D-Micromac's MicroFORMING system.

Model	MicroFORMING
Work envelope	Ø25 mm or Ø50 mm × 40 mm (height)
Laser	50 kHz
Build speed	0.02 mm³/s
Resolution	3–5 μm (lateral)
Min. layer thickness	1 μm
Roughness	Ra = 1.5 μm

5.10.3. Process

The 3D-Micromac systems are able to process powder materials with submicrometer grain sizes in layers as thin as one micrometer. The process is essentially a further development from the well-established laser sintering method. It uses a special recoating device to apply thin powder layers and a pulsed solid-state laser to locally melt the powder. In this manner, microparts can be produced in various metals giving excellent material properties, e.g., a detail resolution down to 30 micrometers is possible.[38] Figure 5.31 shows an illustration of the MicroSINTERING process.

Materials used in the MicroSINTERING process include tungsten, copper, aluminum, gold, titanium, silver, molybdenum composites and metal-ceramic-based composites. As an example, for tungsten material, the density of microparts achievable is between 40% and 70%, or for tungsten/aluminum powder mixtures (composites), more than 95% density.

5.10.4. Applications

The micro-components produced find applications in areas such as micro-mechanics, telecommunications, medical, electronics and computer technologies. Other applications are found in industries such as the automotive, mold construction, hard-carbide tools, tool and mold making and precision engineering industries.

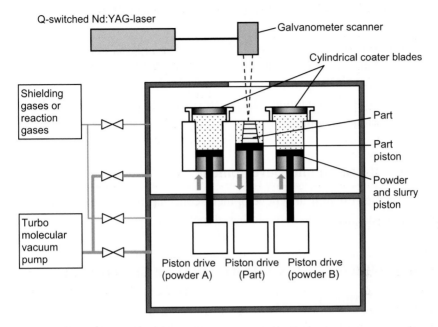

Fig. 5.31. Schematic illustration of the MicroSintering process.

5.10.5. Strengths and Weaknesses

The key strengths of the MicroSINTERING process are as follows:

(1) *Accurate and precisely build parts*: As the MicroSINTERING process is conducted in a vacuum chamber and with the use of a highly accurate Nd:YAG laser, accurate and precisely build micro-structures can be achieve with the micron-size metal powder.
(2) *Fine surface finish*: The use of micro-size powder allows the process to attain surface finishes up to 1.5 μm R_a.
(3) *Wide range of submicron grained metal powder*: The system is able to process a wide range of submicron grained metal powder including: single component metals of tungsten, aluminum, copper, silver, titanium, molybdenum and steel; blend materials of copper/tungsten, aluminum/ tungsten, copper/molybdenum and aluminum/molybdenum.

The main weaknesses of the Sintermask SMS are as follows:

(1) *Need to maintain the vacuum chamber*: The process is carried out in a vacuum chamber that has to be maintained as it has direct impact on the quality of the part built.
(2) *High power consumption*: The power consumed for using the Nd:YAG laser is relatively high.

5.10.6.   Research and Development

Much of the research and development concentration is on developing new materials including ceramic-based materials.

5.11.   THERICS INC's THERIFORM TECHNOLOGY

5.11.1.   Company

Therics Inc, a biopharmaceutical company founded in 1996, is a subsidiary of Tredeger Corporation. Based on the same printing technology for Z Corporation's 3DP developed at Massachusetts Institute of Technology, the TheriForm™ technology is protected by a broad portfolio of strong patents, including an exclusive license from MIT for the worldwide healthcare market. From 1993, Therics has worked closely with MIT to develop successful generations of TheriForm™ fabrication machines that can create both macroshapes and complex microarchitectures that can meet stringent regulatory requirements applicable to healthcare products. Therics Inc. today does not sell their RP machine. Instead, the company focuses on designing, developing and manufacturing a variety of tissue engineering scaffolds and synthetic bone substitute products that promote the growth of bone while eliminating the risk of disease transmission. In July 2005, Therics LLC was formed as an Ohio limited liability company by Randy Theken through the purchase of ownership from Tredegar Corp.

The company produces a line of Beta-Tricalcium Phosphate (ß-TCP) products using the Theriform technology. It manufactures devices from the inside-out. Printheads travel back and forth across a fine layer-bed of ß-TCP, delivering microdroplets of binder solution, which bind the powder particles, creating a cross-sectional layer. This process continues layer by layer until the specified product is complete. Through this method, Therics can control both essential elements of a true scaffolding material — the microarchitecture (internal interconnectivity of the pores) and the macroarchitecture (profile shape). The control of the internal microarchitecture is essential for cell migration and vessel ingrowths, thus producing a solid bone fusion. The address of the Therics LLC headquarters is 115 Campus Drive, Princeton, NJ 08540, USA.

5.11.2. Process

The TheriForm™ manufacturing process (see Fig. 5.32) works in a manner similar to an ink jet printer, creating 3D products composed on a series of 2D layers (see Sec. 5.2). The process contains the following steps[39]:

(1) Products are fabricated by printing microdrops of binders, drugs and other materials and even living cells onto an ultra-thin layer of powdered polymers and biomaterials in a computer-directed sequence.

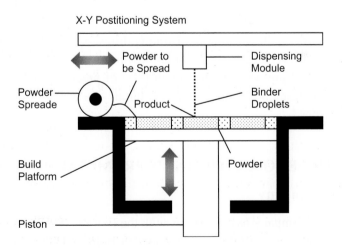

Fig. 5.32. Schematic of the TheriForm™ Process.

(2) These droplets bind with the powder to form a particular 2D layer of the product. After each layer is printed, the build platform descends and a new layer of powder is spread and printing process is repeated.
(3) Thus, the successive layers are built upon and bind to previous layer until entire structure is completed.

5.11.3. Principle

TheriForm™ technology is based on the 3D printing technology developed by MIT. Binder droplets are selectively dispensed to bond the powders in the same layer together as well as to the next layer. The strengths and weaknesses of this process are very similar to those described in Sec. 5.2.5 except for the availability of color.

5.11.4. Applications

TheriForm™ technology is ideally suited for the creation of unique reconstructive implants including bone replacement products from 3D imaging data (e.g., from MRI or CT).[40] It is also used to produce cell-containing tissue that can incorporate drugs and growth factors at specific geometrical locations in functional gradients to enhance the development of functional tissue within the body.[41] TheriForm™'s ability to fabricate structures with architectures and material compositions provides a useful RP system that responds to the challenge of administrating drugs, proteins and biological substances. This is particularly important when conventional manufacturing methods do not provide a high enough level of precision and versatility necessary to fashion a dosage form exhibiting the desired drug release characteristics.

5.12.   THE Ex One COMPANY'S PROMETAL

5.12.1. Company

The Ex One Company has its origins with the Extrude Hone Corporation. The late Lawrence J. Rhoades, first founded Extrude Hone Corporation and when it expanded and grew, sold the technologies relating to abrasive flow machining, Surftran electrochemical deburring and ThermoBurr deburring to Kennametal, Inc. in 2005. He then set up the Ex One Company as a new business to serve as an incubator for inventive, new technologies that have the potential to improve manufacturing techniques. ProMetal® is one of the divisions in the Ex One Company involved with RP. The ProMetal® 3D printing has the ability to build fully functional metal work pieces, casting molds and cores directly from CAD data within days. ProMetal® offers reduced time to market and greater design latitude for both internal and external geometries. The address for E address is 8075 Pennsylvania Avenue, Irwin, PA 15642, USA.

5.12.2. Products

ProMetal® systems are capable of creating multiple parts at one go. The systems available are categorized into Direct Metal (R-1 and R-2) and Rapid Casting (S15, S-Print and SR2), which offer different build volumes. ProMetal® R-1 is used for prototyping small, complex parts for research and educational purposes. ProMetal® R-2 allows a much bigger build volume and is capable of fabricating extremely intricate metal tools and workpieces. ProMetal® S15™, S-Print™ and SR-2 are able to manufacture intricate molds and cores, which can be directly assembled for casting, thus abbreviating the entire casting process chain. ProMetal® S15™ has the largest build envelope and the system uses foundry grade materials. The ProMetal® Direct Metal R2 system is shown in Fig. 5.33. Figure 5.34 shows a photograph of the Rapid Casting S15 system. Specifications of the machines are summarized in Tables 5.11 and 5.12.

5.12.3. Process

Utilizing 3DP technology, the ProMetal® process has the ability to build metal components by selectively binding metal powder layer by layer. The finished structural skeleton is then sintered and infiltrated with bronze to produce a finished part that is 60% steel and 40% bronze. The process consists of the following steps. Figure 5.35 illustrates the ProMetal® process schematically.[42]

Fig. 5.33. The ProMetal® Direct Metal R-2 machine (courtesy The Ex One Company, LLC).

Fig. 5.34. The ProMetal® Rapid Casting S15™ machine (courtesy The Ex One Company, LLC).

(1) A part is first designed on a computer using commercial CAD software.
(2) The CAD image is then transferred to the control unit. A very smooth layer of steel powder is then collected from the metal powder supply and spread onto the part build piston.
(3) The CAD image is printed with an ink jet print head depositing millions of droplets of binder per second. These droplets dry quickly upon deposition.

Table 5.11. Summarized specifications of ProMetal® machines.

	Model		
	R-1 system	R-2 system	SR-2
Build volume, mm	50.8 × 38.1 × 50.8	200 × 200 × 150	200 × 250 × 200
Build rate	1 min 20 s per layer	30–90 s per layer	
Layer thickness, μm	50–200	100–175 with fine resolution layers at 50 μm	200–250
User interface	Input file format is STL PC user interface Data transfer: Ethernet	Windows-based PC with color Touch Screen; STL CAD file input	Input file format is SLC; PC user interface; Data transfer: Ethernet
Machine dimension, mm	965 × 711 × 1066	1750 × 1220 × 1580	702 × 1169 × 1488
Power supply	120 VAC, 4.1 A, 60 Hz	240 VAC, 50 A, 50/60 Hz, 3 phase	120 VAC, 4.1A, 60 Hz

Table 5.12. Summarized specifications of ProMetal RCT™ machines.

	Model	
	S15™	S-Print™
Build volume, mm	1500 × 750 × 700	750 × 380 × 400
Build rate	15000 cm^3/h	7500 cm^3/h
Machine dimensions, mm	3113 × 3354 × 2164	2625 × 2450 × 2150
Weight, kg	4500	2500
Data interface	STL, CLI	STL, CLI
Power supply	5 kW, 400 V/3 phase	5 kW, 400 V/3 phase

(4) The part build piston lowers approximately 120–170 μm. The process is repeated until the part is completely printed.

(5) The resulting "green" part, of about 60% density, is removed from the machine and excess powder is brushed away.

(6) The "green" part is next sintered in a furnace, while burning off the binder. It is then infiltrated with molten bronze via capillary action to obtain full density. This is carried out in an infiltration furnace.

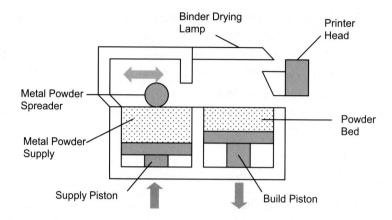

Fig. 5.35. Schematic illustration of the ProMetal® process.

(7) Post-processing include machining, polishing and coating to enhance wear and chemical resistance, e.g., nickel and chrome plating.

5.12.4.   Principle

The working principle of ProMetal's 3DP uses an electrostatic ink-jet printing head to deposit a liquid binder onto the powder metals. The part is built one layer at a time based on sliced cross-sectional data. The metal powder layer is spread on the build piston and a sliced layer is printed onto the powder layer by the ink-jet print head depositing droplets of binder that are in turn dried by the binder drying lamp.[43] The process is repeated until the part build is completed.

5.12.5.   Strengths and Weaknesses

The strengths of the ProMetal® process are as follows:

(1) *Fast*: The ProMetal® machine creates multiple parts simultaneously and not sequentially like laser systems. Interchangeable build chambers allow quick turnaround between jobs. Build rates can be over 4,000 cm^3 per hour.

(2) *Flexible*: Virtually no restriction on design flexibility, complex internal geometries and undercuts can be easily created.

(3) *Reliable*: There is auto-tuning and calibration for maximum performance and also built-in self-diagnostics and status reporting. The ink-jet printing process is simple and reliable.

(4) *Large parts*: Large steel mold parts measuring $1,016 \times 508 \times 254$ mm can be rapid prototyped.

The weaknesses of the ProMetal® process are as follows:

(1) *Large space required*: The machine needs a very large area to house it.

(2) *Limited materials*: The system only prototypes parts with its own metal powder.

5.12.6. Applications

The ProMetal™ 3DP is primarily used to rapidly fabricate complex stainless or steel tooling parts. Applications include injection molds, extrusion dies, direct metal components and blow molding.[44] The technology is also suitable for repairing worn out metal tools.

5.13. VOXELJET TECHNOLOGY GmbH's VX SYSTEM

5.13.1. Company

Voxeljet Technology GmbH was founded in 1999 by Ingo Ederer with the aim of providing new generative processes for production. The special know-how of the company lies in the connection between high-performance ink jet technology and rapid manufacturing.

The company developed a process, which enabled tool making to be revolutionized in the casting area. With the type designation GS 1500 now S15, the machine system was developed as one of the largest commercially available RP system in the world with its unique productivity. Using the equipment, sand casting molds can be generated without tools automatically. The system technology has been implemented successfully worldwide for customers such as BMW AG and Daimler-Chrysler AG.

In mid-2003, the company licensed the technology to the Extrudehone Corporation. Voxeljet then focused itself on process application, in order to be able to offer a fast generating of molds to a broad client base. Today, Voxeljet operates on three fields: in the supply of services, molds for casting, cast parts and plastic parts. In the field of printing and dosing technology, high-performance ink-jet systems are developed and produced for the most diverse applications. Industrial 3D printing systems for the manufacture of plastic components form the latest product line. The company address is Am Mittleren Moos 15, D-86167 Augsburg-Lechhausen, Germany.

5.13.2. Products

The Voxeljet produces two machines, the VX500 and VX800 (shown in Fig. 5.36), that produce thermoplastic models to order from 3D data, without tools automatically. A selectively bonded particle material is applied layer by layer to create models.

Once the building process is completed, the interchangeable container is removed from the system via a rail-guide system. As soon as a second

Fig. 5.36. The Voxeljets' VX800 (courtesy of Voxeljet Technology GmbH).

interchangeable container (optional accessory) is inserted, the system is ready to start operation again.

With the aid of an unpacking station and an industrial vacuum cleaner, the components are removed from the surrounding powder before being dried in a convection oven (both optional accessories). A summary of the VX500 and VX800 machine specifications are summarized in Table 5.13.

5.13.3. Principle

The principle of VX technology from Voxeljet is based on the 3DP (see Sec. 5.2).

5.13.4. Applications

Voxeljet Technology GmbH produces plastic parts, sand molds and cores for use in sand casting.

Table 5.13. Specifications of Voxeljet machines.

	Model	
	VX500	VX 800
Build volume, mm, $W \times D \times H$	$500 \times 400 \times 300$	$850 \times 450 \times 500$
Resolution x, y, mm	0.1×0.1 (250 dpi)	0.1×0.1 (250 dpi)
Layer thickness, mm	0.1–0.15	0.1×0.2
Build Speed, mm/h	12–48	12–48
Parts build material		Modified acrylic glass
Parts accuracy	0.3%; min ± 0.1 mm	0.3%; min ± 0.1 m
Data interface	STL	STL
Network connection	Ethernet	Ethernet
PC	Windows XP	Windows XP
Outer dimensions, mm, $W \times D \times H$	$1790 \times 1852 \times 1660$	$2600 \times 2350 \times 2300$
Weight, kg	Approximate 1200	Approximate 2500
Air pressure	6–10 bar	6–0 bar
Input supply voltage	400 V, 3 phase, 50 Hz	400 V, 3 phase, 50 Hz
Power consumption	16 A	16 A
Power consumption	<5 kW	<5 kW

5.14. OTHER NOTABLE POWDERED-BASED RP SYSTEMS

5.14.1. Soligen's Direct Shell Production Casting (DSPC)

5.14.1.1. *Company*

Soligen Technologies Inc. was founded by Yehoram Uziel, its President and CEO in 1991 and went public in 1993. It first installed its Direct Shell Production Casting (DSPC) System at three "alpha" sites in 1993. It bought the license to MIT's 3D printing patents for metal casting, which was valid till 2006.

5.14.1.2. *Product*

DSPC creates ceramic molds for metal parts with integral coves directly and automatically from CAD files.

5.14.1.3. *Process*

The DSPC technology is derived from a process known as 3DP and was invented and developed at the MIT, USA. The process steps are illustrated in Fig. 5.37 and are comprised of the following steps[45,46]:

(1) A part is first designed on a computer, using commercial computer-aided design (CAD) software.
(2) The CAD model is then loaded into the shell design unit — the central control unit of the equipment. The computer model for the casting mold is prepared by taking into considerations such as scaling the dimensions to compensate for shrinkage, adding fillets, etc. The mold maker then decides on the number of mold cavities on each shell and the type of gating system. Once the CAD mold shells are modified, the shell design unit generates the necessary data files to be transferred to the shell production unit.
(3) The shell production unit begins depositing a thin layer of fine alumina powder over the shell working surface for the first slice of the casting mold. A roller follows the powder, leveling the surface.

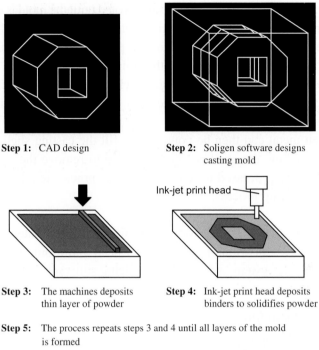

Step 1: CAD design

Step 2: Soligen software designs casting mold

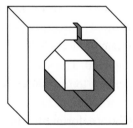

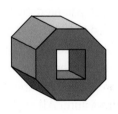

Ink-jet print head

Step 3: The machines deposits thin layer of powder

Step 4: Ink-jet print head deposits binders to solidifies powder

Step 5: The process repeats steps 3 and 4 until all layers of the mold is formed

Step 6: Loose powder is removed from the completed mold

Step 7: Molten metal is poured into mold to create the finished product

Fig. 5.37. Soligen Inc.'s DSPC process.

(4) An ink-jet print head moves over the layer, injecting tiny drops of colloidal silica binder onto the powder surface. The binder solidifies the powder into ceramic on contact and the unbounded alumina remains as support for the following layer. The work area lowers and another layer of powder is distributed.

(5) The process through steps 3 and 4 is repeated until the mold is complete.
(6) After the building process is completed, the casting shell remains buried in a block of loose alumina powder. The unbound excess powder is then separated from the finished shell. The shell can then be removed for post-processing, which may include firing in a kiln to remove moisture or pre-heat to appropriate temperature for casting.
(7) Molten metal can then be poured in to fill the casting shell or mold. After cooling, the shell can be broken up to remove the cast, which can then be processed to remove gatings, sprues, etc., thus completing the casting process.

The hardware of the DSPC system contains a PC computer, a powder holder, a powder distributor, rolls, a print head and a bin. The software includes a CAD system and a Soligen's slicing software.

5.14.1.4. *Principle*

The principle of Soligen's DSPC is based on 3DP technology invented, developed and patented by MIT. The 3DP is licensed to Soligen on a worldwide basis for the field of metal casting. Using this technology, binder from the nozzle selectively binds the ceramic particles together to create the ceramic shell layer by layer until it is completed. The shell is removed from the DSPC machine and fired to ventrification temperatures to harden and remove all moisture. All excess ceramic particles are blown away.

In the process, parameters that influence performance and functionality are the layer thickness, powder's properties, the binders and the pressure of rollers.

5.14.1.5. *Strengths and Weaknesses*

The key strengths of the DSPC process include the production of patternless casting and net-integral molds. In patternless casing, direct tooling is possible and thus eliminating the need to produce any patterns. There are no parting lines, core prints or draft angles required. Integral gatings and chills can also be appropriately added.

The main limitation of the DSPC process is that the DSPC 300 only focuses on making ceramics molds primarily for metal casting.

5.14.2. Fraunhofer's MJS

5.14.2.1. *Company*

The Fraunhofer–Gessellschaft is Germany's leading organization of applied research, maintaining 46 research establishments at 31 locations.[47] Two of these establishments, the Fraunhofer Institute for Applied Materials Research (IFAM) and the Fraunhofer Institute for Manufacturing Engineering and Automation (IPA) cooperated in developing an RP process named MJS.

5.14.2.2. *Process*

The MJS process is the one that is able to produce metallic or ceramic parts. It uses low-melting point alloys or powder-binder mixture, which is squeezed out through a computer-controlled nozzle to build the part layer by layer.[48]

The MJS process comprises two main steps: data preparation and model building.

Data preparation: In the first step, the part is designed on a 3D CAD system. The data are then imported to the MJS system and together with process parameters such as machining speed and materials flow rate added, a controller file for the machine is then generated. This file is subsequently downloaded onto the controller of the machine for the build process.

Model building: The material used is usually a powder-binder mixture, but it can also be a liquefied alloy. At the beginning of the build process, the material is heated to beyond its melting point in a heated chamber. It is then squeezed out through a computer-controlled nozzle by a pumping system and deposited layer by layer onto a platform.[48,49] The melted material solidifies when it comes into contact with the platform or the previous layer as both temperature and pressure decrease and heat is dissipated to the part and the surrounding. The contact of the liquefied material leads to partial remelting of the previous layer and thus bonding between layers results. After one layer is finished, the extrusion jet moves in the Z-direction and the next layer is built. The part is built layer by layer until it is complete.

The main components of the apparatus used for the MJS process comprises a PC, a computer-controlled three-axes positioning system and a heated chamber with a jet and a hauling system. The machine precision of the positioning system is ± 0.01 mm and has a work volume of $500 \times 540 \times 175$ mm. The chamber is temperature stabilized within ± 1°C. The material is supplied as powder, pellets, or bars. Extrusion temperature can reach up to 200°C. Extrusion orifices vary from 0.5 to 2.0 mm.

5.14.2.3. *Principle*

The working principle of the MJS process[50] is shown in Fig. 5.38. The basic concept applies the extrusion of low-viscosity materials through a jet layer by layer, similar to the fused deposition modeling process.

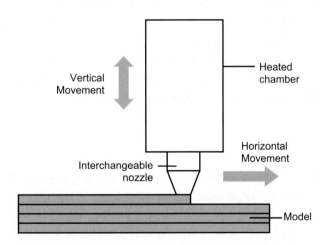

Material is heated above its melting point and squeezed through a nozzle and deposited layer by layer.

Fig. 5.38. Working principle of the MJS process.

The main differences between the two processes are in the raw material used to build the model and the feeding system. For the MJS process, the material is supplied in different phases using power-binder mixture or liquefied alloys instead of using material in the wire form. As the form of the material is different, the feed and nozzle systems are also different. The material used is a wax loaded with up to approximately 50% volume fraction metal powder.

In the MJS process, parameters that influence its performance and functionality are the layer thickness, the feed material, i.e., whether it is liquefied alloys (usually low-melting-point metals) or powder-binder mixture (usually materials with high melting point), the chamber pressure, the machining speed (build speed), the jet specification, the material flow and the operating temperature.

5.14.3. AeroMet Corporation's LASFORM Technology

5.14.3.1. *Company*

In 1997, AeroMet™ was formed as a subsidiary of MTS Systems Corporation (MTS). The initial Lasform™ system installed at its Eden Prairie, Minnesota facility is operated in collaboration with US Army Research Laboratory (ARL) of Aberdeen Proving Ground, Maryland. The AeroMet™ Lasform™ technology is based on research performed jointly by the Applied Physics Laboratory of John Hopkins University, the Applied Research Laboratory of Penn State University and MTS Systems Corporation. AeroMet Corporation ceased operation in 2005.

5.14.3.2. *Products*

The Lasform™ process uses commercially available precursor materials and creates parts that require minimal post-machining or heat treatment prior to use. As the precursor material is in the form of metal powders, it is also possible to produce "graded alloys" across the geometry of a component via real-time mixing of elemental constituents. This is a very unique feature of great interest to RP users and designers.[51] Since the AeroMet™ process takes place in an inert environment, it is possible to use the Lasform process in niobium, rhenium and other reactive materials, which require protective processing atmospheres.

5.14.3.3. *Process*

The Lasform[SM] process is described as follows[52]:

(1) The AeroMet[TM] laser forming process starts with a 3D CAD representation of the part. This is then translated via proprietary software to generate trajectory paths for the laser-forming system. These paths are transmitted as machine instructions to the laser forming system.
(2) The focused laser beam traces out the structural shape pattern of the desired part by moving the titanium target plate beneath the beam in the approximate x–y trajectories.
(3) Titanium pre-alloyed powder is introduced into the molten metal head and provides for the buildup of the desired shape as the molten spot is traversed over a target plate in the desired pattern.
(4) The 3D structure is fabricated by repeating the pattern, layer by layer over the desired geometry and indexing the focal point up one layer for the repeat pattern. This layer-by-layer registry with metallurgical integrity between layers generates the desired integral ribbed structure called a machining pre-form. Post-processes include heat treatment, machining and inspection.

5.14.3.4. *Principle*

The Lasform[SM] process uses gas atomized and hydride–dehydride titanium alloy powders introduced into the focus region of the CO_2 laser beam.[53] The focus region is shifted in the X–Y plane as determined by the CAD slice. This is achieved by driving a numerical controlled manipulator to reproduce the desired shape. A solid titanium deposit layer remains and the process is repeated for the next layer in the Z-direction. The new layer is also fused with the previous one, building layer upon layer until the part is complete. The process is carried out in an argon-filled environment. The production of high-quality titanium shapes via laser direct metal deposition requires the integration of several technologies. These include high power laser beam generation and delivery, metal powder handling, robotics, process sensing and control and environmental management.

5.14.3.5. *Strengths and Weaknesses*

The strengths of the Lasform^{SM} includes the following. It allows very high-quality titanium parts to be rapidly produced. Parts mechanical tested showed that AMS standards for commercially pure Ti, Ti-6Al-4V and Ti-5Al-2.5Sn are met. As it has a very large work volume, large parts can be seamlessly accommodated, while maintaining an oxygen-free inert atmosphere. The laser-formed shapes also require minimal post-machining and heat treating. This provides substantial cost and time savings by eliminating high materials waste, costly manufacturing tooling and long machining times. The process is flexible for its ability to vary the composition of the material throughout the part that could result in the formation of a functionally gradient material where the microstructure and mechanical properties can vary depending on the composition. Thus, the composition of different regions of the part can be customized according to functional and economic requirements.

The Lasform^{SM}, however, requires a very large space to be dedicated to house the system. It also specializes only in the fabrication of titanium and other metals parts. It therefore cannot be used to produce polymer parts or parts to be used as nonfunctional models.

REFERENCES

1. 3D Systems, Products: SLS® Systems [On-line], available at http://www.3dsystems.com/products/sls/index.asp (June 2007).

2. 3D Systems, Product Brochures and Datasheets (Material Specifications) (2006).

3. J. L. Johnson, *Principles of Computer Automated Fabrication*, Chapter 3, Palatino Press (1994), pp. 75–84.

4. M. S. M. Sun, J. C. Nelson, J. J. Beaman and J. J. Barlow, A model for partial viscous sintering, *Proc. Solid Freeform Fabrication Symposium*, University of Texas (1991).

5. W. F. Hug and P. F. Jacobs, Laser technology assessment for stereolithographic systems, *Proc. Second Int. Conf. Rapid Prototyping* (23–26 June 1991), pp. 29–38.

6. J. J. Barlow, M. S. M. Sun and J. J. Beaman, Analysis of selective laser sintering, *Proc. Second Int. Conf. Rapid Prototyping* (23–26 June 1991), pp. 29–38.

7. 3D Systems (DTM Corp.), Horizons Q4 (1999), pp. 6–7.

8. 3D Systems, SLS® Technology Featured at the "Extreme Textiles" Exhibition at the Smithsonian's Cooper-Hewitt, National Design Museum, [On-line], available at http://www.3dsystems.com/newsevents/ newsreleases/pdfs/3D_Systems_SLS_technology_featured_at_ Extreme.pdf (June 2005).

9. 3D Systems, Bertrandt [On-line], available at http://www.3dsystems. com/ appsolutions/casestudies/pdf/CS_Bertrandt.pdf (June 2007).

10. 3D Systems, 4D Concepts for Hekatron [On-line], available at http://www.3dsystems.com/appsolutions/casestudies/pdf/CS_4D_ Concepts.pdf (2006).

11. Z Corporation, Z Corporation Raises 3D Printing Quality and Affordability Standard with ZPrinter 310 Plus [On-line], available at http://www.zcorp.com/Press-Room/Z-Corporation-Raises-3D-Printing/ news.aspx (2005).

12. Solid Caddgroup Inc., Imakenews.com, *Autoformer* 11, 6 February (2002).

13. Z Corporation, Materials options [On-line], available at https://www.zcorp.com/Products/3D-Printers/Material-Options/ spage.aspx (July 2006).

14. Z Corporation, 3D Printing White Paper [On-line], available at https://www. zcorp.com/documents/108_3D%20Printing%20White %20Paper%20FINAL.pdf (April 2006).

15. Z Corporations, Solutions, [On-line], available at https://www.zcorp. com/Solutions/Rapid-Prototypes—CAD/spage.aspx (April 2008).

16. C. Goldsberry, Rapid processes gaining ground, *Mod. Plastics Worldwide* **84**(9), 35–38 (2007).

17. Annonymous, EOS takes fine approach to laser sintering, *Metal Powder Rep.* **56**(3), 18 (2001).

18. A. Grochowski, Rapid prototyping — Rapid tooling, CADCAM Forum (2000), 39–41.

19. J. C. W. Serbin, C. Pretsch and M. Shellabear, STEREOS and EOSINT 1995 — New Developments and State of the Art, *EOS* GmbH (1995).

20. EOS GmbH, Success Stories [On-line], available at http://www.eos. info/jp/newsevents/success-stories/success-stories/select/success-stories/article/21/weltpremiere.html (July 2006).

21. P. Mognol, D. Lepicart and N. Perry, Rapid prototyping: Energy and environment in the spotlight, *Rapid Prototyping J.* **12**(1), 26–34 (2006).

22. W. Hofmeister, M. Wert, J. E. Smugeresky, J. A. Philliber, M. Griffith and M. Ensz, Investigating solidification with the laser-engineered net-shaping (LENS™) process, *J. Mater.* **5**(7) (1999).

23. Z. Xu, R. S. Amano and P. Rohatgi, A study on laser engineered net shaping prototyping technology, *Proc. ASME Int. Design Engineering Technical Conf. and Computers and Information in Engineering Conf.* DETC2007 2 (B) (2008), pp. 717–722.

24. NDS.com, Rapid prototyping to market, *Manufacturing Industry News* (27 August 2001).

25. Y. Xiong, J. E. Smugeresky, L. Ajdelsztajn and J. M. Schoenung, Fabrication of WC-Co cermets by laser engineered net shaping, *Mater. Sci. Eng.* **A 496**(1–2), 261–266 (2008).

26. J. Y. Hwang, A. Neira, T. W. Scharf, J. Tiley and R. Banerjee, Laser-deposited carbon nanotube reinforced nickel matrix composites, *Script. Mater.* **59**(5), 487–490 (2008).

27. M. Hedges, M. Renn and M. Kardos, Mesoscale deposition technology for electronics applications, *Polytronic 2005: 5th Int. Conf. Polymers and Adhesives in Microelectronics and Photonics — Proc. 2005*, art. No. 1596486 (2005), pp. 53–57.

28. V. Zollmer, M. Muller, M. Renn, M. Busse, I. Wirth, D. Godlinski and M. Kardos, Printing with aerosols, *Eur. Coat. J.* **7–9**, 46–50 (2006).

29. V. R. Marinov, Y. A. Atanasov, A. Khan, D. Vaselaar, A. Halvorsen, D. L. Schulz and D. B. Chrisey, Direct-write vapor sensors on FR4 plastic substrates, *IEEE Sensors J.* **7**(6), 937–944 (2007).

30. M. Carter, T. Amundson, J. Colvin and J. Sears, Characterization of soft magnetic nano-material deposited with M3D technology, *J. Mater. Sci.* **42**(5), 1828–1832 (2007).

31. A. B. Arcam, Electron Beam Melting [On-line], available at http://www. arcam.com/technology/tech_ebm.asp (June 2007).

32. G. J. Gibbons and R. G. Hansell, Direct tool steel injection mold inserts through the Arcam EBM free-form fabrication process, *Assembly Automat.* **25**(4), 300–305 (2005).

33. O. Edelmann, LaserCUSING: Improved mold manufacturing and reduced cycle times, *Society of Plastic Engineers — Modern Moldmaking Techniques Conf.* (2004), pp. 131–161.

34. Prototype Magazine, MCP SLM Machines [On-line], available at http://www.prototypemagazine.com/index.php?option=com_content&task= view&id=102&Itemid=2 (July 2007).

35. M. Van Elsen, F. Al-Bender and J. Krith, Application of dimensional analysis to selective laser melting, *Rapid Prototyping J.* **14**(1), 15–22 (2008).

36. Y. Yang, Y. Huang and W. Wu, One-step shaping of NiTi biomaterial by selective laser melting, *Proc. SPIE — The Int. Society for Optical Engineering,* art. No. 68250C (2008).

37. Sintermask Technologies AB, Sintermask Technologies [On-line], available at http://www.sintermask.se/page.php?p=7 (August 2007).

38. T. Petsch, P. Regenfuß, R. Ebert, L. Hartwig, S. Klötzer, T. H. Brabant and H. Exner, Industrial Laser Micro Sintering, *Proc. 23rd Int. Congress on Applications of Lasers and Electro-Optics*, San Francisco, (October 2004), pp. 4–7.

39. Therics Inc., Theriform™ Technology [On-line], available at http://www. therics.com/technology/ (July 2007).

40. T. Livingston, Scaffolds for stem cell-based tissue regeneration, *Biomaterials News* (Spring 2002).

41. P. Sharke, Rapid transit to manufacturing, *Mech. Eng.*, ASME 1–6 (2001).

42. The Editors, Added options for producing "impossible" shapes, rapid traverse — Technology and trends spotted by the Editors of Modern Machine Shop, MMS Online (11 September 2001).

43. G. S. Vasilash, A quick look at rapid prototyping, *Automot. Des. Product.* 3 (September 2001).

44. P. J. Waterman, RP3: Rapid prototyping, pattern making and production, *DE Online* (March 2000).

45. Y. Uziel, Art to part in 10 days, *Machine Design* (10 August 1995), pp. 56–60.

46. A. Gregor, From PC to factory, *Los Angeles Times* (12 October 1994).

47. Fraunhofer–Gesellschaft, Profile of the Fraunhofer–Gesellschaft: Its purpose, capabilities and prospects, Fraunhofer–Gesellschaft (1995).

48. M. Geiger, W. Steger, M. Greul and M. Sindel, Multiphase jet solidification, European action on rapid prototyping, *EARP Newsletter* **3** (1994).

49. Rapid Prototyping Report, Multiphase Jet Solidification (MJS), CAD/CAM Publishing Inc. **4**(6), 5 (June 1994).

50. M. Greulick, M. Greul and T. Pitat, Fast, functional prototypes via multiphase jet solidification, *Rapid Prototyping J.* **1**(1), 20–25 (1995).

51. F. G. Arcella, E. J. Whitney and D. Krantz, Laser forming near shapes in titanium, *ICALEO '95: Laser Materials Processing*, Vol. 80 (1995), pp. 178–183.

52. D. H. Abott and F. G. Arcella, AeroMet implementing novel Ti process, *Metal Powder Rep.* **53**(2), 24 (1998).

53. F. G. Arcella, D. H. Abbot and M. A. House, Rapid laser forming of titanium structures, *Metallurgy World Conf. and Exposition*, Grenada, Spain 18–22 October 1998.

PROBLEMS

1. Using a sketch to illustrate your answer, describe the Selective Laser Sintering process.
2. Discuss the types of materials available for the Sinterstation HiQ Series.
3. Describe the differences between the Selective Laser Sintering process and the three-dimensional printing process.
4. List the advantages and disadvantages of the three-dimensional printing process.
5. Present the types of EOS systems and their capabilities in terms of materials processing.
6. Compare and contrast the laser-based EOS process with the nonlaser-based Direct Shell Production Casting for the production of foundry casting molds.
7. What are the critical factors that influence the performance and applications of the following RP processes?
   (a) 3D Systems' SLS,
   (b) Z-Corporation's 3DP,
   (c) Optomec's LENS,
   (d) Arcam's EBM,
   (e) Theric's Theriform.
8. Name three laser powder-based RP systems and three nonlaser powder-based RP systems.

9. Compare and contrast EOS's EOSINT M system with Optomec's LENS system. What are the advantages and disadvantages for each of these systems?

10. Describe the system and process for Optomec's Maskless Mesoscale Materials Deposition (M^3D) system.

11. Describe Arcam's Electron Beam Melting (EBM) technology.

12. List the advantages and disadvantages for the Concept Laser's LaserCUSING process.

13. Discuss the advantages and disadvantages of powder-based RP systems compared with
    (a) liquid-based RP systems,
    (b) solid-based RP systems.

14. List the significances of MCP-HEK's selective laser melting process?

15. Discuss possible applications for the Phenix Systems' PM Series.

16. Describe the process and principles of the Selective Mask Sintering process by Speed Part.

17. What are the key features of the 3D Micromac AG's MicroSINTERING process?

18. Discuss the differences between Ex One's Prometal™ process and that of the Selective Laser Sintering process for the production of metal parts.

19. List Voxeljet's field of operation for their RP system.

20. Discuss the processes, strengths and limitations of the following RP processes:
    (a) Direct Shell Pattern Casting,
    (b) Multijet Solidification,
    (c) Lasform Technology.

Chapter 6
RAPID PROTOTYPING DATA FORMATS

6.1. STL FORMAT

Representation methods used to describe CAD geometry vary from one system to another. A standard interface is needed to convey geometric descriptions from various CAD packages to rapid prototyping (RP) systems. The STL (STereoLithography) file, as the *de facto* standard, has been used in many, if not all, RP systems.

The STL file,[1-3] conceived by the 3D Systems, USA, is created from the CAD database via an interface on the CAD system. This file consists of an unordered list of triangular facets representing the outside skin of an object. There are two formats of the STL file. One is the ASCII format and the other is the binary format. The size of the ASCII STL file is larger than that of the binary format but is human readable. In an STL file, triangular facets are described by a set of X, Y and Z coordinates for each of the three vertices and a unit normal vector with X, Y and Z to indicate the side of the facet, which is inside or outside the object. An example is shown in Fig. 6.1.

Because the STL file is a facet model derived from precise CAD drawings, it is, therefore, an approximate model of the part. Also, many commercial CAD models are not robust enough to generate the facet model (STL file) and frequently have problems.

Nevertheless, there are several advantages of the STL file. First, it provides a simple method of representing 3D CAD data. Second, it is already a *de facto* standard and has been used by most CAD systems and RP systems. Finally, it can provide small and accurate files for data transfer for certain shapes.

On the other hand, several disadvantages of the STL file exist. First, the STL file is many times larger than the original CAD data file for a given accuracy parameter. The STL file carries much redundant information such as duplicate vertices and edges shown in Fig. 6.2. Second, geometry

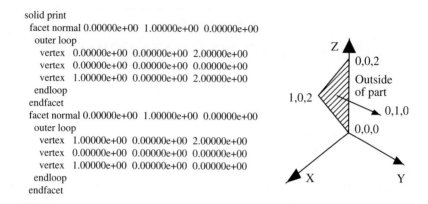

Fig. 6.1. A sample STL file.

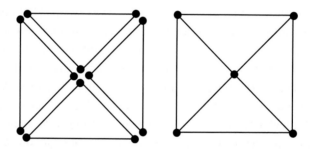

Fig. 6.2. Edge and vertex redundancy in STL format.

flaws exist in the STL file because many commercial tessellation algo-rithms used by CAD vendors today are not robust enough. This gives rise to the need for a "repair software", which slows the production cycle time. Finally, the subsequent slicing of large STL files can take many hours. However, some RP processes can slice while they are building the previous layer and this will alleviate this disadvantage.

6.2.　STL FILE PROBLEMS

Several problems plague STL files and they are due to the very nature of STL files as they contain no topological data. Many commercial tessellation algorithms used by CAD vendors today are also not robust,[4-6] and as a result they tend to create polygonal approximation models, which exhibit the following types of errors:

(1) gaps (cracks, holes, punctures), that is, missing facets;
(2) degenerate facets (where all its edges are collinear);
(3) overlapping facets;
(4) nonmanifold topology conditions.

The underlying problem is due, in part, to the difficulties encountered in tessellating trimmed surfaces, surface intersections and controlling numerical errors. This inability of the commercial tessellation algorithm to generate valid facet model tessellations makes it necessary to perform model validity checks before the tessellated model is sent to the RP equipment for manufacturing. If the tessellated model is invalid, procedures become necessary to determine what the specific problems are, whether they are due to gaps, degenerate facets or overlapping facets, etc.

Early research has shown that repairing invalid models is difficult and not at all obvious.[7] However, before proceeding any further to discuss the procedures that are generated to resolve these difficulties, the following sections shall clarify what the problems, as mentioned earlier, are. In addition, an illustration would be presented to show the consequences brought about by a model having a missing facet, that is, a gap in the tessellated model.

6.2.1. Missing Facets or Gaps

Tessellation of surfaces with large curvature can result in errors at the intersections between such surfaces, leaving gaps or holes along the edges of the part model.[8] A surface intersection anomaly, which results in a gap, is shown in Fig. 6.3.

6.2.2. Degenerate Facets

A geometrical degeneracy of a facet occurs when all of the facets' edges are collinear even though all its vertices are distinct. This might be caused

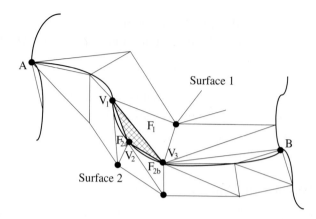

Fig. 6.3. Gaps due to missing facets.[4]

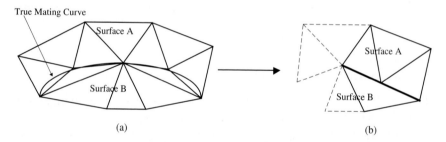

Fig. 6.4. Shell punctures (a) created by unequal tessellation of two adjacent surface patches along their common mating curve and (b) eliminated at the expense of adding degenerate facet.

by stitching algorithms that attempt to avoid shell punctures as shown in Fig. 6.4(a).[9]

The resulting facets generated, as shown in Fig. 6.4(b), eliminate the shell punctures. However, this is done at the expense of adding a degenerate facet. While degenerate facets do not contain valid surface normals, they do represent implicit topological information on how two surfaces mate. This important information is consequently stored prior to discarding the degenerate facet.

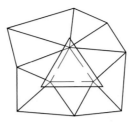

Fig. 6.5. Overlapping facets.

6.2.3. Overlapping Facets

Overlapping facets may be generated due to numerical round-off errors occurring during tessellation. The vertices are represented in 3D space as floating point numbers instead of integers. Thus, the numerical round-off can cause facets to overlap if tolerances are set too liberally. An example of an overlapping facet is illustrated in Fig. 6.5.

6.2.4. Nonmanifold Conditions

There are three types of nonmanifold conditions, namely a nonmanifold edge, a nonmanifold point and a nonmanifold face.

These may be generated because tessellations of the fine features are susceptible to round-off errors. An illustration of a nonmanifold edge is shown in Fig. 6.6(a). Here, the nonmanifold edge is actually shared by four different facets as shown in Fig. 6.6(b). A valid model would be one whose facets have only an adjacent facet each, that is, one edge shares two facets only. Hence, the nonmanifold edges must be resolved such that each facet has only one neighboring facet along each edge, that is, reconstructing a topologically manifold surface.[4] Shown in Figs. 6.6(c) and 6.6(d) are two other types of nonmanifold conditions.

All problems that have been mentioned previously are difficult for most slicing algorithms to handle and cause fabrication problems for RP processes, which essentially require valid tessellated solids as input. Moreover, these problems arise because tessellation is a first-order

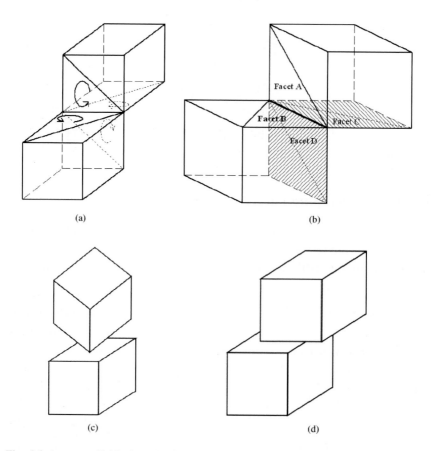

(a)   (b)

(c)   (d)

Fig. 6.6. A nonmanifold edge (a) whereby two imaginary minute cubes share a common edge and (b) whereby four facets share a common edge after tessellation. (c) Nonmanifold point and (d) nonmanifold face.

approximation of more complex geometric entities. Thus, such problems have become almost inevitable as long as representation of the solid model is done using the STL format, which inherently has these limitations.

6.3.   CONSEQUENCES OF BUILDING VALID AND INVALID TESSALLATED MODELS

The following sections present an example each of the outcomes of a model built using a valid and an invalid tessellated model as an input to the RP systems.

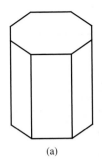
(a)

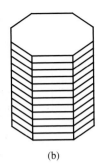
(b)

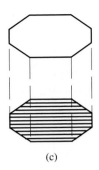
(c)

Fig. 6.7. (a) A valid 3D model, (b) a 3D model sliced into 2D planar layers and (c) conversion of 2D layers into 1D scan lines.

6.3.1. A Valid Model

A tessellated model is said to be valid if there are no missing facets, degenerate facets, overlapping facets, or any other abnormalities. When a valid tessellated model (see Fig. 6.7(a)) is used as an input, it will first be sliced into 2D layers, as shown in Fig. 6.7(b). Each layer would then be converted into unidirectional (or 1D) scan lines for the laser or other RP techniques to commence building the model as shown in Fig. 6.7(c). The scan lines would act as on/off points for the laser beam controller so that the part model can be built accordingly without any problems.

6.3.2. An Invalid Model

However, if the tessellated model is invalid, a situation may develop as shown in Fig. 6.8.

A solid model is tessellated nonrobustly and results in a gap as shown in Fig. 6.8(a). If this error is not corrected and the model is subsequently sliced, as shown in Fig. 6.8(b), in preparation for it to be built layer by layer, the missing facet in the geometrical model would cause the system to have no pre-defined stopping boundary on the particular slice, thus the building process would continue right to the physical limit of the RP machine, creating a stray physical solid line and ruining the part being produced, as illustrated in Fig. 6.8(c).

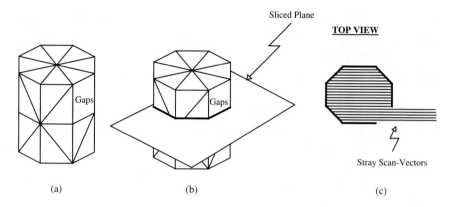

Fig. 6.8. (a) An invalid tessellated model, (b) an invalid model being sliced and (c) a layer of an invalid model being scanned.

Therefore, it is of paramount importance that the model be "repaired" before it is sent for building. Thus, the model validation and repair problem are stated as follows:

> Given a facet model (a set of triangles defined by their vertices), in which there are gaps, i.e., missing one or more sets of polygons, generate "suitable" triangular surfaces which "fill" the gaps.[4]

6.4.   STL FILE REPAIR

The STL file repair can be implemented using a generic solution and dedicated solutions for special cases.

6.4.1.   Generic Solution

In order to ensure that the model is valid and can be robustly tessellated, one solution is to check the validity of all the tessellated triangles in the model. This section presents the basic problem of missing facet and a proposed generic solution to solve the problem with this approach.

In existing RP systems, when a punctured shell is encountered, the course of action taken usually requires a skilled technician to manually repair the shell. This manual shell repair is frequently done without any knowledge of the designer's intent. The work can be very time-consuming and tedious, thus negating the advantages of RP as the cost would increase and the time taken might be longer than that taken if traditional prototyping processes were used.

The main problem of repairing the invalid tessellated model would be that of matching the solution to the designer's intent when it may have been lost in the overall process. Without the knowledge of the designer's intent, it would indeed be difficult to determine what the "right" solution should be. Hence, an "educated" guess is usually made when faced with ambiguities of the invalid model.

The algorithm for a generic solution to solve the "missing facets" problem aims to match, if not exceed, the quality of repair done manually by a skilled technician when information of the designer's intent is not available. The basic approach of the algorithm would be to detect and identify the boundaries of all the gaps in the model. Once the boundaries of the gap are identified, suitable facets would then be generated to repair and "patch up" these gaps. The size of the generated facets would be restricted by the gap's boundaries while the orientation of its normal would be controlled by comparing it with the rest of the shell. This is to ensure that the generated facet orientation is correct and consistent throughout the gap closure process.

The orientation of the shell's facets can be obtained from the STL file, which lists its vertices in an ordered manner following Mobius' rule. The algorithm exploits this feature so that the repair carried out on the invalid model, using suitably created facets, would have the correct orientation.

Thus, this generic algorithm can be said to have the ability to make an inference from the information contained in the STL file so that the following two conditions can be ensured:

(1) The orientation of the generated facet is correct and compatible with the rest of the model.
(2) Any contoured surface of the model would be followed closely by the generated facets due to the smaller facet generated. This is in contrast to manual repair whereby, in order to save time, fewer facets are

generated to close the gaps desired, resulting in large generated facets that do not follow closely to the contoured surfaces.

Finally, the basis for the working of the algorithm is due to the fact that in a valid tessellated model, there must only be two facets sharing every edge. If this condition is not fulfilled, then this indicates that there are some missing facets. With the detection and subsequent repair of these missing facets, the problems associated with an invalid model can then be eliminated.

6.4.1.1. *Solving the "Missing Facets" Problem*

The following procedure illustrates the detection of gaps in the tessellated model and its subsequent repair. It is carried out in four steps:

Step 1:   Checking for approved edges with adjacent facets

The checking routine executes as follows for Facet A as seen in Fig.6.9:

(a) (i)   Read in first edge {vertex 1–2} from the STL file.
     (ii)   Search file for a similar edge in the opposite direction {vertex 2–1}.

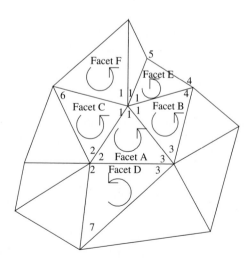

Fig. 6.9. A representation of a portion of a tessellated surface without any gaps.

(iii) If edge exists, store this under a temporary file (e.g., file B) for approved edges.

(b) (i) Read in second edge {vertex 2–3} from the STL file.
  (ii) Search file for a similar edge in the opposite direction {vertex 3–2}.
  (iii) Perform as in a(iii).

(c) (i) Read in third edge {vertex 3–1} from the STL file.
  (ii) Search file for a similar edge in the opposite direction {vertex 1–3}.
  (iii) Perform as in a(iii).

This process is repeated for the next facet until all the facets have been searched.

Step 2: Detection of gaps in the tessellated model

With reference to Fig. 6.10, the detection routine executes as follows:

(a) (i) For Facet A, read in edge {vertex 2–3} from the STL file.
  (ii) Search file for a similar edge in the opposite direction {vertex 3–2}.

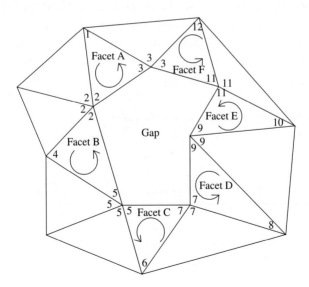

Fig. 6.10. A representation of a portion of a tessellated surface with a gap present.

(iii) If edge does not exist, store edge {vertex 3–2} in another temporary file (e.g., file C) for suspected gap's bounding edges and store vertex 2–3 in file B1 for existing edges without adjacent facets (this would be used later for checking the generated facet orientation).

(b) (i) For Facet B, read in edge {vertex 5–2} from the STL file.
    (ii) Search file for a similar edge in the opposite direction {vertex 2–5}.
    (iii) If it does not exist, perform as in a(iii) above.

(c) (i) *Repeat for edges*:   5–2; 7–5; 9–7; 11–9; 3–11.
    (ii) *Search for edges*:   2–5; 5–7; 7–9; 9–11; 11–3.
    (iii) Store all the edges in that temporary file B1 for edges without any adjacent facet and store all the suspected bounding edges of the gap in temporary file C. File B1 can appear as in Table 6.1.

Step 3:   Sorting of erroneous edges into a closed loop

When the checking and storing of edges (both with and without adjacent facets) are completed, a sort would be carried out to group all the edges without adjacent facets to form a closed loop. This closed loop would represent the gap detected and be stored in another temporary file (e.g., file D) for further processing. The following is a simple illustration of what could be stored in file C for edges that do not have an adjacent edge:

Assuming all the "erroneous" edges are stored according to the detection routine (see Fig. 6.10 for all the erroneous edges), then file C can appear as in Table 6.2.

As can be seen in Table 6.2, all the edges are all unordered. Hence, a sort would have to be carried out to group all the edges into a closed loop.

Table 6.1. File B1 contains existing edges without adjacent facets.

Vertex	Edge					
	First	Second	Third	Fourth	Fifth	Sixth
First	2	7	3	5	9	11
Second	3	5	11	2	7	9

Table 6.2. File C containing all the "erroneous" edges that would form the boundary of each gap.

Vertex					Edge				
	First	Second	Third	Fourth	Fifth	Sixth	Seventh	Eighth	Ninth
First	3	5	*	11	2	7	*	9	*
Second	2	7	*	3	5	9	*	11	*

* represents all the other edges that would form the boundaries of other gaps.

Table 6.3. File D containing sorted edges.

	First edge	Second edge	Third edge	Fourth edge	Fifth edge	Sixth edge
First vertex	3	2	5	7	9	11
Second vertex	2	5	7	9	11	3

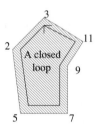

Fig. 6.11. A representation of a gap bounded by all the sorted edges.

When the edges have been sorted, it would then be stored in a temporary file, say file D. Table 6.3 is an illustration of what could be stored in file D.

Figure 6.11 is a representation of the gap, with all the edges forming a sorted closed loop.

Step 4:   Generation of facets for the repair of the gaps

When the closed loop of the gap is established with its vertices known, facets are generated one at a time to fill up the gap. This process is summarized in Table 6.4 and illustrated in Fig. 6.12.

Table 6.4. Process of facet generation.

		V3	V2	V5	V7	V9	V11
Generation	F1	1	2	—	—	—	3
of	F2	E	1	—	—	2	3
facets	F3	E	1	2	—	3	E
	F4	E	E	1	2	3	E

V, vertex; F, facet; and E, eliminated from the process of facet generation.

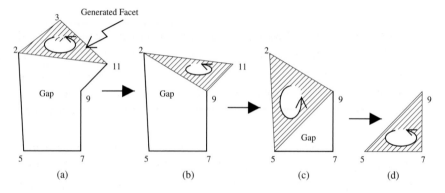

Fig. 6.12. (a) First facet generated, (b) second facet generated, (c) third facet generated and (d) fourth facet generated.

With reference to File D,

(a) *Generating the first facet*: First two vertices (V3 and V2) in the first two edges of file D will be connected to the first vertex in the last edge (V11) in file D and the facet is stored in a temporary file E (see Table 6.5 on how the first generated facet would be stored in file E). The facet is then checked for its orientation using the information stored in file B1. Once its orientation is checked to be correct, the first vertex (V3) from file D will be temporarily removed.

(b) *Generating the second facet*: Of the remaining vertices in file D, the previous second vertex (V2) will become the first edge of file D.

Table 6.5. Illustration of how data could be stored in file E.

Generated facet	First edge		Second edge		Third edge	
	First vertex	Second vertex	First vertex	Second vertex	First vertex	Second vertex
First	V3	V2	V2	V11	V11	V3
Second	V2	V9	V9	V11	V11	V2
Third	V2	V5	V5	V9	V9	V2
Fourth	V5	V7	V7	V9	V9	V5

The second facet is formed by connecting the first vertex (V2) of the first edge with that of the last two vertices in file D (V9, V11) and the facet is stored in temporary file E. It is then checked to confirm if its orientation is correct. Once it is determined to be correct, the vertex (V11) of the last edge in file D is then removed temporarily.

(c) *Generating the third facet*: The whole process is repeated as it was done in the generation of facets 1 and 2. The first vertex of the first two edges (V2, V5) is connected to the first vertex of the last edge (V9) and the facet is stored in temporary file E. Once its orientation is confirmed, the first vertex of the first edge (V2) will be removed from file D temporarily.

(d) *Generating the fourth facet*: The first vertex in the first edge will then be connected to the first vertices of the last two edges to form the fourth facet and it will again be stored in the temporary file E. Once the number edges in file D is less than 3, the process of facet generation will be terminated. After the last facet is generated, the data in file E will be written to file A and its content (file E's) will be subsequently deleted. Table 6.5 shows how file E may appear.

The above procedures work for both types of gaps whose boundaries consist of either odd or even number of edges. Figure 6.13 and Table 6.6 illustrate how the algorithm works for an *even* number of edges or vertices in file D.

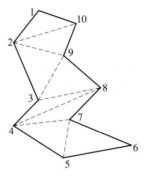

Fig. 6.13. Gaps with even number of edges.

Table 6.6. Process of facet generation for gaps with even number of edges.

Facets	Vertices									
	V1	V2	V3	V4	V5	V6	V7	V8	V9	V10
F1	1	2								3
F2	E	1							2	3
F3	E	1	2						3	E
F4	E	E	1					2	3	E
F5	E	E	1	2				3	E	E
F6	E	E	E	1			2	3	E	E
F7	E	E	E	1	2		3	E	E	E
F8	E	E	E	E	1	2	3	E	E	E

With reference to Table 6.6,

First facet generated:

Edge 1 → V1, V2
Edge 2 → V2, V10
Edge 3 → V10, V1

Second facet generated:

Edge 1 → V2, V9
Edge 2 → V9, V10
Edge 3 → V10, V1

and so on until the whole gap is covered. Similarly, Fig. 6.14 and Table 6.7 illustrate how the algorithm works for an *odd* number of edges or vertices in file D.

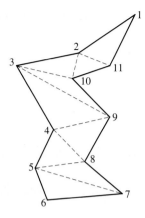

Fig. 6.14. Gaps with odd number of edges.

Table 6.7. Process of facet generation for gaps with odd number of edges.

Facets	V1	V2	V3	V4	V5	V6	V7	V8	V9	V10	V11
					Vertices						
F1	1	2									3
F2	E	1								2	3
F3	E	1	2							3	E
F4	E	E	1						2	3	E
F5	E	E	1	2					3	E	E
F6	E	E	E	1				2	3	E	E
F7	E	E	E	1	2			3	E	E	E
F8	E	E	E	E	1		2	3	E	E	E
F9	E	E	E	E	1	2	3	E	E	E	E

The process of facet generation for *odd* vertices is also done in the same way as *even* vertices. The process of facet generation has the following pattern:

F1→ First and second vertices are combined with the last vertex. Once completed, eliminate first vertex. The remainder is 10 vertices.

F2→ First vertex is combined with last two vertices. Once completed, eliminate the last vertex. The remainder is nine vertices.

F3→   First and second vertices are combined with the last vertex. Once completed, eliminate first vertex. The remainder is eight vertices.

F4→   First vertex is combined with last two vertices. Once completed, eliminate the last vertex. The remainder is seven vertices.

This process is continued until all the gaps are patched.

6.4.1.2. *Solving the "Wrong Orientation of Facets" Problem*

In the case when the generated facet's orientation is wrong, the algorithm should be able to detect it and take corrective action to rectify this error. Figure 6.15 shows how a generated facet with a wrong orientation can be corrected.

It can be seen that facet Z (vertices 1, 2, 11) is oriented in a clockwise direction and this contradicts the right-hand rule adopted by the STL format. Thus, this is not acceptable and needs correction.

This can be done by shifting the last record in file D of Table 6.8 to the position of the first edge in file D of Table 6.9. All the edges, including the original first edge will be shifted one position to the right (assuming that the records are stored in the left to right structure). Once this is done, Step 4 of facet generation can be implemented.

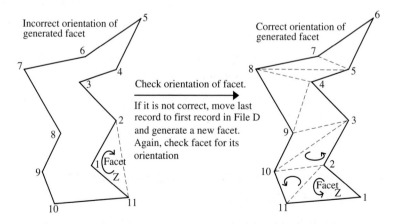

Fig. 6.15. Incorrect orientation of generated facet and its repair.

Table 6.8. Illustration showing how file D is manipulated to solve orientation problems.

	Before the shift										
	Edge										
Vertex	First	Second	Third	Fourth	Fifth	Sixth	Seventh	Eighth	Ninth	Tenth	Eleventh
First	1	2	3	4	5	6	7	8	9	10	11
Second	2	3	4	5	6	7	8	9	10	11	1

Table 6.9. Illustration showing the result of the shift to correct the facet orientation.

	After the shift										
	Edge										
Vertex	First	Second	Third	Fourth	Fifth	Sixth	Seventh	Eighth	Ninth	Tenth	Eleventh
First	11	1	2	3	4	5	6	7	8	9	10
Second	1	2	3	4	5	6	7	8	9	10	11

As can be seen from the above example, vertices 1 and 2 are used initially as the first edge to form a facet. However, this resulted in a facet having a clockwise direction. After the shift, vertices 1 and 11 are used as the first edge to form a facet.

Facet Z, as shown on the right-hand side of Fig. 6.15, is again generated (vertices 1, 2, 11) and checked for its orientation. When its orientation is correct (i.e., in the anti-clockwise direction), it is saved and stored in temporary file E.

All subsequent facets are then generated and checked for their respective orientations. If any of its subsequent generated facets has an incorrect orientation, the whole process would be restarted using the initial temporary file D. If all the facets are in the right orientation, it will then be written to the original file A.

6.4.1.3. *Comparison with an Existing Algorithm for Facet Generation*

An illustration of an existing algorithm that might cause a very narrow facet (shaded) to be generated is shown in Fig. 6.16.

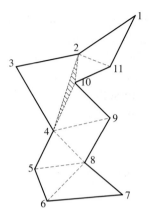

Fig. 6.16. Generation of facets using an algorithm that uses the smallest angle between edges.

This results from using an algorithm that uses the smallest angle to generate a facet. In essence, the problem is caused by the algorithm's search for a time-local rather than a global-optimum solution.[10] Also, calculation of the smallest angle in 3D space is very difficult.

Figure 6.16 is similar to Fig. 6.14. However, in this case, the facet generated (shaded) can be very narrow. In comparing the algorithms, the result obtained from the first algorithm would match, if not, exceed the algorithm that uses the smallest angle to generate a facet.

6.4.2. Special Algorithms

The generic solution presented could only cater to gaps (whether simple or complex) that were isolated from one another. However, should any of the gaps meet at a common vertex, the algorithm may not be able to work properly. In this section, the algorithm is expanded to include solving some of these special cases. These special cases include

(1) two or more gaps formed from a coincidental vertex;
(2) degenerate facets;
(3) overlapping facets.

The special cases are classified as such because these errors are not commonly encountered in the tessellated model. Hence, it is not advisable to include this expanded algorithm in the generic solution as it can be very time-consuming to apply in a during a normal search. However, if there are still problems in the tessellated model after the generic solution's repair, the expanded algorithm can then be used to detect and solve the special case problems.

6.4.2.1. *Two or More Gaps Formed from a Coincidental Vertex*

The first special case deals with problems where two or more gaps are formed from a coincidental vertex. Appropriate modifications to the general solution may be made according to the solutions discussed as follows. As can be seen from Fig. 6.17, there exists two gaps that are connected to vertex 1.

The algorithm for the generic solution would have a problem identifying which vertex to go to when the search reaches vertex 1 (either vertex 2 in gap 1 or vertex 5 in gap 2). Table 6.10 illustrates what file C would look like, given the two gaps.

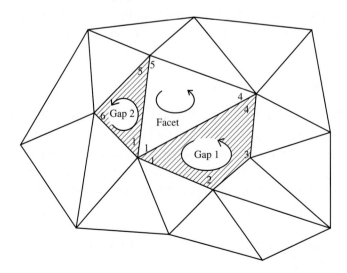

Fig. 6.17. Two gaps sharing one coincidental vertex.

Table 6.10. File C containing "erroneous" edges that would form the boundaries of gaps.

				File C			
				Edge			
Vertex	First	Second	Third	Fourth	Fifth	Sixth	Seventh
First	V3	V5	V4	*V1*	V6	V2	*V1*
Second	V4	V6	*V1*	*V5*	V1	V3	*V2*

When the search starts to find all the edges that would form a closed loop, the previous algorithm might mistakenly connect edges: 3–4, 4–1 and 1–5, instead of 1–2. This is clearly an error as the edge that is supposed to be included in file D should be 1–2 and *not* 1–5. It is therefore pertinent that for every edge searched, the second vertex (e.g., V1 of the third edge, shaded, in Table 6.10) of that edge should be searched against the first vertex of subsequent edges (e.g., V1 of edge 1–5 in the fourth edge) and this should not be halted the first time, the first vertex of subsequent edges is found. The search should continue to check if there are other edges with V1 (e.g., edge 1–2, in the seventh edge). Every time, say, for example, vertex 1 is found, there should be a count. When the count is more than two, that would indicate that there is more than one gap sharing the same vertex; this may be called the coincidental vertex. Once this happens, the following procedure would then be used:

Step 1:   Conducting of normal search for the boundary of gap 1

At the start of the normal search, the first edge of file C, vertices 3 and 4, as seen as in Table 6.11(a), is saved into a temporary file C1.

The second vertex in the first edge (V4) of file C1, is searched against the first vertex of subsequent edge in file C (refer to Table 6.11(a), shaded box). Once it is found to be the same vertex (V4) and that there are no other edges sharing the same vertex, the edge (i.e., vertex 4–1) is stored as the second edge in file C1 (refer to Table 6.11(b)).

Table 6.11(a). Representation of how files C and C1 would look like.

			File C				
				Edge			
Vertex	First	Second	Third	Fourth	Fifth	Sixth	Seventh
First	V3	V5	V4	V1	V6	V2	V1
Second	V4	V6	V1	V5	V1	V3	V2

	File C1		
		Edge	
Vertex	First	Second	Third
First	V3	?	?
Second	V4	?	?

Step 2: Detection of more than one gap

The second vertex of the second edge (V1) in file C1 is searched for an equivalent first vertex of subsequent edges in file C (refer to Table 6.11(b), shaded box containing vertex). Once it is found (V1, first vertex of fourth edge), a count of one is registered and at the same time, that edge is noted.

The search for the same vertex is continued to determine if there are other edges sharing the same vertex 1. If there is an additional edge sharing the same vertex 1 (Table 6.11(b), seventh edge), another count is registered, making a total of two counts. Similarly, the particular record in which it happens again is noted. The search is continued until there are no further edges sharing the same vertex. When completed, the following is carried out for the third edge in file C1.

For the first count, reading from file C, all the edges that would form the first alternative closed loop is sorted. These first alternatives in file C2 (refer to Table 6.11(c) and Fig. 6.18(a) for a graphical representation of

Table 6.11(b). Representation on how files C and C1 would look like during the normal search (special case).

			File C			
			Edge			
Vertex	First	Second	Third	Fourth	Fifth	Sixth	Seventh
---	---	---	---	---	---	---	---
First	V3	V5	V4	V1	V6	V2	V1
Second	V4	V6	V1	V5	V1	V3	V2
				↑			↑
				Count 1			Count 2

	File C1		
		Edge	
Vertex	First	Second	Third
First	V3	V4	?
Second	V4	V1	?

Table 6.11(c). First alternative closed loop that may represent a gap boundary.

			File C2			
			Edge			
Vertex	First	Second	Third	Fourth	Fifth	Sixth	Seventh
---	---	---	---	---	---	---	---
First	V3	V4	V1	V5	V6	V1	V2
Second	V4	V1	V5	V6	V1	V2	V3

how the first alternatives might look like) are then saved. A closed loop is established once the second vertex of the last edge (V3) is the same as the first vertex of the first edge (V3).

For the second count, the edges are sorted to form a second alternative of a closed loop that will represent the boundary of the gap. These

Table 6.11(d). Second alternative closed loop that may represent a gap boundary.

| Vertex | File C3 | | | |
| | Edge | | | |
	First	Second	Third	Fourth
First	V3	V4	V1	V2
Second	V4	V1	V2	V3

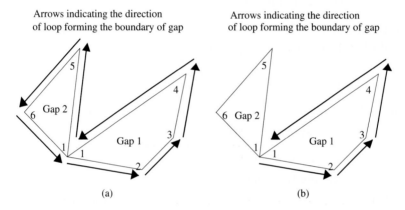

(a)   (b)

Fig. 6.18. Graphical representation of the two gaps sharing (a) a coincidental vertex (first alternative) and (b) the same vertex (second alternative).

edges are then saved in another temporary file C3 (refer to Table 6.11(d) and Fig. 6.18(b) for a graphical representation of how the second alternatives may look like). Once a closed loop is established, the search is stopped.

Step 3: Comparison of alternatives for least record to form
gap boundary

From file C2 (first alternative) and file C3 (second alternative), it can be seen that the second alternative has the least record to form the boundary of gap 1. Hence the second alternative data would be written to file D for the next stage of facet generation and the two temporary files C2 and C3 would be discarded.

Once gap 1 is repaired, gap 2 can be repaired by using the generic solution.

The algorithm can cater to more than two gaps sharing the same vertex, as shown in Fig. 6.19(a), or even three gaps arranged differently, as shown in Fig. 6.19(b).

Step 1:   Conducting of normal search

For the case shown Fig. 6.19(b), Tables 6.12(a) and 6.12(b) show how file C and file C1 may appear, respectively.

Step 2:   Detection of multiple gaps

Referring to Table 6.12(a), it can be seen that for V1, there are two counts and for V2, there are two counts.

For V1 → there are two counts.
For V2 → there are two counts.

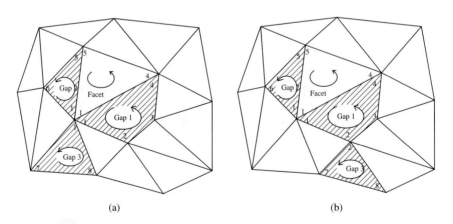

(a)   (b)

Fig. 6.19. Three facets sharing (a) one coincidental vertex and (b) two coincidental vertices.

Table 6.12(a). Illustration on how file C may appear given three gaps sharing same two vertices.

	File C									
	Edge									
Vertex	First	Second	Third	Fourth	Fifth	Sixth	Seventh	Eighth	Ninth	Tenth
First	V3	V5	V4	V1	V6	V2	V1	V2	V8	V7
Second	V4	V6	V1	V5	V1	V3	V2	V7	V2	V8

			↑		↑	↑			
			Count 1 for V1		Count 1 for V1	Count 2 for V1	Count 2 for V2		

Table 6.12(b). File C1 during normal search.

	File C1			
	Edge			
Vertex	First	Second	Third	Fourth
First	V3	V4	?	?
Second	V4	V1	?	?

Table 6.12(c). First alternative of closed loop that may represent a gap boundary.

	Edge									
Vertex	First	Second	Third	Fourth	Fifth	Sixth	Seventh	Eighth	Ninth	Tenth
First	V3	V4	V1	V5	V6	V1	V2	—	—	—
Second	V4	V1	V5	V6	V1	V2	V3	—	—	—

For Count 1 for V1 and Count 1 for V2, from file C (Table 6.12(a)) the first alternative closed loop can be generated and the file is shown in Table 6.12(c). Figure 6.20(a) illustrates a graphical representation of the gap's boundary.

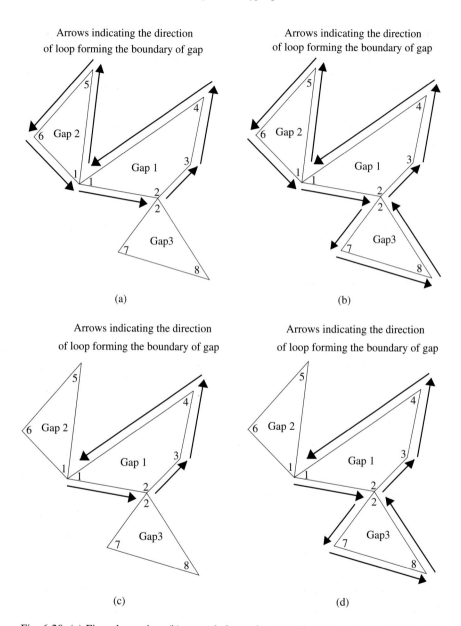

Fig. 6.20. (a) First alternative, (b) second alternative, (c) third alternative and (d) fourth alternative — graphical representation of the three gaps sharing two coincidental vertices.

Table 6.12(d). Second alternative of closed loop that may represent a gap boundary.

Vertex					Edge					
	First	Second	Third	Fourth	Fifth	Sixth	Seventh	Eighth	Ninth	Tenth
First	V3	V4	V1	V5	V6	V1	V2	V7	V8	V2
Second	V4	V1	V5	V6	V1	V2	V7	V8	V2	V3

Table 6.12(e). Third alternative of closed loop that may represent a gap boundary.

Vertex					Edge					
	First	Second	Third	Fourth	Fifth	Sixth	Seventh	Eighth	Ninth	Tenth
First	V3	V4	V1	V2	—	—	—	—	—	—
Second	V4	V1	V2	V3	—	—	—	—	—	—

Table 6.12(f). Fourth alternative of closed loop that may represent a gap boundary.

Vertex					Edge					
	First	Second	Third	Fourth	Fifth	Sixth	Seventh	Eighth	Ninth	Tenth
First	V3	V4	V1	V2	V7	V8	V2	—	—	—
Second	V4	V1	V2	V7	V8	V2	V3	—	—	—

For Count 1 for V1 and Count 2 for V2, the second alternative of a closed loop can be sorted and is shown in Table 6.12(d). Figure 6.20(b) illustrates a graphical representation of the gap's boundary.

For Count 2 for V1 and Count 1 for V2, the third alternative of a closed loop can again be sorted and is shown in Table 6.12(e). Figure 6.20(c) illustrates a graphical representation of the gap's boundary.

For Count 2 for V1 and Count 2 for V2, the fourth alternative of a closed loop can also be sorted and is shown in Table 6.12(f). Figure 6.20(d) illustrates a graphical representation of the gap's boundary.

Step 3: Comparison of the four alternatives

As can be seen from the four alternatives (see Fig. 6.20), the third alternative, as shown in Fig. 6.20(c), is considered the best solution and the correct solution to fill gap 1. This correct solution can be found by comparing which alternative uses the least edges to fill gap 1 up. Once the solution is found, the edges would be saved to file D for the next stage, that of facet generation.

After gap 1 is filled, gaps 2 and 3 can then be repaired using the basic generic solution.

6.4.2.2. *Degenerate Facets*

When dealing with a degenerate facet such as Facet A, shown in Fig. 6.21, which shares the same common edge (a–c) with two different facets B and C, vector algebra is applied to solve it.

The following steps are taken:

Step 1:   Edge a–c is converted into vectors,

$$(c_1\mathbf{i} + c_2\mathbf{j} + c_3\mathbf{k}) - (a_1\mathbf{i} + a_2\mathbf{j} + a_3\mathbf{k}). \tag{6.1}$$

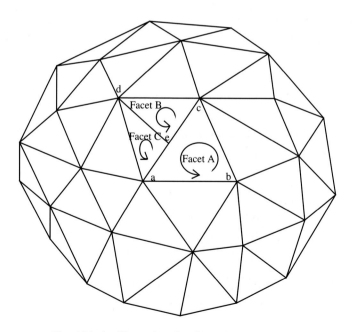

Fig. 6.21. An illustration showing a degenerate facet.

Step 2:   Collinear vectors are checked.

$$\text{Let } x = ac \Rightarrow x_1\mathbf{i} + x_2\mathbf{j} + x_3\mathbf{k}, \tag{6.2}$$

$$y = ce \Rightarrow y_1\mathbf{i} + y_2\mathbf{j} + y_3\mathbf{k}. \tag{6.3}$$

By mathematical definition, the two vectors x and y are said to be collinear if there exists scalars s and t, both nonzero, such that

$$sx + ty = 0. \tag{6.4}$$

However, when applied in computers, it is only necessary to have

$$sx + ty \leq \varepsilon, \tag{6.5}$$

where ε is a definable tolerance.

Step 3:   If the two vectors are found to be collinear vectors, the position
             of vertex e is generated.
Step 4:   Facet A is split into two facets (see Fig. 6.22). The two facets are
             generated using the three vertices in Facet A and vertex e:

$$\text{First facet vertices} \quad \rightarrow \quad a, b, e,$$
$$\text{Second facet vertices} \quad \rightarrow \quad b, c, e.$$

The orientation of each of the facets is checked and the new facets are stored in a temporary file.

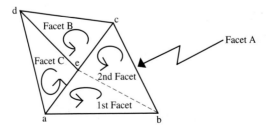

Fig. 6.22. Illustration on how a degenerate facet is solved.

Step 5: Search and delete Facet A from the original file.

Step 6: The two new facets are stored in the original file and the data are deleted from the temporary file.

6.4.2.3. *Overlapping Facets*

The condition of overlapping facets can be caused by errors introduced by inconsistent numerical round-off. This problem can be resolved through vertex merging where vertices within a pre-determined numerical round-off tolerance of one another can be merged into just one vertex. Figure 6.23 illustrates one example of how this solution can be applied. Figure 6.24 illustrates another example of an overlapping facet.

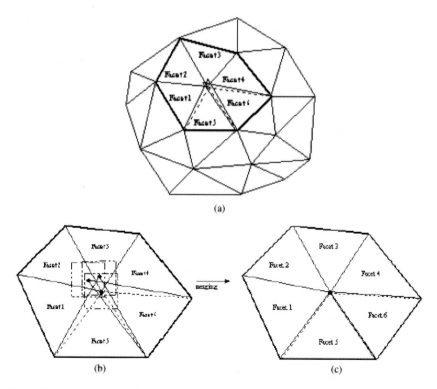

Fig. 6.23. (a) Overlapping facets, (b) numerical roundoff equivalence region and (c) vertices merged.

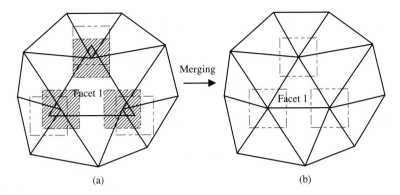

Fig. 6.24. (a) An overlapping facet and (b) facet's vertices merged with vertices of neighboring facets.

It is recommended that this merging of vertices be done before the searching of the model for gaps. This will eliminate unnecessary detection of erroneous edges and save substantial computational time expended in checking whether the edges can be used to generate another facet.

6.4.3. Performance Evaluation

Computational efficiency is an issue whenever CAD-model repair of solids that have been finely tessellated is considered. This is due to the fact that for every unit increase in the number of facets (finer tessellation), the additional increase in the number of edges is 3. Thus, the computational time required for the checking of erroneous edges would correspondingly increase.

6.4.3.1. *Efficiency of the Detection Routine*

Assuming that there are 12 triangles in the cube (Fig. 6.25), the number of edges = $12 \times 3 = 36$. The number of searches is computed as follows:

(1) Read first edge, search 35 edges and remove 2 edges.
(2) Read second edge, search 33 edges and remove 2 edges.
(3) Read third edge, search 31 edges and remove 2 edges and so on.

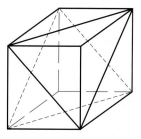

Fig. 6.25. A cube tessellated into 12 triangles.

$$\text{Number of searches} = 1 + 3 + \cdots + 35.$$

In general,

$$\text{Number of searches} = 1 + \cdots + (n-1)$$
$$= n^2/4, \qquad (6.6)$$

where n is the number of edges.

Although this result does not seem satisfactory, it is both an optimum and a robust solution that can be obtained given the inherent nature of the STL format (such as its lack of topological information).

However, if topological information is available, the efficiency of the routine used to detect the erroneous edges can definitely be increased significantly. Some additional points worth noting are as follows: First, binary files are far more efficient than ASCII files because they are only 20–25% as large and thus reduce the amount of physical data that needs to be transferred and because they do not require the subsequent translation into a binary representation. Consequently, one easily saves several minutes per file by using binary instead of ASCII file formats.[10]

As for vertex merging, the use of 1D AVL trees can significantly reduce the search-time for sufficiently identical vertices.[10] The AVL trees, which are usually 12 to 16 levels deep, reduce each search from $0(n)$ to $0(\log n)$ complexity and the total search-time from close to an hour to less than a second.

6.4.3.2. *Estimated Computational Time for Shell Closure*

The computational time required for shell closure is relatively fast. The estimated time can range from a few seconds to less than a minute and is arrived at based on the processing time obtained by Bohn[10] that uses a similar shell closure algorithm.

6.4.3.3. *Limitations of Current Shell Closure Process*

The shell closure process developed thus far does not have the ability to detect or solve the problems posed by any of the nonmanifold conditions. However, the detection of nonmanifold conditions and their subsequent solutions would be the next focus in ongoing research.

A limitation of the algorithm involves the solving of coplanar (see Fig. 6.26(a)) and non coplanar facets (see Fig. 6.26(b)) whose intersections result in another facet. The reasons for such errors are related to the application that generated the faceted model, the application that generated the original 3D CAD model and the user.

Another limitation involves the incorrect triangulation of parametric surface (see Fig. 6.27). One of the overlapping triangles, $T_b = BCD$, should not be present and should thus be removed while the other triangle, $T_a = ABC$, should be split into two triangles so as to maintain the correct

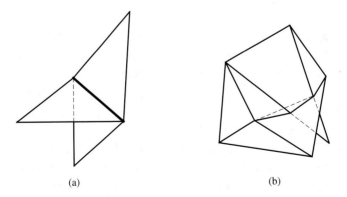

(a)   (b)

Fig. 6.26. (a) Incorrect triangulation (coplanar facet) and (b) non coplanar whereby facets are split after being intersected.

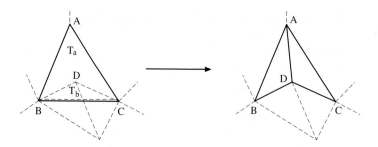

Fig. 6.27. Incorrect triangulation of parametric surface.

contoured surface. The proposed algorithm is presently unable to solve this problem.[11]

Finally, as mentioned earlier, the efficiency of $n^2/4$ (when n is the number of edges) is a major limitation especially when the number of facets in the tessellated model becomes very large (e.g., greater than 40,000). There are differences in the optimum use of computational resources. However, further work is being carried out to ease this problem by using topological information, which is available in the original CAD model.

6.5. OTHER TRANSLATORS

6.5.1. IGES File

IGES (Initial Graphics Exchange Specification) is a standard used to exchange graphics information between commercial CAD systems. It was set up as American National Standard in 1981.[12,13] The IGES file can precisely represent CAD models. It includes not only the geometry information (Parameter Data Section) but also topological information (Directory Entry Section). In the IGES, surface modeling, constructive solid geometry (CSG) and boundary representation (B-rep) are introduced. Especially, the ways of representing the regularized operations for union, intersection and difference have also been defined.

The advantages of the IGES standard are its wide adoption and comprehensive coverage. Since IGES was set up as American National Standard, virtually every commercial CAD/CAM system has adopted

IGES implementations. Furthermore, it provides the entities of points, lines, arcs, splines, NURBS surfaces and solid elements. Therefore, it can precisely represent CAD models.

However, several disadvantages of the IGES standard in relation to its use as an RP format include the following objections:

(1) Because IGES is the standard format to exchange data between CAD systems, it also includes much redundant information that is not needed for RP systems.
(2) The algorithms for slicing an IGES file are more complex than the algorithms slicing an STL file.
(3) The support structures needed in RP systems such as SLA cannot be created according to the IGES format.

IGES is a generally used data transfer medium, which interfaces with various CAD systems. It can precisely represent CAD models. Advantages of using IGES over the current approximate method include precise geometry representations, few data conversions, smaller data files and simpler control strategies. However, the problems are the lack of transfer standards for a variety of CAD systems and system complexities.

6.5.2. HP/GL File

HP/GL (Hewlett-Packard Graphics Language) is a standard data format for graphic plotters.[1,2] Data types are all two-dimensional, including lines, circles, splines, texts, etc. The approach, as seen from a designer's point of view, would be to automate a slicing routine which generates a section slice, invoke the plotter routine, to produce a plotter output file and then loop back to repeat the process.

The advantages of the HP/GL format are that a lot of commercial CAD systems have the interface to output HP/GL format and it is a 2D geometry data format that does not need to be sliced.

However, there are two distinct disadvantages of the HP/GL format. First, because HP/GL is a 2D data format, the files would not be appended, potentially leaving hundreds of small files needing to be given logical names and then transferred. Second, all the support structures required must be generated in the CAD system and sliced in the same way.

6.5.3. CT Data

Computerized tomography (CT) scan data is a particular approach for medical imaging.[1,14] This is not standardized data. Formats are proprietary and somewhat unique from one CT scan machine to another. The scan generates data as a grid of 3D points, where each point has a varying shade of gray indicating the density of the body tissue found at that particular point. Data from CT scans have been used to build skull, femur, knee and other bone models on Stereolithography systems. Some of the reproductions were used to generate implants, which have been successfully installed in patients. The CT data consist essentially of raster images of the physical objects being imaged. It is used to produce models of human temporal bones.

There are three approaches to making models out of CT scan information: (1) via CAD systems, (2) STL-interfacing and (3) direct interfacing. The main advantage of using CT data as an interface of RP is that it is possible to produce structures of the human body by the RP systems. However, disadvantages of CT data include, first, the increased difficulty in dealing with image data as compared with STL data and second, the need for a special interpreter to process CT data.

6.6. NEWLY PROPOSED FORMATS

As seen from above, the STL file — a collection of coordinate values of triangles — is not ideal and has inherent problems. As a result, researchers including the inventor of STL, 3D Systems Inc., USA, have in recent years proposed several new formats and these are discussed in the following sections. However, none of these has been accepted yet as a replacement for STL. STL files are still widely used today.

6.6.1. SLC File

The SLC (StereoLithography Contour) file format is being developed at 3D Systems, USA.[15] It addresses a number of problems associated with the STL format. An STL file is a triangular surface representation of a CAD model. Since the CAD data must be translated to this faceted representation, the surface of the STL file is only an approximation of the real surface of an object. The facets created by STL translation are sometimes noticeable on RP parts (such as the AutoCAD Designer part). When the number of STL triangles is increased to produce smoother part surfaces, STL files become very large and the time required for a RP system to calculate the slices can increase.

SLC attempts to solve these problems by taking 2D slices directly from a CAD model instead of using an intermediate tessellated STL model. According to 3D Systems, these slices eliminate the facets associated with STL files because they approximate the contours of the actual geometry.

Three problems may arise from this new approach. First, in slicing a CAD model, it is not always necessarily more accurate as the contours of each slice are still approximations of the geometry. Second, slicing in this manner requires much more complicated calculations (and is, therefore, very time-consuming) when compared with the relatively straightforward STL files. Third, a feature of a CAD model, which falls between two slices, may simply disappear, if it is just under the tolerances set for inclusion on either of the adjacent slices.

6.6.1.1. SLC File Specification

The SLC file format is a "2½D" contour representation of a CAD model. It consists of successive cross sections taken at ascending Z intervals in which solid material is represented by interior and exterior boundary polylines. SLC data can be generated from various sources, either by conversion from CAD solid or surface models, or more directly from systems, which produce data arranged in layers, such as CT-scanners.

6.6.1.2. Definition of Terms

(1) *Segment*: A segment is a straight line connecting two X/Y vertice points.
(2) *Polyline*: A polyline is an ordered list of X/Y vertice points connected continuously by each successive line segment. The polyline

must be closed whereby the last point must equal the first point in the vertice list.

(3) *Contour boundary*: A boundary is a closed polyline representing interior or exterior solid material. An exterior boundary has its polyline list in counter-clockwise order. The solid material is inside the polyline. An interior boundary has its polyline list in clockwise order and solid material is outside the polyline. Figure 6.28 shows a description of the contour boundary.

(4) *Contour Layer*: A contour layer is a list of exterior and interior boundaries representing the solid material at a specified Z cross section of the CAD model. The cross-section slice is taken parallel to the X/Y plane and has a specified layer thickness.

6.6.1.3. *Data Formats*

Byte	8 bits,
Character	1 Byte,
Unsigned integer	4 Bytes,
Float	4 Bytes IEEE Format.

Most significant byte of FLOAT is specified in the highest addressed byte. The byte ordering follows the Intel PC Little Indian/Big Indian scheme.

Address	0	1	2	3
	Low Word		High Word	
	LSB MSB		LSB MSB	

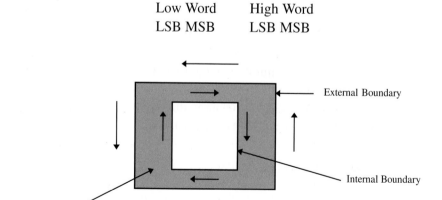

Fig. 6.28. Description of contour boundary.

Most UNIX RISC Workstations are Big Indian/Little Indian, therefore, they need to byte swap all Unsigned Integers and Floats before outputting to the SLC file.

6.6.1.4. *Overview of the SLC File Structure*

The SLC file is divided into a header section, a 3D reserved section, a sample table section and the contour data section.

(1) *Header section*: The header section is an ASCII character string containing global information about the part and how it was prepared. The header is terminated by a carriage return, line feed and control-Z character ($0 \times 0d$, $0 \times 0a$, $0 \times 1a$) and can be a maximum of 2,048 bytes including the termination characters. The syntax of the header section is a keyword followed by its appropriate parameter.

Header Keywords

(i)   "-SLCVER ⟨X.X⟩" specifies the SLC file format version number. The version number of this specification is 2.0.

(ii) "-UNIT ⟨INCH/MM⟩" indicates which units the SLC data is represented in.

(iii) "-TYPE ⟨PART/SUPPORT/WEB⟩" specifies the CAD model type. PART and SUPPORT must be closed contours. WEB types can be open polylines or line segments.

(iv) "-PACKAGE ⟨vendor specific⟩" identifies the vendor package and version number, which produced the SLC file. A maximum of 32 bytes.

(v)   "-EXTENTS ⟨minx,maxx miny,maxy minz,maxz⟩" describes the X, Y and Z extents of the CAD model.

(vi) "-CHORDDEV ⟨value⟩" specifies the cordal deviation, if used, to generate the SLC data.

(vii) "-ARCRES ⟨value in degrees⟩" specifies the arc resolution, if used, to generate the SLC data.

(viii) "-SURFTOL ⟨value⟩" specifies the surface tolerance, if used, to generate the SLC data.

(ix) "-GAPTOL ⟨value⟩" specifies the gap tolerance, if used, to generate the SLC data.

(x) "-MAXGAPFOUND ⟨value⟩" specifies the maximum gap size found when generating the SLC data.

(xi) "-EXTLWC ⟨value⟩" specifies if any line width compensation has been applied to the SLC data by the CAD vendor.

(2) *3D reserved section*: This 256 byte section is reserved for future use.

(3) *Sample table section*: The sample table describes the sampling thicknesses (layer thickness or slice thickness) of the part. There can be up to 256 entries in the table. Each entry descibes the Z start, the slice thickness and what line width compenstation is desired for that sampling range.

Sampling Table Size	1 Byte
Sampling Table Entry	4 Floats
Minimum Z Level	1 Float
Layer Thickness	1 Float
Line Width Compensation	1 Float
Reserved	1 Float

The first sampling table entry Z start value must be the very first Z contour layer. For example, if the cross sections were produced with a single thickness of 0.006 in. and the first Z level of the part is 0.4 in. and a line width compensation value of 0.005 is desired, then the sampling table will look like the following:

Sample Table Size	1
Sample Table Entry	0.4 0.006 0.005 0.0

If, for example, the part was sliced with two different layer thickness the Sample Table could look like the following:

Sample Table Size	2
Sample Table Entry 1	0.4 0.005 0.004 0.0
Sample Table Entry 2	2.0 0.010 0.005 0.0

Slice thicknesses must be even multiples of one other to avoid processing problems.

(4) *Contour data section*: The contour data section is a series of successive ascending Z cross-sections or layers with the accompaning contour data. Each Contour Layer contains the minimum Z layer value, number of boundaries followed by the list of individual boundary data. The boundary data contains the number of x, y vertices for that boundary, the number of gaps and finally the list of floating point vertice points.

The location of a gap can be determined when a vertice point repeats itself. To illustrate, given the contour layer in Fig. 6.28, the contour section could be as follows:

Z layer	0.4
Number of boundaries	2
Number of vertices for the 1st boundary	5
Number of gaps for the 1st boundary	0
Vertex list for 1st boundary	0.0, 0.0
	1.0, 0.0
	1.0, 1.0
	0.0, 1.0
	0.0, 0.0

Note that the direction of the vertice list is counter-clockwise indicating the solid material is inside the polylist. Also, note that the polylist is closed because the last vertice is equal to the first vertice:

Number of vertices for the 2nd boundary	5
Number of gaps for the 2nd boundary	0
Vertex list for 2nd boundary	0.2, 0.2
	0.2, 0.8
	0.8, 0.8
	0.8, 0.2
	0.2, 0.2

Note that the direction of the vertice list is clockwise indicating the solid material is outside the polylist. Also, note that the polylist is closed because the last vertice is equal to the first vertice.

The Contour Layers are stacked in an ascending order until the top of the part. The last layer or the top of the part is indicated by the Z level and a termination unsigned integer (0 × FFFFFFFF).

Contour Layer Section Description

Contour Layer
  Minimum Z Level   Float
    Number of Boundaries   Unsigned Integer
     Number of Vertices   Unsigned Integer
     Number of Gaps   Unsigned Integer
     Vertices List (X/Y)   Number of Vertices * 2 Float
    Repeat Number of   1
      Boundaries
  Repeat Contour Layer until
   Top of Part
Top of Part
  Maximum Z Level   1 Float
  Termination Value   Unsigned Integer (0 × FFFFFFFF)

Minimum Z Level for a given Contour Layer

A one centimeter cube based at the origin 0,0,0 can be represented by only one contour layer and the Top of Part Layer data.

Suppose the cube was to be imaged in 0.010 layers. The sample table would have a single entry with its starting Z level at 0.0 and layer thickness at 0.01. The contour layer data section could be as follows:

Z layer	0.0
Number of boundaries	1
Number of vertices for the 1st boundary	5
Number of gaps for the 1st boundary	0
Vertex list for 1st boundary	0.0, 0.0
	1.0, 0.0
	1.0, 1.0
	0.0, 1.0

	0.0, 0.0
Z layer	1.0
Termination value	$0 \times$ FFFFFFFF

Note only one contour was necessary to describe the entire part. The initial contour will be imaged until the next minimum contour layer or the top of the part at the specified layer thickness described in the sampling table. Now, this part could have 100 identical contour layers but that would have been redundant. This is why the contour Z value is referred to as the minimum Z value. It gets repeated until the next contour or top of the part.

6.6.2. CLI File

The Common Layer Interface (CLI) format is being developed in a Brite Euram project[2,16] with the support of major European car manufacturers. The CLI format is meant as a vendor-independent format for layer-by-layer manufacturing technologies. In this format, a part is built by a succession of layer descriptions. The CLI file can be in binary or ASCII format. The geometry part of the file is organized in layers in ascending order. Every layer is started by a layer command, giving the height of the layer.

The layers consist of a series of geometric commands. The CLI format has two kinds of entities. One is the polyline. The polylines are closed, which means that they have a unique sense, either clockwise or anti-clockwise. This directional sense is used in the CLI format to state whether a polyline is on the outside of the part or surrounding a hole in the part. Counter-clockwise polylines surround the part, whereas clockwise polylines surround holes. This allows correct directions for beam offset. The other is the hatching to distinguish between the inside and outside of the part. As this information is already present in the direction of the polyline and hatching takes up considerable file space, hatches have not been included in output files.

The advantages of the CLI format are given as follows:

(1) Since the CLI format only supports polyline entities, it is a simpler format compared with the HP/GL format.

(2) The slicing step can be avoided in some applications.
(3) The error in the layer information is more easily corrected than that in the 3D information. Automated recovery procedures can be used and if required, editing is also not difficult.

However, there exist several disadvantages of the CLI format. They are given as follows:

(1) The CLI format only has the capability of producing polylines of the outline of the slice.
(2) Although the real outline of the part is obtained, by reducing the curve to segments of straight lines, the advantage over the STL format is lost.

The CLI format also includes the layer information like the HP/GL format but the CLI format only has polyline entities, while HP/GL supports arcs and lines. The CLI format is simpler than the HP/GL format and has been used by several RP systems. It is hoped that the CLI format will become an industrial standard such as STL.

6.6.3. RPI File

The rapid prototyping interface (RPI) format is being designed by the Rensselaer Design Research Center,[4,7] Rensselaer Polytechnic Institute. It is derived from the currently accepted STL format data. The RPI format is capable of representing facet solids, but includes additional information about the facet topology. Topological information is maintained by representing each facet solid entity with indexed lists of vertices, edges and faces. Instead of explicitly specifying the vertex coordinates for each facet, they can be referred to by each facet using index numbers. This contributes to the goal of overall redundant information reduction.

The format is being developed in ASCII to facilitate cross-platform data exchange and debugging. An RPI format file is composed of the collection of entities, each of which internally defines the data it contains. Each entity conforms to the syntax defined by the syntax diagram, shown in Fig. 6.29. Each entity is composed of an entity name, a record

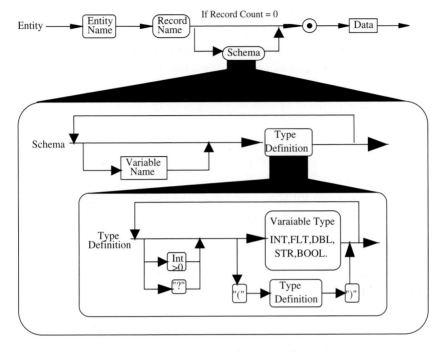

Fig. 6.29. RPI format entity syntax diagram.

count, a schema definition, schema termination symbol and the corresponding data. The data are logically subdivided into records, which are
made up of fields. Each record corresponds to one variable type in the
type definition.

The RPI format includes the following four advantages:

(1) Topological information is added in the RPI format. As the result of
    this, flexibility is achieved. It allows users to balance storage and
    processing costs.
(2) Redundancy in the STL is removed and the size of the file is
    compacted.
(3) Format extensibility is made possible by interleaving the format
    schema with data as shown in Fig. 6.29.
(4) Representation of CSG primitives is provided, as are capabilities to
    represent multiple instances of both facet and CSG solids.

Two disadvantages of the RPI format are given as follows:

(1) An interpreter, which processes a format as flexible and extensible as the RPI format, is more complex than that for STL format.
(2) Surface patches suitable for solid approximation cannot be identified in the RPI format.

The RPI format offers a number of features unavailable in the STL format. The format can represent CSG primitive models as well as facet models. Both can be operated on by the Boolean union, intersection and difference operators. Provisions for solid translation and multiple instancing are also provided. Process parameters, such as process types, scan methods, materials and even machine operator instructions, can be included in the file. Facet models are more efficiently represented as redundancy is reduced. The flexible format definition allows storage and processing costs to be balanced.

6.6.4. LEAF File

The Layer Exchange ASCII Format (LEAF) is being generated by Helsinki University of Technology.[11] To describe this data model, concepts from the object-oriented paradigm are borrowed. At the top level, there is an object called Layer Manufacture Technology file (LMT-file) that can contain parts, which in turn are composed of other parts or by layers. Ultimately, layers are composed of 2D primitives and currently the only ones that are planned for implementation are polylines.

For example, an object of a given class is created. The object classes are organized in a simple tree shown in Fig. 6.30. Attached to each object class is a collection of properties. A particular instance of an object specifies the values for each property. Objects inherit properties from their parents. In LEAF, the geometry of an object is simply one among several other properties.

Fig. 6.30. The object tree.

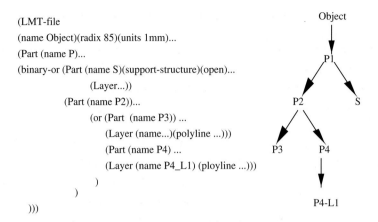

(LMT-file
(name Object)(radix 85)(units 1mm)...
(Part (name P)...
(binary-or (Part (name S)(support-structure)(open)...
        (Layer...))
    (Part (name P2))...
        (or (Part (name P3)) ...
            (Layer (name...)(polyline ...)))
        (Part (name P4) ...
        (Layer (name P4_L1) (ployline ...)))
)
)
    )))

Fig. 6.31. An instance tree.

In this example, the *object* is an LMT-file. It contains exactly one child, the object P1. P1 is the combination of two parts, one of which is the support structure and the other one is P2, again a combination of two others. The objects at leaves of the tree — P3, P4 and S — must have been, evidently, sliced with the same z-values so that the required operations, in this case *or* and *binary-or*, can be performed and the layers of P1 and P2 constructed.

In LEAF, the properties, support-structure and open, can also be attached to layers or even polyline objects, allowing the sender to represent the original model and the support structures as one single part. In Fig. 6.31, all parts inherit the properties of *Object*, their ultimate parent. Similarly, all layers of the object S inherit the open property indicating that the contours in the layers are always interpreted as open, even if they are geometrically closed.

Among the many advantages of the LEAF format are as follows:

(1) It is easy to implement and use.
(2) It is not ambiguous.
(3) It allows for data compression and for a human-readable representation.
(4) It is machine independent and LMT process independent.
(5) Slices of CSG models can be represented almost directly in LEAF.

(6) The part representing the support structures can be easily separated from the original part.

The disadvantages of the LEAF format include the following items:

(1) A new interpreter is needed for connecting the RP systems.
(2) The structure of the format is more complicated than that of the STL format.
(3) The STL format cannot be changed into this format.

The LEAF format is described at several levels, mainly at a logical level using a data model based on object-oriented concepts and at a physical level using a LISP-like syntax. At the physical level, the syntax rules are specified by several translation phases. Thus defined, it allows one to choose at which level interaction with LEAF is desirable and at each level there is a clear and easy-to-use interface. It is doubtful that LEAF currently supports the needs of all processes currently available but it is hopefully a step forward in this direction.

6.7. STANDARD FOR REPRESENTING LAYERED MANUFACTURING OBJECTS

For the past several years, research and development has been going on for ISO 10303 on Rapid Prototyping and Layered Manufacturing (RPLM). ISO 10303 stands for International Standard for the Exchange of Product (STEP) model data, which is used for computer-interpretable representation and exchange of product data for engineering purposes.[17] Under the manufacturing environment, standards are categorized under two groups: formal standards and informal standards.[18]

STL format is grouped under informal standards commonly called industry or *de facto* standards. STL format is the interface between CAD and RP systems. It is widely used in most RP systems available in the market due to its simplicity and benefits. Experienced RP systems users know the problems created by the STL format the time and effort required to correct the data.[18] Furthermore, RP applications are expanding widely with time, creating even more complex 3D shapes and more problems are generated as the STL format is only an approximation of 3D shape. More importantly, nonshape information and model accuracy in RPLM are issues regarding the future of RP which several researchers are developing alternative data formats for.

STEP is grouped under the formal category, which is approved internationally and supported by a public standards-making authority. STEP itself covers a broad range of product types and product life cycle stages. STEP internally consists of several application protocols (APs), which are used to describe a particular life cycle stage of a particular product type. The exchange of the actual information indicating description and the method used is purely based on APs. To implement the description, a set of Integrated Resources (IRs) are constructed to form the APs and the construction of the IRs is significantly applied for specific purposes.

Developers have long noticed the importance of nonshape information and model accuracy in STL toward the future in RP whereby alternative data formats need to be developed for the replacement of the STL format. Since the proposal for the development of a new STEP AP for RPLM process, discussion and feedback have been carried out with manufacturing users, RP vendors, CAD vendors, RP service bureaus, or RP researchers.[18] Currently, there is still no STEP AP developed specially for RPLM. Main resources such as exact representation and modeling in geometry, representation of material type, tolerance specification and function specification are needed for the new development. Replacement of the STL format will be a long-term solution, which consists of questions and decisions in order to determine the new AP for RPLM, requiring more interest and stronger participation.[19] Additional improvement methods like direct slicing STEP-based NURBS models for layered manufacturing,[20] segmented object manufacturing,[21] and conversion of 2D CAD model directly to sliced data[22] have been proposed for enhancement. Meanwhile, the STL format will still be the solution as it is so widely used by all systems. More support and greater interest from RP vendors and developer are needed to push on further to speed up the progress.

Fabrication of heterogeneous objects by layered manufacturing is the other research area being focused on after the proposal of ISO 10303 STEP

AP for RPLM. A solid object with the material attributes of a heterogeneous object is referred to as a Heterogeneous Solid Model (HSM).
Functionally Graded Materials (FGM) refers to a composition material
variation along the geometry. With the variation of material in material
composition, it creates more potential applications especially for industry
and biomedical applications. An integrated system for the design and fabrication of heterogeneous objects has been introduced with description of
utilized material information.[23] Design and fabrication of heterogeneous
objects are drawn up with the representation of heterogeneous object and
process planning method for RPLM together with the processing method.
Fabricated parts modeled with the integrated system consist of errors like
incorrect mixture of materials, which may occur. Indeed with the errors
encountered from the results, more reliable parts with actual material resolution will be presented soon.

REFERENCES

1. P. F. Jacobs, Rapid prototyping and manufacturing, Society of
   Manufacturing Engineers (1992).

2. R. Famieson and H. Hacker, Direct Slicing of CAD Models for Rapid
   Prototyping. *Rapid Prototyping J.* (1995).

3. R. J. Donahue, CAD model and alternative methods of information
   transfer for rapid prototyping systems, *Proc. Second Int. Conf. Rapid
   Prototyping* (1991), pp. 217–235.

4. M. J. Wozny, Systems issues in solid freeform fabrication, *Proc. Solid
   Freeform Fabrication Symposium 1992*, Texas, USA (3–5 August
   1992), pp. 1–15.

5. K. F. Leong, C. K. Chua and Y. M. Ng, A study of stereolithography
   file errors and repair, Part 1 — Generic solutions, *Int. J. Adv. Manufact.
   Technol.* **12**(6), 407–414 (1996).

6. K. F. Leong, C. K. Chua and Y. M. Ng, A study of stereolithography
   file errors and repair, Part 2 — Special cases, *Int. J. Adv. Manufact.
   Technol.* **12**(6), 415–422 (1996).

7. S. J. Rock and M. J. Wozny, A flexible format for solid freeform fabrication, *Proc. Solid Freeform Fabrication Symposium 1991*, Texas, USA (12–14 August 1991), pp. 1–12.

8. R. H. Crawford, Computer aspects of solid freeform fabrication: Geometry, process control and design, *Proc. Solid Freeform Fabrication Symposium 1993*, Texas, USA (9–11 August 1993), pp. 102–111.

9. J. H. Bohn and M. J. Wozny, Automatic CAD-model repair: Shell-closure, *Proc. Solid Freeform Fabrication Symposium 1992*, Texas, USA (3–5 August 1992), pp. 86–94.

10. J. H. Bohn, *Automatic CAD-Model Repair*, UMI, Ann Arbor, MI, USA (1993).

11. A. Dolenc and I. Malela, A data exchange format for LMT processes, *Proc. Third Int. Conf. Rapid Prototyping* (1992), pp. 4–12.

12. K. Reed, D. Harrvd and W. Conroy, *Initial Graphics Exchange Specification (IGES) version 5.0*, CAD-CAM Data Exchange Technical Centre (1990).

13. J. Li, Improving stereolithography parts quality — Practical solutions, *Proc. Third Int. Conf. Rapid Prototyping* (1992), pp. 171–179.

14. B. Swaelens and J. P. Kruth, Medical applications of rapid prototyping techniques, *Proc. Fourth Int. Conf. Rapid Prototyping* (1993), pp. 107–120.

15. W. Vancraen, B. Swawlwns and J. Pauwels, Contour interfacing in rapid prototyping — Tools that make it work, *Proc. Third European Conf. Rapid Prototyping and Manufacturing* (1994), pp. 25–33.

16. G. Smith-Moritz, 3D Systems, *Rapid Prototyping Rep. Rapid Prototyping Manufact.* **4**(12), 3 (1994).

17. L. Patil, D. Dutta, A. D. Bhatt, K. Jurrens, K. Lyons, M. J. Pratt and R. D. Sriram, A proposed standards-based approach for representing heterogeneous objects for layered manufacturing, *Rapid Prototyping J.* **8**(3), 134–146 (2002).

18. M. Pratt, New Work Item Proposal, ISO/TC 184/SC 4 N (June 2000).

19. M. J. Pratt, A. D. Bhatt, D. Dutta, K. W. Lyons, L. Patil and R. D. Sriam, Progress towards an international standard for data transfer in rapid prototyping and layered manufacturing, *Comput. Aid. Des.* **34**(14), 1111–1121 (2002).

20. B. Starly, A. Lau, W. Sun, W. Lau and T. Bradbury, Directly slicing of STEP based NURBS models for layered manufacturing, *Comput. Aid. Desi.* **37**(4), 387–397 (2005).

21. K. P. Karunakaran, S. Agrawa, P. D. Vengurlekar, O. S. Sahahrabudhe, V. Pushpa and R. H. Ely, Segmented object manufacturing, *IIE Trans.* **37**(4), 291–302 (2005).

22. G. H. Liu, Y. S. Wong, Y. F. Zhang and H. T. Loh, Modelling cloud data for prototype manufacturing, *J. Mater. Process. Technol.* **138**, 53–57 (2003).

23. K. H. Shin, H. Natu, D. Dutta and J. Mazumder, A method for the design and fabrication of hetergenous object, *Mater. Des.* **24**(5), 339–353 (2003).

PROBLEMS

1. What is the common format used by RP systems? Describe the format and illustrate with an example. What are the pros and cons of using this format?
2. Referring to Fig. 6.32, write a sample STL file for the shaded triangle.
3. Based on the STL format, how many triangles and coordinates would a cube contain?
4. What causes *missing facets or gaps* to occur?
5. Illustrate, with diagrams, the meaning of *degenerate facets*.
6. Explain *overlapping facets*.
7. What are the three types of nonmanifold conditions?

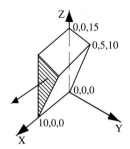

Fig. 6.32. Sample STL file.

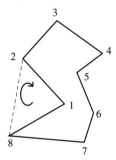

Fig. 6.33. Incorrectly generated facet's orientation.

8. What are the consequences of building a valid and invalid tessellated model?

9. What problems can the generic solution solve?

10. Describe the algorithm to solve the *missing facets'* problem.

11. In Fig. 6.33, facet *X* is incorrectly orientated. Describe how the problem can be resolved. Draw the newly generated facet *X* with the corrected orientation.

12. Describe the algorithm used to solve special case 1 where two or more gaps are formed from a coincidental vertex.

13. Prove and illustrate how a generate facet can be repaired by the use of vector algebra.

14. How can the problem of overlapping facets be solved?

15. What is the efficiency of the detection routine? Illustrate using the example of a cube.
16. What are some of the limitations of the solutions, both generic and special cases, described to solve STL-related problems?
17. Name some other translators used in place of STL.
18. What problems does the SLC file format seek to address?
19. Some newly proposed formats are CLI, RPI and the LEAF files. Describe them briefly and contrast their strengths and weaknesses.

Chapter 7
APPLICATIONS AND EXAMPLES

7.1. APPLICATION–MATERIAL RELATIONSHIP

Areas of applications are closely related to the purposes of prototyping and consequently, the materials used. As such, the closer the rapid prototyping (RP) materials are to traditional prototyping materials in physical and behavioral characteristics, the wider will be the range of applications. Unfortunately, there are marked differences in these areas between current RP materials and traditional materials in manufacturing. The key to increasing the applicability of RP technologies, therefore, lies in widening the range of materials.

In the early developments of RP systems, the emphasis of the tasks at hand was oriented towards the creation of "visualization" and "touch-and-feel" models to support design, i.e., creating 3D objects with little or no regard to their function and performance. These are broadly classified as "Applications in Design". It is a result that is influenced and in many cases limited by the materials available in these RP systems. However, as the initial costs of the machines are high, vendors are constantly on the look-out for more areas of applications, with the logical search for functional evaluation and testing applications and eventually tooling. This not only calls for improvements in RP technologies in terms of the process to create stronger and more accurate parts, but also in terms of developing an even wider range of materials, including metals and ceramic composites. Applications of RP prototypes were first extended to "Applications in Engineering, Analysis and Planning" and later extended further to "Applications in Manufacturing and Tooling". These typical application areas are summarized in Fig. 7.1 and discussed in the following sections.

The major breakthrough in RP technologies in manufacturing has been their abilities in enhancing and improving product development

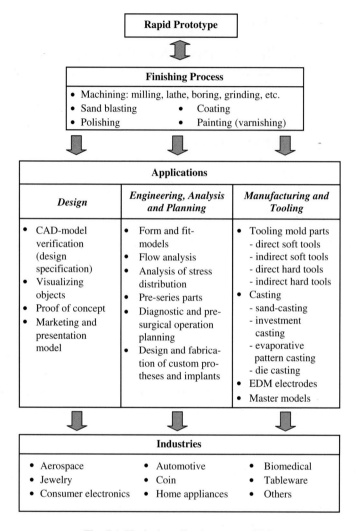

Fig. 7.1. Typical application areas of RP.

while at the same time reducing costs and time required to take the product from conception to market. Sections 7.6–7.12 contain examples of applications in the aerospace, automotive, biomedical, jeweler, coin and tableware industries. These examples are by no means exhaustive, but they do represent their applications in a wide cross section of the industry.

7.2. FINISHING PROCESSES

As there are various influencing factors such as shrinkage, distortion, curling and accessible surface smoothness, it is necessary to apply some post-RP finishing processes to the parts just after they have been produced. These processes can be carried out before the RP parts are used in their desired applications. Furthermore, additional processes may be necessary in specific cases, e.g., when creating screw threads.

7.2.1. Cutting Processes

In most cases, the resins or other materials used in the RP systems can be subjected to conventional cutting processes, such as milling, boring, turning and grinding.

These processes are particularly useful in the following cases:

(1) Deviations in geometrical measurements or tolerances due to unpredictable shrinkage during the curing or bonding stages of the RP process.
(2) Incomplete generation of selected form features. This could be due to fine or complex-shaped features that are difficult to achieve.
(3) Clean removal of necessary support structures or other remainder materials attached to the RP parts.

In all these cases, it is possible to achieve economic surface finishing of the objects generated with a combination of NC machining and computer-aided NC programming.

7.2.2. Sandblasting and Polishing

Sandblasting or abrasive jet deburring can be used as an additional cleaning operation or process to achieve better surface quality. However, there is a trade-off in terms of accuracy. If better finishing is required, additional polishing by mechanical means with super-fine abrasives or simple rotary sanding can also be used after sandblasting.

7.2.3. Coating

Appropriate surface coatings can be used to further improve the physical properties of the surface of plastic RP parts. One example is galvano-coating, a coating which provides very thin metallic layers to plastic RP parts.

7.2.4. Painting

Painting is applied fairly easily on RP parts made of plastics or paper. It is carried out mainly to improve the aesthetic appeal or for presentation purposes, e.g., for marketing or advertising presentations.

Once the RP parts are appropriately finished, they can then be used for the various areas of application as shown in Fig. 7.1.

7.3.   APPLICATIONS IN DESIGN

7.3.1. CAD Model Verification

This is the initial objective and strength of RP systems, in that designers often need the physical part to confirm the design that they have created in the CAD system. This is especially important for parts or products designed to fulfill aesthetic functions or are intricately designed to fulfill functional requirements.

7.3.2. Visualizing Objects

Designs created on CAD systems need to be communicated not only amongst designers within the same team, but also to other departments like manufacturing and marketing. Thus, there is a need to create objects from the CAD designs for visualization so that all these people will be referring to the same object in any communication. Tom Mueller in his paper entitled "Application of Stereolithography in injection molding"[1] characterizes this necessity by saying:

> *"Many people cannot visualize a part by looking at print. Even engineers and toolmakers who deal with print everyday require*

several minutes or even hours of studying a print. Unfortunately, many of the people who approve a design (typically senior management, marketing analysts and customers) have much less ability to understand a design by looking at a drawing."

7.3.3. Proof of Concept

Proof of concept relates to the adaptation of specific details to an object environment or aesthetic aspects, e.g., verifying that a car telephone design is suitable and blends in well within a specific car. It also relates to the specific details of the design on the functional performance of a desired task or purpose, e.g., how a lever arm of a CD player opens the door of the CD tray.

7.3.4. Marketing and Commercial Applications

Frequently, the marketing or commercial departments require a physical model for presentation and evaluation purposes, especially for assessment of the project as a whole. The mock-up or presentation model can even be used to produce promotional brochures and related materials for marketing and advertising even before the actual product becomes available.

7.4. APPLICATIONS IN ENGINEERING, ANALYSIS AND PLANNING

Other than creating a physical model for visualization or proofing purposes, designers are also interested in the engineering aspects of their designs. This invariably relates to the functions of the design. RP technologies become important as they are able to provide the information necessary to ensure sound engineering and function of the product. What makes it more attractive is that it also helps to save development time and reduce costs. Based on the improved performance of processes and materials available in current RP technologies, some applications for functional models are presented in the following sections.

7.4.1. Scaling

RP technology allows easy scaling down (or up) of the size of a model by scaling the original CAD model. In a case of designing bottles for perfumes with different holding capacities, the designer can simply scale the CAD model appropriately for the desired capacities and view the renderings on the CAD software. With the selected or preferred capacities determined, the CAD data can be modified accordingly to create the corresponding RP model for visualization and verification purposes (see Fig. 7.2).

7.4.2. Form and Fit

Other than dealing with sizes and volumes, forms have to be considered from the aesthetics and functional standpoints as well. How a part fits into a design and its environment are important aspects which have to be addressed. For example, the wing mirror housing for a new car design has to be of a form that augments well with the general appearance of the exterior design. This will also include how it fits to the car door. The model will be used to evaluate how it satisfies both aesthetic and functional requirements.

Form and fit models are used not just in the automotive industries. They can also be used for industries involved in aerospace and others like consumer electronic products and appliances.

7.4.3. Flow Analysis

Designs of components that affect or are affected by air or fluid flow cannot be easily modified if produced by the traditional manufacturing

Fig. 7.2. Perfume bottles with different capacities.

routes. However, if the original 3D design data can be stored in a computer model, then any change of object data based on some specific tests can be realized with computer support. The flow dynamics of these products can be computer simulated with software. Experiments with 3D physical models are frequently required to study product performance in air and liquid flow. Such models can be easily built using RP technology. Modifications in design can be done on computer and rebuilt for retesting very much faster than using traditional prototyping methods. Flow analyses are also useful for studying the inner sections of inlet manifolds, exhaust pipes, replacement heart valves,[2] or similar products that at times can have rather complex internal geometries. If required, transparent parts can also be produced using rapid tooling methods to aid visualization of internal flow dynamics. Typically, flow analyses are necessary for products manufactured in the aerospace, automotive, biomedical and shipbuilding industries.

7.4.4. Stress Analysis

In stress analysis using mechanical or photo-optical methods or otherwise, physical replicas of the part being analyzed is necessary. If the material properties or features of the RP technologies generated objects are similar to those of the actual functional parts, they can be used in these analytical methods to determine the stress distribution of the product.

7.4.5. Mock-Up Parts

"Mock-up" parts, a term first introduced in the aircraft industry, are used for final testing of parts from different aspects. Generally, mock-up parts are assembled into the complete product and functionally tested at predetermined conditions, e.g., for fatigue. Some RP techniques are able to generate "mock-ups" very quickly to fulfill these functional tests before the design is finalized.

7.4.6. Pre-Production Parts

In cases where mass-production will be introduced once the prototype design has been tested and confirmed, pilot-production runs of 10 or more parts are usual. The pilot-production parts are used to confirm tooling design and specifications. The necessary accessory equipment, such as fixtures, chucks, special tools and measurement devices required for the mass-production process are prepared and checked. Many of the RP methods are able to quickly produce pilot-production parts, thus helping to shorten the process development time, thereby accelerating the overall time-to-market process.

7.4.7.   Diagnostic and Surgical Operation Planning

In combining engineering prototyping methodologies with surgical procedures, RP models can complement various imaging systems, such as magnetic resonance imaging and computed tomography (CT) scanning, to produce anatomical models for diagnostic purposes. These RP models can also be used for surgical and reconstruction operation planning. This is especially useful in surgical procedures that have to be carried out by different teams of medical specialists and inter-departmental communication is of essence. Several related examples and case studies can be found in Chap. 8.

7.4.8.   Design and Fabrication of Custom Prosthesis and Implant

RP can be applied to the design and fabrication of customized prostheses and implants. A prosthesis or implant can be made from anatomical data inputs from imaging systems, e.g., laser scanning and CT. In cases, such as producing ear prostheses, a scan profile can be taken of the good ear to create a computer-mirrored exact replica replacement using RP technology. These models can be further refined and processed to create the actual prostheses or implants to be used directly on a patient. The ability to efficiently customize and produce such prostheses and implants is important, as standard sizes are not always an ideal fit for the patient. Also, a less than ideal fit, especially for artificial joints and weight bearing implants, can often result in accumulative problems and damage to the surrounding tissue structures.

7.5. APPLICATIONS IN MANUFACTURING AND TOOLING

Central to the theme of rapid tooling is the ability to produce multiple copies of a prototype with functional materials properties in short lead-times. Apart from mechanical properties, material can also include functionalities such as color dyes, transparency, flexibility and the likes. Two issues are to be addressed here: tooling proofs and process planning. Tooling proofs refer to getting the tooling right so that there will not be a need to do a tool change during production because of process problems. Process planning is meant for laying down the process plans for the manufacture as well as assembly of the product based on the prototypes produced.

Rapid tooling can be classified as soft and hard tooling, direct and indirect tooling,[3] as schematically shown in Fig. 7.3. Soft tooling, typically made of silicon rubber, epoxy resins, low-melting point alloys and foundry sands, generally allows for only single casts or for small batch production runs. Hard tooling, on the other hand, usually made from tool steels, generally allows for longer production runs.

Direct tooling is referred to when the tool or die is created directly by the RP process. As an example in the case of injection molding, the main cavity and cores, runner, gating and ejection systems, can be produced directly using the RP process. In indirect tooling, on the other hand, only the master pattern is created using the RP process. A mold, made of silicone rubber, epoxy resin, low-melting point metal, or ceramic, is then created from the master pattern.

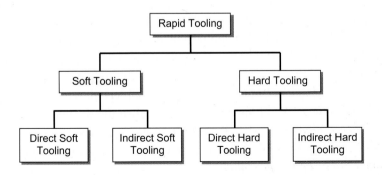

Fig. 7.3. Classification of rapid tooling.

7.5.1.   Direct Soft Tooling

This is where the molding tool is produced directly by the RP systems. Such tooling can be used for liquid metal sand casting, in which the mold is destroyed after a single cast. Other examples, such as composite molds, can be made directly using stereolithography. These are generally used in injection molding of plastic components and can withstand between 100 and 1,000 shots. As these molding tools can typically only support a single cast or small batch production before breaking down, they are classified as soft tooling. The following section list several examples of direct soft tooling methods.

7.5.1.1.   *Selective Laser Sintering® of Sand Casting Molds*

Sand casting molds can be produced directly using the selective laser sintering (SLS®) process. Individual sand grains are coated with a polymeric binder. Laser energy is applied to melt this binder thereby coating the individual sand grains and bonding the grains of sand together in the shape of a mold.[4] Accuracy and surface finish of the metal castings produced from such molds are similar to those produced by conventional sand casting methods. Functional prototypes can be produced this way and if modifications are necessary, a new prototype can be produced within a few days.

7.5.1.2.   *Direct ACES Injection Molding*

This is a rapid tooling method developed by 3D CAD/CAM systems that uses the SLA to produce resin molds which allows direct injection of thermoplastic materials. Known as the Direct ACES injection molding (AIM),[5] this method is able to produce parts with high levels of accuracy. However, build times using this method is relatively slow on the standard stereolithography (SLA) machine. Also, because the mechanical properties of these molds are generally poor, tool damage can occur during the ejection of the part. This is more evident when producing geometrically complex parts using these molds.

7.5.1.3. *SL Composite Tooling*

This method builds molds with thin shells of resin with the required surface geometry which is then backed-up with aluminum powder-filled epoxy resin to form the rest of the mold tooling.[6] This method is advantageous in that higher mold strengths can be achieved when compared to those produced by the Direct AIM method which builds a solid SLA resin mold. To further improve the thermal conductivity of the mold, aluminum shot can be added to back the thin shell, thus promoting faster build times for the mold tooling. Other advantages of this method include higher thermal conductivity of the mold and lower tool development costs when compared to molds produced by the Direct AIM method.

7.5.2. Indirect Soft Tooling

In this rapid tooling method, a master pattern is first produced using RP. From the master pattern, a mold tooling can be built out of an array of materials such as silicone rubber, epoxy resin, low-melting point metals and ceramics.

7.5.2.1. *Arc Spray Metal Tooling*

Using metal spraying on the RP model, it is possible to create very quickly an injection mold that can be used to mold a limited number of prototype parts. A typical metal spray process for creating injection mold is shown in Fig. 7.4. The metal spraying process is operated manually, with a hand-held gun. An electric arc is introduced between two wires, which melt the wires into tiny droplets.[7] Compressed air blows out the droplets in small layers of approximately 0.5 mm of metal.

Using the master pattern produced by any RP process, this is mounted onto a base and bolster, which are then layered with a release agent. A coating of metal particles using the arc spray is then applied to the master pattern to produce the female form cavity of the desired tool. Depending on the type of tooling application, a reinforcement backing is selected and applied to the shell. Types of backing materials include filled epoxy

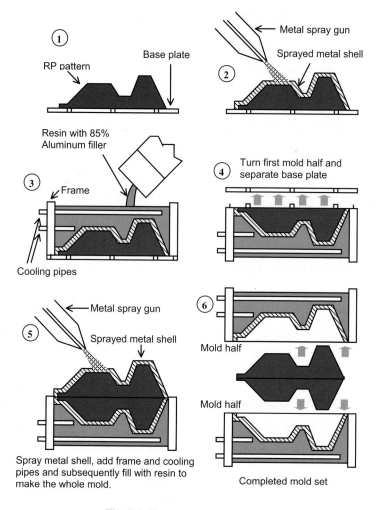

Fig. 7.4. The metal arc spray system.

resins, low-melting point metal alloys and ceramics. This method of producing soft tooling is cost and lead-time saving.

7.5.2.2. *Silicone Rubber Molds*

In manufacturing functional plastic, metal and ceramic components, vacuum casting with the silicone rubber mold has been the most flexible rapid tooling process and the most used to date. They have the following advantages:

- Extremely high resolution of master model details can be easily copied to the silicone cavity mold.
- Gross reduction of backdraft problems (i.e., die lock, or the inability to release the part from the mold cavity because some of the geometry is not within the same draw direction as for the rest of the part).

The master pattern, attached with a system of sprue, runner, gating and air vents, is suspended in a container. Silicone rubber slurry is poured into the container engulfing the master pattern. The silicone rubber slurry is baked at 70°C for 3 h and upon solidification, a parting line is cut with a scalpel. The master pattern is removed from the mold thus forming the tool cavity. The halves of the mold are then firmly taped together. Materials, such as polyurethane, are poured into the silicone tool cavity under vacuum to avoid asperities caused by entrapped air. Further baking at 70°C for 4 h is carried out to cure the cast polymer part. The vacuum casting process is generally used with such molds. Each silicone rubber mold can produce up to 20 polyurethane parts before it begins to break apart.[8] These problems are commonly encountered when using hard molds, making it necessary to have expensive inserts and slides. They can be cumbersome and take a longer time to produce. These are virtually eliminated when the silicone molding process is used.

RP models can be used as master patterns for creating these silicon rubber molds. Figures 7.5(a)–7.5(f) describe the typical process of creating a silicon rubber mold and the subsequent urethane-based part.

A variant of this is a process developed by Shonan Design Co Ltd. This process, referred to as the "Temp-less" (temperature-less) process, makes use of similar principles of preparing the silicone mold and casting the liquid polymer except that no baking is necessary to cure the materials. Instead, ultraviolet rays are used for curing of the silicone mold and urethane parts. The advantages this gives are higher accuracy of replicating the master model because no heat is used, less equipment are required and it takes only about 30% of the time to produce the parts as compared to the standard silicone molding processes.[9]

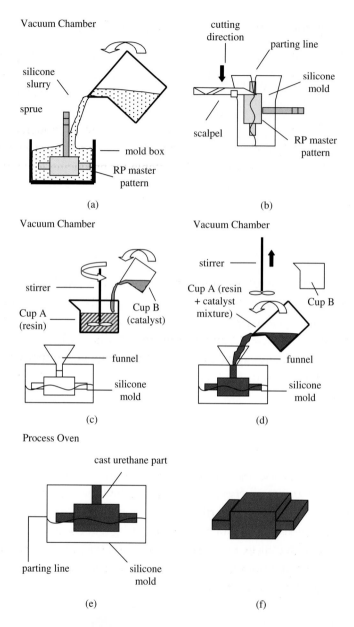

Fig. 7.5. Vacuum casting with silicone molding. (a) Producing the silicone mold, (b) removing the RP master pattern, (c) mixing the resin and catalyst, (d) casting the polymer mixture, (e) cast urethane part cured in a baking oven and (f) the final rapid tooled urethane part.

7.5.2.3. *Spin Casting with Vulcanized Rubber Molds*

Spin casting, as its name implies, applies spinning techniques to produce sufficient centrifugal forces in order to assist in filling the cavities. Circular tooling molds made from vulcanized rubber are produced in much the same way as in silicone rubber molding. The tooling cavities are formed closer to the outer parameter of the circular mold to increase centrifugal forces. Polyurethane or zinc-based alloys can be cast using this method.[10] This process is particularly suitable for producing low volumes of small zinc prototypes that will ultimately be mass-produced by die-casting.

7.5.2.4. *Castable Resin Molds*

Similar to the silicone rubber molds, the master pattern is placed in a mold box with the parting line marked out in plasticine.[11] The resin is painted or poured over the master pattern until there is sufficient material for one half of the mold. Different tooling resins may be blended with aluminum powder or pellets so as to provide different mechanical and thermal properties. Such tools are able to typically withstand up to 100–200 injection molding shots.

7.5.2.5. *Castable Ceramic Molds*

Ceramic materials that are primarily sand-based can be poured over a master pattern to create the mold.[12] The binder systems can vary with preference of binding properties. For example, in colloidal silicate binders, the water content in the system can be altered to improve shrinkage and castability properties. The ceramic-binder mix can be poured under vacuum conditions and vibrated to improve the packing of the material around the master pattern.

7.5.2.6. *Plaster Molds*

Casting into plaster molds have been used to produce functional prototypes.[13] A silicon rubber mold is first created from the master pattern and a plaster mold is then made from this. Molten metal is then poured into the plaster mold which is broken away once the metal has solidified. Silicon rubber is used as an intermediate stage because the pattern can be easily separated from the plaster mold.

7.5.2.7. *Casting*

In the metal casting process, a metal, usually an alloy, is heated until it is in a molten state, whereupon it is poured into a mold or die that contains a cavity. The cavity will contain the shape of the component or casting to be produced. Although there are numerous casting techniques available, three main processes are discussed here: the conventional sand casting, investment casting and evaporative casting processes. RP models render themselves well to be the master patterns for the creation of these metal dies.

Sand casting molds are similarly created using RP master patterns. RP patterns are first created and placed appropriately in the sand box. Casting sand is then poured and packed very compactly over the pattern. The box (cope and drag) is then separated and the pattern carefully removed leaving behind the cavity. The box is assembled together again and molten metal is cast into the sand mold. Sand casting is the cheapest and most practical for casting of large parts. Figure 7.6 shows a cast metal mold resulting from a RP pattern made by LOM.

Another casting method, the investment casting process, is probably the most important molding process for casting metal. Investment casting molds can be made from RP master patterns. The pattern is usually wax,

Fig. 7.6. Cast metal (left) and RP pattern for sand casting (courtesy Helysis, Inc.).

foam, paper, or other materials that can be easily melted or vaporized. The pattern is dipped in slurry ceramic compounds to form a relatively strong coating, or investment shell, over it.[14] This is repeated until the shell builds up thickness and strength. The shell is then used for casting, with the pattern being melted away or burned out of the shell, resulting in a ceramic cavity. Molten metal can then be poured into the mold to form the object. The shell is then cracked open to release the desired object in the mold. The investment casting process is ideal for casting miniature parts with thin sections and complex features. Figure 7.7 shows schematically the investment casting process from an RP-produced wax master pattern while Fig. 7.8 shows an investment casting mold resulting from a RP pattern.

The third casting process discussed in this book is the evaporative pattern casting. As its name implies, this uses an evaporative pattern, such as polystyrene foam, as the master pattern. This pattern can be produced using the SLS process along with CastForm™ polystyrene material. The master pattern is attached to sprue, riser and gating systems to form a "tree". This polystyrene "tree" is then surrounded by foundry sand in a container and vacuum compacted to form a mold. Molten steel is then poured into the container through the sprue. As the metal fills the cavity, the polystyrene evaporates with very low-ash content.[15] The part is cooled before the casting is removed. A variety of metals, such as titanium, steel, aluminum, magnesium and zinc can be cast using this method. Figure 7.9 shows schematically how an RP master pattern is used with the evaporative pattern casting process.

7.5.3. Direct Hard Tooling

Hard tooling produced by RP systems has been a major topic for research in recent years. Although several methods have been demonstrated, much research is still being carried out in this area. The advantages of hard tooling produced by RP methods are fast turnaround times to create highly complex-shaped mold tooling for high volume production. Fast response to modifications in generic designs can be almost immediate. The following are some examples of direct hard tooling methods.

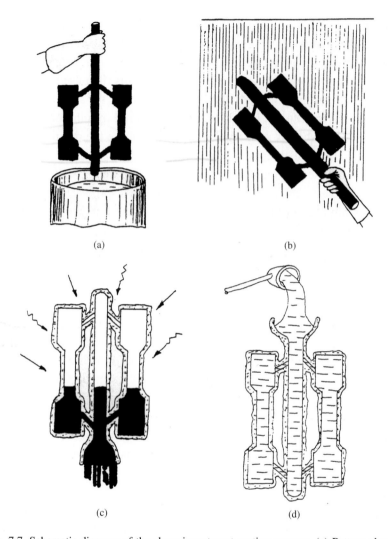

(a)   (b)

(c)   (d)

Fig. 7.7. Schematic diagram of the shess investment casting process. (a) Pattern clusters are dipped in ceramic slurry and (b) refractory grain is sifted onto the coated patterns. Steps (a) and (b) are repeated several times to obtain desired shell thickness, (c) after the mold material has set and dried, the patterns are melted out of the mold and (d) hot molds are filled with metal by gravity, pressure vaccum, or centrifugal force.

7.5.3.1. *RapidTool*™

RapidTool™ is a technology invented by the DTM Corporation to produce metal molds for plastic injection molding directly from the

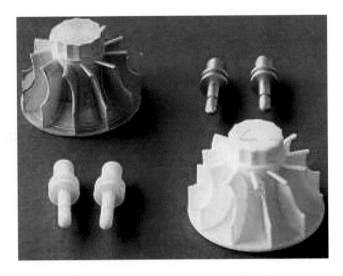

Fig. 7.8. Investment casting of fan impeller from RP pattern.

SLS Sinterstation. The molds are capable of being used in the conventional injection molding machines to mold the final product with the functional material.[16] The CAD data is fed into the Sinterstation™ which bonds polymeric binder coated metal beads together using the SLS process. Next, debinding takes place and the green part is cured and infiltrated with copper to make it solid. The furnace cycle is about 40 h with the finished part having similar properties equivalent to aluminum. The finished mold can be easily machined. Shrinkage is reported to be no more than 2%, which is compensated for in the software.

Typical time frames allow relatively complex molds to be produced in 2 weeks as compared to 6–12 weeks using conventional techniques. The finished mold is capable of producing up to tens of thousands of injection-molded parts before breaking down.

7.5.3.2. *Laminated Metal Tooling*

This is another method that may prove promising for RT applications. The process applies metal laminated sheets with the laminated object manufacturing (LOM) method. The sheets can be made of steel or any other material which can be cut by the appropriate means, for example by CO_2

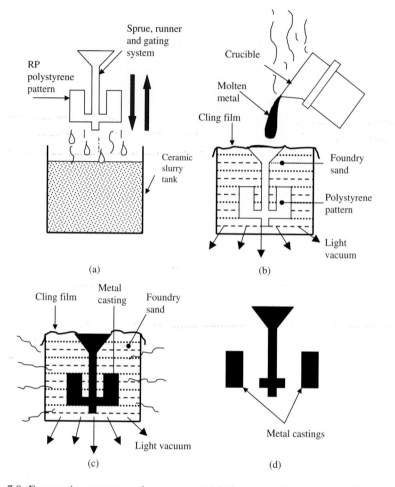

Fig. 7.9. Evaporative pattern casting process. (a) Polystyrene RP pattern "tree" is coated by dipping with a ceramic slurry and air-dried, (b) coated RP pattern is packed with foundry sand in a container. The container is sealed with cling film and vacuumed to compact the sand further, (c) the polystyrene pattern evaporates as the molten metal is cast into the mold. The casting is then left to cool and (d) after solidification, the final cast parts are removed from the sprue, runner and gating system.

laser, water jet, or milling, based on the LOM principle.[17] The CAD 3D data provides the sliced 2D information for cutting the sheets layer by layer. However, instead of bonding each layer as it is cut, the layers are all assembled after cutting and either bolted or bonded together.

7.5.3.3. *Direct Metal Laser Sintering Tooling*

The direct metal laser sintering technology was developed by EOS. The process uses a very high-powered laser to sinter metal powders directly. The powders available for use by this technology are the bronze-based and steel-based materials. Bronze is used for applications where strength requirements are not crucial. Upon sintering of the bronze powder, an organic resin, such as epoxy, is used to infiltrate the part. For steel powders, the process is capable of producing direct steel parts of up to 95% density so that further infiltration is not required. Several direct applications are produced with this technology including mold inserts and other metal parts.[18]

7.5.3.4. *ProMetal™ Rapid Tooling*

Similar to MIT's Three-Dimensional Printing (3DP) process, the ProMetal™ Rapid Tooling System is capable of creating steel parts for tooling of plastic injection molding parts, lost foam patterns and vacuum forming. This technology uses an electrostatic ink-jet print head to eject liquid binders onto the powder, selectively hardening slices of an object a layer at a time. A fresh coat of metal powder is spread on top and the process repeats until the part is complete. Loose powders act as supports for the object to be built. The RP part is then infiltrated at furnace temperatures with a secondary metal to achieve full density. Toolings produced by this technology for use in injection molding have reported withstanding pressures up to 30,000 psi (200 MPa) and survived 100,000 shots of glass-filled nylon.[19]

7.5.4. Indirect Hard Tooling

There are numerous indirect RP tooling methods that fall under this category and this number continues to grow. However, many of these processes remain largely similar in nature except for small differences, e.g., binder system formulations or type of system used. Processes include the Rapid Solidification Process, Ford's (UK) Sprayform, Cast Kirksite Tooling, CEMCOM's Chemically Bonded Ceramics and Swift Technologies Ltd

"SwiftTool", just to name a few. This section will only cover selected processes that can also be said to generalize all the other methods under this category. In general, indirect methods for producing hard tools for plastic injection molding generally make use of casting of liquid metals or steel powders in a binder system. For the latter, debinding, sintering and infiltration with a secondary material are usually carried out as post-processes.

7.5.4.1. *3D Keltool*

The 3D Keltool process has been developed by 3D Systems to produce a mold in fused powdered steel.[20] The process uses an SLA model of the tool for the final part that is finished to a high quality by sanding and polishing. The model is placed in a container where silicone rubber is poured around it to make a soft silicone rubber mold that replicates the female cavity of the SLA model. This is then placed in a box and silicone rubber is poured around it to produce a replica copy of the SLA model. This silicone rubber is then placed in a box and a proprietary mixture of metal particles, such as tool steel and a binder material is poured around it, cured and separated from the silicone rubber model. This is then fired to eliminate the binder and sinter the green metal particles together. The sintered part is about 70% steel and 30% void is then infiltrated with copper to give a solid mold, which can be used in injection molding.

An alternative to this process is described as the reverse generation process. This uses a positive SLA master pattern of the mold and requires one step less. This process claims that the CAD solid model to injection-molded production part can be completed in 4–6 weeks. Cost savings of around 25–40% can be achieved when compared to that of conventional machined steel tools.

7.5.4.2. *EDM Electrodes*

A method successfully tested in research laboratories, but so far not widely applied in industry is the possible manufacturing of copper electrodes for electro-discharge machining (EDM) processes using RP technology.

To create the electrode, the RP-created part is used to create a master for the electrode. An abrading die is created from the master by making a cast using an epoxy resin with an abrasive component. The resulting die is then used to abrade the electrode. A specific advantage of the SLS procedure (see Sec. 5.1) is the possible usage of other materials. Using copper in the SLS process, it is possible to generate the electrodes used in EDM quickly and affordably.

7.5.4.3. *Ecotool*

This is a development between the Danish Technological Institute (DTI) in Copenhagen, Denmark and the TNO Institute of Industrial Technology of Delft in Holland. The process uses a new type of powder material with a binder system to rapidly produce tools from RP models. As its name implies, the binder is friendly to the environment in that it uses a water-soluble base. An RP master pattern is used and a parting line block produced. The metal powder-binder mixture is then poured over the pattern and parting block and left to cure for an hour at room temperature. The process is repeated to produce the second half of the mold in the same way. The pattern is then removed and the mold baked in a microwave oven.

7.5.4.4. *Copy Milling*

Although not broadly applied nowadays, RP master patterns can be provided by manufacturers to their vendors for use in copy milling, especially if the vendor for the required parts is small and does not have the more expensive, but accurate CNC machines. In addition, the principle of generating master models only when necessary, allows some storage space to be saved. The limitation of this process is in that only simple geometrical shapes can be made.

7.6. AEROSPACE INDUSTRY

With the various advantages that RP technologies promise, it is only natural that high value-added industries like the aerospace industry have taken special interest in it even though initial investment costs may be high. There are abundant examples of the use of RP technology in the aerospace industry. The following are a few examples.

7.6.1. Design Verification of an Airline Electrical Generator

Sundstrand Aerospace, which manufactures inline electrical generators for military and commercial aircraft, needed to verify its design of an integrated drive generator for a large jetliner.[21] It decided to use Helisys's LOM to create the design-verification model. The generator is made up of an external housing and about 1,200 internal parts. Each half of the housing measures about 610 mm in diameter is 300 mm tall and has many intricate internal cavities into which the subassemblies must fit.

Such complex designs are difficult to visualize from 2D drawings. A physical model of the generator housing and many of its internal components is a good way to identify design problems before the expensive tooling process. But time and expense needed to construct the models by traditional means are prohibitive. Thus, Sundstrand decided to turn to RP technologies. Initial designs for the generator housing and internal subassemblies were completed on a CAD system and the subsequent. STL files were sent to a service bureau. Within two weeks, Sundstrand was able to receive the parts from the service bureau and began its own design verification.

Sundstrand assembled the various parts and examined them for form, fit and limit function. Clearances and interferences between the housing and the many subassemblies were checked. After the initial inspection, several problematic areas were found which would have otherwise been missed. These were corrected and incorporated into the CAD design and in some cases, new RP models were made. Apart from design verification, Sundstrand was able to use the physical models to help toolmakers plan and design casting patterns. The models were also used for manufacturing process design, tool checking and assembly sequence design. Though the approximate cost for the RP models was US$16,500, the savings realized from removing engineering and design changes were immeasurable and the time saved (estimated to be about eight to ten weeks) was significant.

7.6.2. Engine Components for Fanjet Engine

In an effort to reduce the developmental time of a new engine, AlliedSignal Aerospace used 3D Systems' QuickCast™ to produce a turbofan jet engine for a business aviation jet.[22] Basically, RP is used for the generation of the casting pattern for an impeller compressor shroud engine component. This part is the static component that provides the seal for the high-pressure compressor in the engine. Three different designs were required for cold rig, hot rig and first engine to test. Using QuickCast™, the 3D Technology Center was able to produce directly patterns for investment castings using the stereolithography technology. The patterns produced were durable, had improved accuracy, good surface finish and were single large piece patterns. In fact, the patterns created were accurate enough for a design revision error in the assembly fixture to be easily detected and corrected. With the use of these RP techniques, production time was slashed by eight to ten weeks and a saving of US$50,000 for tooling in the three design iterations was realized.

7.6.3. Prototyping Air Inlet Housing for Gas Turbine Engine

Sundstrand Power Systems, a manufacturer of auxiliary engines for military and commercial aircraft, needed prototypes of an air inlet housing for a new gas turbine engine.[23,24] It first needed mock-ups of the complex design and also several fully functional prototypes to test on the development engines. The part, which measures about 250 mm in height and 300 mm in diameter, has wall thickness as thin as 1.5 mm (see Fig. 7.10). It would have been difficult and costly to build using traditional methods.

To realize the part, Sundstrand used the SLS® system (see Sec. 5.1) at a service bureau to build the evaluation models of the housing and then generate the necessary patterns for investment casting, ultimately the method used for the manufacture. The SLS® system was chosen primarily because the air inlet housing has several overhanging structures from which removal of supports would have been extremely difficult.

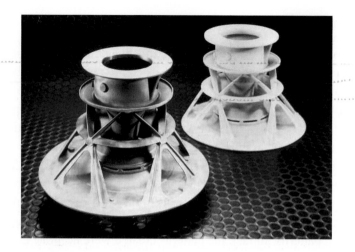

Fig. 7.10. Polycarbonate investment casting pattern (right) and the steel air inlet housing (left) for a jet turbine engine (courtesy DTM Corporation).

Sundstrand designed several iterations of the housing as solid models on its CAD system. These models were converted to the STL format and sent to build the nylon evaluation models. As the program progressed, Sundstrand wanted to test the part. As the designs were finalized, new SLS® versions of the part were created as tooling for investment casting. Polycarbonate patterns were created, sealed with wax and sent for casting. The patterns were first coated with a thin layer of polyurethane to fill any remaining surface pores and provide the necessary surface finish. Then the patterns were used to cast the part in Inconel 718 steel, which were sent back to Sundstrand for testing. In all, Sundstrand saved more than four months of tooling and prototyping time and saved more than US$88,000.

7.6.4.   Fabrication of Flight-Certified Production Castings

Bell Helicopter has successfully used stereolithography, first to verify parts design, then to aid with fit and functional testing and finally to produce investment casting patterns for the manufacture of Federal Aviation Authority (FAA)-certifiable production parts.[25] About 50 of the parts that made up the new helicopter's flight control system were developed with stereolithography. The largest support structure for the hydraulic system, measured approximately $500 \times 500 \times 200$ mm and the smallest, $25 \times 25 \times 1.1$ mm. In production, all parts will be investment cast, most in aluminum while others will be in steel alloys.

Initially, half-scale models were used for design verification, as they were large enough to confirm design intent and were much quicker to fabricate on the SLA machines. Once a design was finalized, full-size SLA models were fabricated for use in "virtual installation".[25] In virtual installation, full-sized SLA parts were assembled with other components and installed on the actual production helicopter in order to test the fit and kinematics of the assembly. Parts used for virtual installation included all the features that would normally be machined into rough production castings. Problems associated with interferences and clearances were identified and rectified before they could arise in later stages, which by then would be more costly to rectify.

After virtual installation, Bell made QuickCast™ investment casting patterns of each part. These patterns were sent for casting, with the resulting parts being sent for FAA flight certification. In previous projects, Bell would have machined parts to simulate production castings and send them for certification. When the castings became available in about 45 weeks, the parts would have to be recertified. With QuickCast™ patterns, Bell could produce production-grade metal investment castings in as little as three weeks and did not need recertification when wax tooling eventually became available. The overall development time was shortened with the use SLA models and QuickCast™ for creating investment casting and came to about six months, resulting in substantial cost savings and a better product offered to the market.

7.7. AUTOMOTIVE INDUSTRY

7.7.1. Prototyping Complex Gearbox Housing for Design Verification

Volkswagen has utilized the LOM to speed up the development of a large, complex gearbox housing for its Golf and Passat car lines.[26] The CAD model for the housing was extremely complex and difficult to visualize.

VW wanted to build a LOM part to check the design of the CAD model and then use the part for packaging studies.

Using traditional methods, such a prototype would have been costly and time consuming to build; and it might not have been always possible to include all fine details of the design. Fabrication of the model based on drawings was often subjected to human interpretation and consequently error-prone, thus further complicating the prototyping process. All these difficulties were avoided by using RP technology as the fabrication of the model was based entirely on the CAD model created.

The gearbox housing was too large for the build volume of the LOM machine. The CAD model was thus split into five sections and reassembled after fabrication. It took about ten days to make and finish all five sections and once they were completed, patternmakers glued them together to complete the final model. The LOM model was first used for verifying the design and subsequently, to develop sand-casting tooling for the creation of metal prototypes. The RP process had shrunk the prototype development time from eight weeks to less than two and considerable time and cost savings were achieved.

7.7.2. Prototyping Advanced Driver Control System with Stereolithography

At General Motors, in many of its divisions, RP is becoming a necessary tool in the critical race to be the first to market.[27] For example, Delco Electronics, its automotive electronics subsidiary, was involved in the development of the Maestro project. Designed to blend an advanced Audio System, a hands-free cellular phone, Global Positioning System navigation, Radio Data System information and climate control into a completely integrated driver control system, the Maestro was to be a marvel.

With many uniquely-shaped push-buttons, two active-matrix LCD screens and a local area network allowing for future expansion, the time needed to develop the system was the most critical factor. The system was to be launched at Convergence '94, the world's premier automotive electronics conference held only once every two years, for maximum impact in the automotive market.

Working with Modern Engineering, an engineering service company, Delco Electronics developed the first renderings and concept drawings for the Maestro project. In order to speed up the project in time for the conference, the designers needed the instrument panel with its myriad of push-buttons working early in the design cycle. Unfortunately, the large number of buttons meant a corresponding large number of rubber molds with all the problems associated with the conventional molding process. From the stylist's concepts, models for each button face were manually machined. Once the designs were confirmed, the machined models were laser scanned, generating the CAD data needed for the creation of SLA models. The final prototype buttons needed to be accurate enough to ensure proper fit and function, as well as translucent, so that they could be back-lit.

The SLA models generated on 3D Systems' SLA machine were accurate enough to be finished, painted and installed in the actual prototype vehicle, eliminating the need for rubber molds. The result was that in less than four months, Delco Electronics was able to complete the functional instrument panel, with all 108 buttons built using the SLA. As such, the Maestro System was launched successfully at Convergence '94, confirming Delco Electronics' position as an automotive technology leader.

7.7.3. Creating Cast Metal Engine Block with RP Process

As new engine design and development is an expensive and time consuming process, the ability to test a new engine and all its auxiliary components before committing to tooling is important in ensuring costs and time savings.[28] The Mercedes-Benz Division of Daimler-Benz AG has initiated a program of physical design verification on prototype engines using SLA parts for initial form and fit testing. After initial design reviews, metal components were produced rapidly using the QuickCast™ process.

Their first project was the design and prototyping of a 4-cylinder engine block for the new Mercedes-Benz "A-Class" car. The aim was to cast the engine block directly from a stereolithography QuickCast™ pattern. The engine block was designed on Mercedes-Benz's own CAD system and the data were transferred to 3D Systems Technology Center at Darmstadt, where the one-piece pattern of the block was built on the SLA machine. The full scale investment casting pattern was generated in 96 h.

The pattern was then sent for shell investment casting, resulting in the $300 \times 330 \times 457$ mm engine block being cast in A356-T6 aluminum in just five weeks. The completed engine block incorporated cast-in water jacket, core passage ways and exhibited Grade B radiographic quality in all areas evaluated. The entire prototyping process using RP technology lasted only six weeks (compared to 15–18 weeks using traditional methods) and the cost savings were approximately US$150,000 as compared to traditional methods. These are both significant, especially in the need for short time-to-market requirement.

7.7.4.   Using Stereolithography to Produce Production Tooling

Ford Motor Company has used 3D Systems' QuickCast™ to create production tool of a rear wiper-motor cover for the 1994 Explorer sport utility vehicle.[29,30] The part measured approximately 200×150 mm by 75 mm and was to be injection molded with polypropylene during production. Traditional methods would have provided the necessary tools for molding in three months.

Ford first built the SLA model of the cover and fitted it over the wiper motor to verify the design (see Fig. 7.11). Dimensional and assembly problems were identified and rectified before the design was confirmed. From the CAD model data, originally created on the CAD software Pro Engineer, the Pro/MOLDDESIGN® software was used to create "negative" mold halves. Shrink factors were then applied to compensate for the photo-curable resin, A2 steel and polypropylene. The QuickCast™ process was then used to build the SLA patterns of the actual tool inserts (in halves). They were then investment cast out in A2 tool steel. Once cast, the tool inserts were fitted onto an injection molding machine and used to produce the plastic wiper-motor covers. With the application of such "rapid tooling" techniques, Ford was able to start durability and water flow testing 18 months ahead of schedule, with costs reduction of 45% and time savings of more than 40%.

Fig. 7.11. QuickCast™ generated patterns and the investment cast inserts for the rear wiper-motor cover.

7.8. JEWELRY INDUSTRY

The jewelry industry has traditionally been regarded as one which is heavily craft-based and automation is generally restricted to the use of machines in the various individual stages of jewelry manufacturing. The use of RP technology in jewelry design and manufacture offers a significant breakthrough in this industry. In an experimental computer-aided jewelry design and manufacturing system jointly developed by Nanyang Technological University and Gintic Institute of Manufacturing Technology in Singapore, the SLA (from 3D Systems) was used successfully to create fine jewelry models.[31] These were used as master patterns to create the rubber molds for making wax patterns that were later used in investment casting of the precious metal end product (see Fig. 7.12). In an experiment with the design of rings, the overall quality of the SLA models were found to be promising, especially in the generation of intricate details in the design. However, due to the nature of the step-wise building of the model, steps at the "gentler" slope of the model were visible. With the use of better resin and finer layer thickness, this problem was reduced but not fully eliminated. Further processing was found to be necessary and abrasive jet deburring was identified to be most suitable.[32]

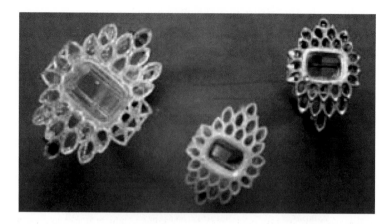

Fig. 7.12. An investment cast silver alloy prototype of a brooch (right), the full-scale wax pattern produced from the silicone rubber molding (center) and the two-times scaled SLA model to aid visualization (left).

Though post-processing of SLA models is necessary in the manufacture of jewelry, the ability to create models quickly (a few hours compared to days or even weeks, depending on the complexity of the design) and its suitability for use in the manufacturing process offer great promise to improve the design and manufacture in the jewelry industry.

7.9. COIN INDUSTRY

Similar to the jewelry industry, the mint industry has traditionally been regarded as very labor-intensive and craft-based. It relies primarily on the skills of trained craftsmen in generating the "embossed" or relief designs on coins and other related products. In another experimental coin manufacturing system using CAD/CAM, CNC and RP technologies developed by Nanyang Technological University and Gintic Institute of Manufacturing Technology in Singapore, the SLA (from 3D Systems) was used successfully with a Relief Creation Software to create tools for coin manufacture.[33] In the system involving RP technology, its working methodology consists of several steps.

Firstly, 2D artwork is read into ArtCAM, the CAD/CAM system used in the system, utilizing a Sharp JX A4 scanner. Figure 7.13 shows the 2D

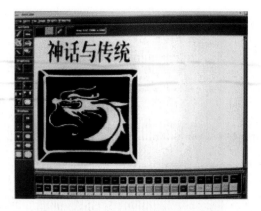

Fig. 7.13. Two-dimensional artwork of a series of Chinese characters and a roaring dragon.

artwork of a series of Chinese characters and a roaring dragon. In the ArtCAM environment, the scanned image is reduced from a color image to a monochrome image with the fully automatic "Gray Scale" function. Alternatively, the number of colors in the image can be reduced using the "Reduce Color" function. A color palette is provided for color selection and the various areas of the images are colored, either using different sizes and types of brushes or the automatic flood fill function.

The second step is the generation of surfaces. The shape of a coin is generated to the required size in the CAD system for model building. A triangular mesh file is produced automatically from the 3D model. This is used as a base onto which the relief data is wrapped and later combined with the relief model to form the finished part.

The third step is the generation of the relief. In creating the 3D relief, each color in the image is assigned a shape profile. There are various fields that control the shape profile of the selected colored region, namely, the overall general shape for the region, the curvatures of the profile (convex or concave), the maximum height, base height, angle and scale. The relief detail generated can be examined in a dynamic Graphic Window within the ArtCAM environment itself. Figure 7.14 illustrates the 3D relief of the roaring dragon artwork.

The fourth step is the wrapping of the 3D relief onto the coin surface. This is done by wrapping the 3D relief onto the triangular mesh file

Fig. 7.14. Three-dimensional relief of artwork of the roaring dragon.

generated from the coin surfaces. This is a true surface wrap and not a simple projection. The wrapped relief is also converted into triangular mesh files. The triangular mesh files can be used to produce a 3D model suitable for color shading and machining. The two sets of triangular mesh files, of the relief and the coin shape, are automatically combined. The resultant model file can be color-shaded and used by the SLA to build the prototype.

The fifth step is to convert the triangular mesh files into the STL file format. This is to be used for building the RP model. After the conversion, the STL file is sent to the SLA to create the 3D coin pattern which will be used for proofing of design.[33]

7.10. TABLEWARE INDUSTRY

In another application to a traditional industry, the tableware industry, CAD and RP technologies are used in an integrated system to create better designs in a faster and more accurate manner. The general methodology used is similar to that used in the jewelry and coin industries. Additional computer tools with special programs developed to adapt decorative patterns to different variations of size and shape of tableware are needed for this particular industry.[34] Also, a method for generating motifs along a circular arc has been developed to supplement the capability of such a system.[35]

The general steps involved in the art to part process for the tableware include the following:

(1) Scanning of the 2D artwork.
(2) Generation of surfaces.
(3) Generation of 3D decoration reliefs.
(4) Wrapping of reliefs on surfaces.
(5) Converting triangular mesh files to STL file.
(6) Building of model by the RP system.

Two RP systems are selected for experimentation in the tableware system. One is 3D Systems' SLA and the other is Helysis' LOM. The SLA has the advantages of being a pioneer and a proven technology with many excellent case studies available. It is also advantageous to use in tableware design as the material is translucent and thus allows designers to view the internal structure and details of tableware items like tea pot and gravy bowls. On the other hand, the use of LOM has its own distinct advantages. Its material cost is much lower and because it does not need support in its process (unlike the SLA); it saves a lot of time in both pre-processing (deciding where and what supports to use) and post-processing (removing the supports). Examples of dinner plates produced using the systems are shown in Fig. 7.15.

Fig. 7.15. Dinner plate prototype built using SLA (left) and LOM (right).

In an evaluation test of making the dinner plate prototype, it was found that the LOM prototype was able to recreate the floral details more accurately. The dimensional accuracy was slightly better in the LOM prototype. In terms of built-time, including pre- and post-processing, the SLA was about 20% faster than the LOM process. However, with sanding and varnishing, the LOM prototype appeared to be a better model which could be used later to create the plaster of Paris molds for the molding of the ceramic table ware (see Fig. 7.16 for a tea-pot built using LOM). Apart from these technical issues, the initial investment, operating and maintenance costs of the SLA were considerably higher than that of the LOM, estimated to be about 50–100% more.

In the ceramic tableware production process, the LOM model can be used directly as a master pattern to produce the block mold. The mold is made of plaster of Paris. The result of this trial is shown in Fig. 7.17. The trials highlighted the fact that plaster of Paris is an extremely good material for detailed reproduction. Even slight imperfections left after hand finishing the LOM model are faithfully reproduced in the block mold and pieces cast from these molds.

Whichever RP technology is adopted, such a system saves time in designing and developing tableware, particularly in building a physical

Fig. 7.16. LOM model of a tea pot (courtesy Champion Machine Tools, Singapore).

Fig. 7.17. Block mold cast from the LOM model of the dinner plate (courtesy Oriental Ceramics Sdn Bhd, Malaysia).

prototype. It can also improve designs by simply amending the CAD model and the overall system is easy and friendly to use.

7.11. GEOGRAPHIC INFORMATION SYSTEM APPLICATIONS

RP has been applied to create physical models of 3D Geographic Information System (GIS) objects to replace 2D representation of geographical information. Figure 7.18 shows a 3D physical model of a city.[36] To be able to do this, Contex Scanning Technology introduced its PUMA HS 36 color scanner, which features iJET Technology and the 18 flatbed color scanner, which is designed to scan all types of originals, including rare and valuable documents up to A2/C size.

7.11.1. 3D Physical Map

The first experiences concerning the reception of the models were very encouraging and surprising: nearly everybody tried to touch the models with his hands, to get a feeling in the literal sense.[36] The sensual capability

Fig.7.18. A physical model of a city (courtesy CONTEX Scanning Technology A/S).

creates a possibility to provide maps for blind and visually impaired persons. The haptic experience could be utilized as an additional stimulus for transmitting a cartographic message and to induce insight.

7.11.2.   3D Representation of Land Prices

A 3D map visualizing the average cost of land in a country or city could be plotted. The height of the prisms would be proportional to the average price of the land. In contrast to a 2D choropleth map depicting value classes, the absolute differences in height could be perceived immediately.[36]

7.11.3.   3D Representation of Population Data

The volume of the prisms could proportional to the number of inhabitants in each unit area. The height of the prisms would then be proportional to the population density (see Fig. 7.19).

7.12.   ARTS AND ARCHITECTURE

7.12.1.   Architecture

Architectural models are used for visualizing the perception and elements of a building that are considered most important to the decision makers (see Fig. 7.20). Architectural model making has revolutionized recently in the past few years. RP technology has been considered to replace

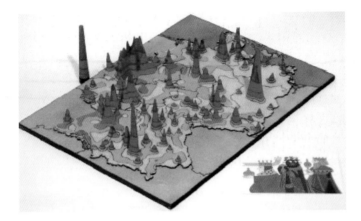

Fig. 7.19. Smooth surface depicting population density (courtesy Federal Office for Building and Spatial Planning (BBR)).

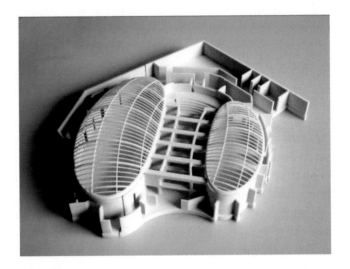

Fig. 7.20. An architecture model built from 3DP (courtesy CONTEX Scanning Technology A/S).

traditional techniques as it has the following advantages: (a) The ability to create scaled models directly from 3D CAD data in a few hours. It is currently used by large architectural firms for the creation of proposed and as-built designs. Design changes in particular segments of the construction

Fig. 7.21. Torus Knot (courtesy Stewart Dickson).

can be readily rebuilt based on the reproducibility of the RP models and (b) the ability to create multiple models at reasonable cost. For larger projects involving more than one key main contractor, it is advantageous to build multiple copies of the model for coordination and communication.

7.12.2. Arts

Today, RP technology is being used by an increasing number of artists to build a wide variety of sculptural objects. Some of these works are realistic and representational while others are abstract. The abstract objects can be the result of pure imagination and artistic free will, or may be derived solely from mathematics or computation. Some of the works created with RP may not have been possible to be made by any other way.[37]

A Trefoil Torus-Knot with computer-generated texture map rendered in 3DP. Color 3D printers now build computer-specified color texture directly modeled-in to the 3D part (Fig. 7.21). It is possible to reproduce surface coloration which defies reproduction by hand — i.e., spaces where a paint brush cannot fit.

REFERENCES

1. T. Mueller, Applications of stereolithography in injection molding, *Proc. Second Int. Conf. Rapid Prototyping*, 23–26 June, Dayton, USA (1991), pp. 323–329.

2. J. H. Wang, C. S. Lim and J. H. Yeo, CFD investigations of steady flow in bi-leaflet heart valve, *Critical Rev. Biomed. Eng.* **28**(1–2), 61–68 (2000).

3. C. K. Chua, K. H. Hong and S. L. Ho, Rapid tooling technology (Part 1) — A comparative study, *Int. J. Adv. Manuf. Technol.* **15**(8), 604–608 (1999).

4. C. Wilkening, Fast production of rapid prototypes using direct laser sintering of metals and foundry sand, *Second National Conference on Rapid Prototyping and Tooling Research*, 18–19 November, UK (1996), pp. 153–160.

5. H. B. Tsang and G. Bennett, Rapid tooling — Direct use of SLA moulds for investment casting, *First National Conference on Rapid Prototyping and Tooling Research*, 6–7 November, UK (1995), pp. 237–247.

6. D. Atkinson, *Rapid Prototyping and Tooling: A Practical Guide* (Strategy Publications, U.K., 1997).

7. C. K. Chua, K. H. Hong and S. L Ho, Rapid tooling technology (Part 2) — A case study using arc spray metal tooling, *Int. J. Adv. Manuf. Technol.* **15**(8), 609–614 (1999).

8. A. D. Venus, S. J. Crommert and S. O. Hagan, The feasibility of silicone rubber as an injection mould tooling process using rapid prototyped pattern, *Second National Conference on Rapid Prototyping and Tooling Research*, 18–19 November UK (1996), pp. 105–110.

9. Shonan Design Co. Ltd, Temp-Less 3.4.3 — UV RTV process for quick development and fast to market (1996).

10. L. Schaer, Spin casting fully functional metal and plastic parts from stereolithography models, *The Sixth Int. Conf. on Rapid Prototyping*, Dayton, USA (1995), pp. 217–236.

11. J. C. Male, N. A. B. Lewis and G. R. Bennett, The accuracy and surface roughness of wax investment casting patterns from resin and silicone rubber tooling using a stereolithography master, *Second National*

Conf. on Rapid Prototyping and Tooling Research, 18–19 November, UK (1996), pp. 43–52.

12. J. S. Bettay and R. C. Cobb, A rapid ceramic tooling system for prototyping plastic injection moldings, *First National Conference on Rapid Prototyping and Tooling Research,* 6–7 November, UK (1995), pp. 201–210.

13. M. C. Warner, Rapid prototyping methods to manufacture functional metal and plastic parts, *Proc. Rapid Prototyping and Manufacturing Conference,* 11–13 May, Dearborn, Michigan (1993), pp. 137–144.

14. C. S. Lim, L. Siaminwe and A. J. Clegg, Mechanical property enhancement in an investment cast aluminium alloy and metal-matrix composite, *9th World Conference on Investment Casting,* 1–18 October 1996, San Francisco, USA (1996), p. 17.

15. A. J. Clegg, *Precision Casting Processes* (Pergamon Press Plc., Oxford, England, 1991).

16. A. D. Venus and S. J. Crommert, Direct SLS nylon injection, *Second National Conference on Rapid Prototyping and Tooling Research,* 18–19 November, UK (1996), pp. 111–118.

17. R. C. Soar, A. Arthur and P. M. Dickens, Processing and application of rapid prototyped laminate production tooling, *Second National Conference on Rapid Prototyping and Tooling Research,* 18–19 November, UK (1996), pp. 65–76.

18. Industrial Technology, Laser sintering for rapid production [On-line], available at http://www.industrialtechnology.co.uk/2001/may/eos.html (June 2002).

19. Wohlers Report 2000, Rapid Prototyping and Tooling State of the Industry, *Wohlers Associates Inc.,* USA (2000).

20. P. Eyerer, Rapid tooling — Manufacturing of technical prototypes and small series, *Mech. Eng.* **118**(6), 45–47 (1996).

21. Rapid Prototyping Report, Sundstrand Aerospace uses laminated object manufacturing to verify large complex assembly, **5**(6), *CAD/CAM Publishing Inc.* (June 1995), pp. 1–2.

22. 3D Systems, User Focus: AlliedSignal Aerospace, Stereolithography and QuickCastTM provide the winning combination for meeting critical deadline in AlliedSignal's development of the TFE 731-20 Turbo Fanjet Engine (1993).

23. Rapid Prototyping Report, Sundstrand Power Systems uses selective laser sintering to create large investment casting patterns, **5**(1), *CAD/CAM Publishing Inc.* (January 1995), pp. 1–2.

24. DTM Corporation Press Release, SLS-generated Polycarbonate Patterns Speed Casting of Intricate Aircraft Engine Part for Sundstrand Power Systems, DTM Corporation (1994).

25. Rapid Prototyping Report, Bell Helicopter uses QuickCast to fabricate flight-certified production casting, **5**(7), *CAD/CAM Publishing Inc.* (July 1995), pp. 1–2.

26. Rapid Prototyping Report, Volkswagen uses laminated object manufacturing to prototype complex gearbox housing, **5**(2), *CAD/CAM Publishing Inc.* (February 1995), pp. 1–2.

27. 3D Systems, Stereolithography provides a symphony of benefits, *The Edge* **6**(2), 6–7 (1995).

28. 3D Systems, The Winner: Mercedes-Benz AG, 1994 European Stereolithography Excellence Awards (1995).

29. Rapid Prototyping Report, Ford uses stereolithography to cast production tooling, **4**(7), *CAD/CAM Publishing Inc.* (July 1995), pp. 1–3.

30. 3D Systems, The Winner: Ford Motor Company, 1994 Stereolithography Excellence Awards (1995).

31. H. B. Lee, M. S. H. Ko, R. K. L. Gay, K. F. Leong and C. K. Chua, Using computer-based tools and technology to improve jewellery design and manufacturing, *Int. J. Comput. Appl. Technol.* **5**(1), 72–88 (1992).

32. K. F. Leong, C. K. Chua and H. B. Lee, Finishing techniques for jewellery models built using the stereolithography apparatus, *J. Inst. Eng. Singapore* **34**(4), 54–59 (1994).

33. C. K. Chua, R. K. L. Gay, S. K. F. Cheong, L. L. Chong and H. B. Lee, Coin manufacturing using CAD/CAM, CNC and rapid prototyping technologies, *Int. J. Comput. Appl. Technol.* **8**(5–6), 344–354 (1995).

34. C. K. Chua, W. Hoheisel, G. Keller and E. Werling, Adapting decorative patterns for ceramic tableware, *Comput. Control Eng. J.* **4**(5), 209–217 (1993).

35. C. K. Chua, R. K. L. Gay and W. Hoheisel, A method of generating motifs aligned along a circular arc, *Comput Graph: An Int. J. Syst. Appl. Comput. Graph.* **18**(3), 353–362 (1994).

36 W-D. Rase, Physical models of GIS objects by rapid prototyping, *Symposium on Geospatial Theory, Processing and Applications* (Ottawa, 2002).

37. S. Dickson, Computer-aided rapid mechanical prototyping or automatic fabrication [On-line], available at http://www.emsh.calarts.edu/-mathart/R_Proto_ref.html (2003).

PROBLEMS

1. How is application of RP models related to the purpose of prototyping? How does it also relate to the materials used for prototyping?
2. List the types of industries that RP can be used in. List specific industrial applications.
3. What are the finishing processes that are used for RP models and explain why are they necessary?
4. What are the typical RP applications in design? Briefly describe each of these applications and illustrate them with examples.
5. What are the typical RP applications in engineering and analysis? Briefly describe each of them and illustrate them with examples.
6. How would you differentiate between the following types of rapid tooling processes: (a) direct soft tooling, (b) indirect soft tooling, (c) direct hard tooling and (d) indirect hard tooling.
7. Explain how a RP pattern can be used for vacuum casting with silicon molding. Use appropriate examples to illustrate your answer.

8. What are the ways RP pattern can be used to create injection mold for plastic parts? Briefly describe the processes.
9. Compare and contrast the use of RP patterns for the following:
   (i) casting of die inserts,
   (ii) sand casting and
   (iii) investment casting.
10. What are the RP systems that are suitable for sand casting? Briefly explain why and how they are suitable for sand casting.
11. Compare the relative merits of using LOM parts with SLA parts for investment casting.
12. Explain whether RP technology is more suitable for "high technology" industries like aerospace than it is for consumer product industries like electronic appliances. Give examples to substantiate your answer.
13. Explain how RP systems can be applied to traditional industries like jewelry, coin and tableware.
14. Briefly describe how RP systems can be applied in GIS.
15. What the three advantages of employing RP technology in the field of architecture?

Chapter 8
MEDICAL AND BIOENGINEERING
APPLICATIONS

8.1. PLANNING AND SIMULATION OF COMPLEX SURGERY

When facing complex operations, such as craniofacial and maxillofacial surgeries, surgeons usually have difficulty in figuring out visually the exact location of a tumor or the precise profile of a defect. Precision and speed of a surgical operation depends significantly on the surgeon's prior knowledge of the case and experience. Rapid prototyping (RP) enables surgeons to practice on a precise model and master the essential details before the actual operation. The hands-on experience helps surgeons to achieve the better success rates. Hence, RP models are frequently used and are vital in surgical planning. The following sections describe some of these applications.

8.1.1. Cranioplasty of Large Cranial Defect

Cranioplasty is a surgical correction of a defect in cranial bone by implanting a metal or plastic replacement to restore the missing part.[1] In Fig. 8.1, a patient was severely impacted on the right side of his head. A portion of the skull was crushed and had to be removed. To cover the large hole left in his skull, an implantable prosthesis with good fit was required. Surgeons first performed computed tomography (CT) scans on the patient and then transformed the data to an STL file, which was subsequently sent to an RP machine where the wax prosthesis was produced. The wax pattern was then used as a mold and a biocompatible material (e.g., poly(methyl-methacrylate)) cast into it. After sterilization, the

403

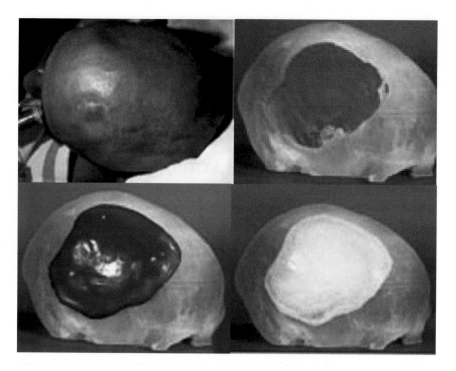

Fig. 8.1. A large cranial prosthesis was modeled using RP technology to close up a defect in the cranium (courtesy Materialise NV).

prosthesis was implanted into the patient. The surgery took very little time as the prosthesis exactly fitted the defect. To experienced surgeons, it was similar to a "plug and play". A good fit has two main advantages: first, the smoothness continues across the defect outline and gives the patient an improved symmetrical cosmesis; second, it reduces redundant or improper protrusions that may damage brain tissue severely, thus preventing post-surgery injury.

8.1.2.   Congenital Malformation of Facial Bones

Restoration of facial anatomy is important in cases of congenital abnormalities, trauma or post-cancer reconstruction. In one case, the patient had a deformed jaw at birth and a surgical operation was necessary to cut out the shorter side of the jaw and alter its position.[2] The difficult part of the operation was the evasion of the nerve canal that ran inside the jawbone. Such an operation was impossible in the conventional procedure because there was no way of visualizing the inner nerve canal.

Using a CAD model reconstructed from the CT images, the position of the canal was identified and a simulation of the amputating process on the workstation was carried out to determine the actual line of cut. Furthermore, the use of a semi-transparent resin prototype of the jawbone allowed the visualization of the internal nerve canal and facilitated the determination of the amputation line prior to the surgery. The end result was a more efficient surgery and improved post-surgery results.

In another case study, a laser digitizer was used instead of the CT[3] to capture the external surface profile of a patient with a harelip problem. The triangulated surface of the patient's face was reconstructed in CAD as seen in Fig. 8.2. Figure 8.3 shows the SLA prototype derived from the CAD data. In this case study, the prototype model provided the validation for the laser scan measurements. In addition, it facilitated an accurate predictive surgical outcome and post-operative assessment of changes in the facial surgery.

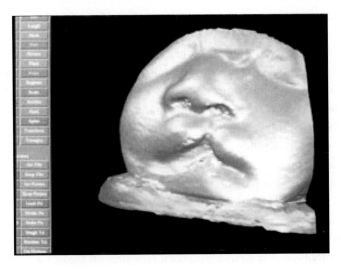

Fig. 8.2. CAD model from laser scanner data of a patient's facial details.

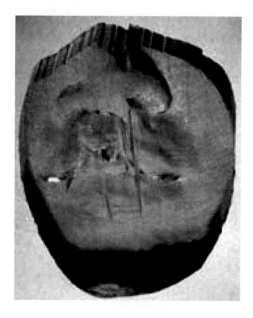

Fig. 8.3. SLA model of a patient's facial details.

8.1.3.   Cosmetic Facial Reconstruction

Due to a traffic accident, a patient had a serious bone fracture of the upper and lateral orbital rim in the skull.[4] In the first reconstructive surgery, the damaged part of the skull was transplanted with the shoulder bone. However, shortly after the surgery, the transplanted bone dissolved. Another surgery was required to retransplant another artificial bone that would not dissolve this time. The conventional procedure of such a surgery would be for surgeons to manually carve the transplanted bone during the operation virtually by "trial and error" until it fitted properly. This operation would have required a lot of time due to the difficulty in carving the bone and it has to be done during the surgery. Using RP, an SLA prototype of the patient's skull was made and then used to prepare an artificial bone that would fit the hole caused by the dissolution. This preparation not only greatly reduced time required for the surgery, but also improved its accuracy.

In another case at Keio University Hospital of Japan,[5] a five-month-old baby had a symptom of scaphocephary in the skull. This is a condition that can lead to serious brain damage because it would only permit the skull to grow in the front and rear directions. The procedure required was to take the upper half of the skull apart and surgically reconstruct it completely, so that the skull would not suppress the baby's brain as it grows. Careful planning was essential for the success of such a complex operation. This was aided by producing a replica skull prototype with the amputation lines drawn into the model. Next, a surgical rehearsal was carried out on the model with the amputation of the skull prototype according to the drawn lines followed by the reconstruction of the amputated part. In this case, the RP model of the skull provided the surgical procedure with: (1) a good 3D visualization support for the planning process, (2) an application as training material and (3) a guide for the real surgery.

8.1.4. Separation of Conjoined Twins

On 24th July 2001, two twin sisters were born in a rural village in Guatemala, healthy in every way except for the fact that they were joined at the head[6] (see Fig. 8.4). X-ray pictures showed that the two brains were separated by a membrane and were otherwise normal in size and had complete structure. This meant that no brain tissue would have to be cut through during the separation surgery. The arteries that carried oxygenated blood to their brains were also separate, but the veins that drained the blood were interwoven and fed into each other's circulatory systems. The most complex part of the operation was to sort out these veins and reroute each girl's blood supply correspondingly. In this, RP had an essential role to play.

A series of CT scans of the two girls were taken. Complicating the task was the fact that the two girls, while connected, could not be arranged in the CT system so that a single scan of their heads could be made. Instead, three sets of scan data were collected at different angles and then combined into a single 3D model. It took about three days to process the CT data and create STL files for RP. Objet Tempo RP system, made by the Israeli company Objet Geometries Ltd, was used to construct the skull models of each girl (Fig. 8.5). The Tempo built parts by selectively jetting tiny droplets of acrylate photopolymer and then curing the drops, layer by

Fig. 8.4. Conjoined twins[6] (courtesy Cyon Research Corporation).

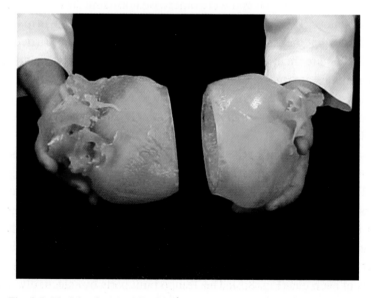

Fig. 8.5. Models of each girl's skull[6] (courtesy Cyon Research Corporation).

Fig. 8.6. A model of the maze of blood vessels where the two skulls connected[6] (courtesy Cyon Research Corporation).

layer, with light. It required semi-hardened supports (like a gel) which could be removed simply by jetting water on it. The model of the intersection of the two skulls helped surgeons to plan how they would reroute the necessary blood vessels (Fig. 8.6). The operation took about 22 h to complete. Similar procedures in the past took as long as 97 h. This significant reduction in time can no doubt be attributed to the RP model.

8.1.5. Tumor in the Jaw

Ameloblastoma is a rare, benign tumor that appears in the jaw. Though these tumors are rarely malignant or metastatic (that is, they rarely spread to other parts of the body), the resulting lesions can cause severe abnormalities in the face and jaw. Additionally, because abnormal cell growth easily infiltrates and destroys surrounding bony tissues, wide surgical excision is required to treat this disorder.

A 51-year-old male was seen in the university clinic of Kiev Medical Academy with ameloblastoma of the posterior part of the left alveolar ridge of the mandible.[7] A series of CT scans were performed and a virtual model was built using a computer. Figure 8.7(a) shows the front view of 3D-reconstruction of CT dataset. Figure 8.7(b) shows a zoomed in picture of the 3D-reconstruction. An ovoid tumor cavity with precise contours is clearly seen. X-ray density is about −41/73 HU, dimensions: $12 \times 25 \times 22$ mm. Based on the CT dataset, an STL file was created and an RP model subsequently built. Figure 8.7(c) shows the model created from segmentation masks (bottom view) whilst Fig. 8.7(d) shows the view from the top (plan view). The size of the tumor and the exact location were identified from the RP model and the surgical procedure was planned in detail. The patient was successfully operated on by Professor A. Timofeev in the CMF Surgery Clinic, Kiev Medical Academy of Postgraduate Education, Kiev, Ukraine.

The planning for removing of the facial bone tumor was enhanced by merging different procedures: 3D-reconstruction, advanced segmentation with different thresholds and direct control on stereolithography models. This procedure increased the quality of the planning, allowing surgeons

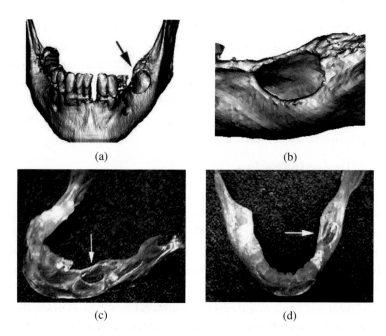

(a)   (b)

(c)   (d)

Fig. 8.7. Ameloblastoma of left mandible using SLA (courtesy Materialise NV).

to practice for real situation before hand and prepare the exact shape of bone graft.

8.1.6. Cancerous Brain

In another case, a patient had a cancerous bone tumor in his temple area requiring the surgeon to have to access the growth via the front through the right eye socket. The operation was highly dangerous as damage to the brain was likely which would result in the impairment of some motor functions. In any case, the patient would have lost the function of the right eye.[8] However, before proceeding with the surgery, the surgeon wanted another examination of the tumor location, but this time using a 3D plastic replica of the patient's skull. By studying the model, the surgeon realized that he could reroute his entry through the patient's jawbone, thus avoiding the risk of harming the eye and motor functions. Eventually, the patient lost only one tooth and of course, the tumor. The plastic RP model used by the surgeon was fabricated by SLA from a series of CT scans of the patient's skull.

Other case studies relating to bone tumor have also been reported to have improved success.[9,10] In all cases, not only was the patient spared physical disability as well as the emotional and financial price tags associated with that, the surgeon also gained valuable insights into his patient through the RP model. From here, nonintuitive alternative improved strategies for the surgery were created by the enhancement to the surgeon's pre-surgical planning stage.

8.1.7. Dental Precision Planning

Treating an impacted maxillary canine requires identifying its exact position; this can pose a challenge to both orthodontists and oral surgeons. By the conventional method of using 2D slices of CT images, it is very difficult to determine the exact location of an impacted tooth.

One case presented a new method for diagnosis and treatment planning of maxillary canine impaction by using CT combined with RP.[11] CT image files of a patient with tooth 13 impaction were edited to produce, by means of RP, an anatomic model of the maxillary teeth and a single attachment model that was to be later used to fabricate a metal attachment for bonding to the impacted tooth. The dental model was used in the diagnosis and orthodontic treatment planning and to communicate with the patient and his parents. The model showed the exact anatomical relationship between the impacted tooth and the other teeth; it was the main aid in the intraoperative navigation during surgery to expose the tooth. The metal attachment built from the prototype was bonded to tooth 13 during surgery. Thus, RP is also an important tool for fabricating brackets and other precision accessories for specific dental needs. In this case, a series of dental models made with RP was the diagnostic procedure of choice for evaluating impacted maxillary canines. Several of the RP manufacturers have introduced dedicated RP machines for the dental industry, e.g., the 3D Systems' ProJet™ DP 3000 System, EnvisionTec GmbH's Digital Dental Printer and Solidscape's D76 3D Dental Printer (See Chap. 4).

8.1.8. Biomodeling as an Aid to Spinal Instrumentation

Stereotactic surgery or stereotaxy is a minimally invasive form of surgical intervention which makes use of a 3D coordinates system to locate small targets inside the body and to perform on them some action such as removal, biopsy, lesion, injection, stimulation and implantation. Previously, frameless stereotaxy was used in spinal surgery, but this has significant limitations. In one study, a novel stereotactic technique using biomodels was developed.[12] Biomodeling was found to be helpful for complex skeletal surgery and had advantages over frameless stereotaxy. In that study, 20 patients with complex spinal disorders requiring instrumentation were recruited. 3D CT scans of their spine were performed and the data was used to generate acrylate biomodels of each spine using RP. The biomodels were used to simulate surgery. Simulation was performed using a standard power drill to place trajectory pins into the spinal biomodel. Acrylate drill guides were manufactured using the biomodels and trajectory pins as templates. The biomodels were found to be highly accurate and of great assistance in the planning and execution of the surgery. The ability to drill optimum screw trajectories into the biomodel and then accurately replicate the trajectory was judged to be especially helpful. Accurate screw placements were confirmed with post-operative CT scanning. The design of the first two templates was suboptimal as the contact surface geometry was too complex. Approximately 20 minutes were spent before the surgery preparing each biomodel and template. Operating time was reduced, as less reliance on intraoperative radiograph was necessary.

8.2. CUSTOMIZED IMPLANTS AND PROSTHESIS

For hip replacements and other similar surgeries, these were previously carried out using standardized replacement parts selected from a set range provided by manufacturers based on available anthropomorphic data and the market needs. This works satisfactorily for some types of procedures and patients, but not all. For those patients outside the standard range, in-between sizes, or with special requirements caused by disease or genetics, the surgical procedure may become significantly more complex and expensive. For imperfect fits, these implants may even cause poor gait outcomes and further wear or injury to other joints. RP has made it possible to manufacture a custom prosthesis that precisely fits a patient at reasonable cost. The following examples are drawn from a range of applications that span the human anatomy.

8.2.1. Cranium Implant

A patient suffered from a large frontal cranium defect after complications from a previous meningioma tumor surgery. This left the patient with a missing cranial section, which caused the geometry of the head to look deformed. Conventionally, a titanium-mesh plate would be hand-formed during the operation by the surgeon. This often resulted in inaccuracies and time wasted on trial and error. Using RP, standard preparations of the patient were made and a CT scan of the affected area and surrounding regions was taken during the pre-operation stage. The 3D CT data file was transferred to a CAD system and the missing section of the cranium topography was generated. After some software repair and cleaning up were carried out on the newly generated section, an inverted mold was produced on CAD. This 3D solid model of the mold was saved in .STL format and transferred to the RP system, such as the SLS, for building the mold. The SLS mold was produced and used mechanically to press the titanium-mesh plate to the required 3D profile of the missing cranium section. During the operation, the surgeon cleared the scalp tissue of the defect area and fixated the perfectly pre-profiled plate onto the cranium using self-tapping screws. The scalp tissue was then replaced and sutured. At post-operation recovery, results observed showed improved surgical results, reduced operation time and a reduced probability of complications.

8.2.2. Hip Implant

In one case, a 30-year-old female was diagnosed as having bilateral pseudohypo-chondroplasia with multiple epiphyseal dysplasia.[13] She had a poor range of movement and constant pain in her hips. A total hip arthroplasty with a custom designed femoral implant was recommended. Conventional X-ray (see Fig. 8.8(a)) showed very distorted femoral cavities which were wide proximally and extremely narrow in mid-shaft. In order to design a custom stem, CT scans and a RP model (see Fig. 8.8(b)) were required, which showed a slot type femoral canal. An implant was designed from the CT scan and the model (see Fig. 8.8(c)). Post-operative X-ray in Fig. 8.8(d) shows that the implant produced a line-to-line fit with the cavity. One-year post-operation during a clinical follow-up showed that the hips were pain free and functioning well.

In another case, a 46-year-old female was diagnosed with malunion of a femoral fracture.[14] She had previous multiple femoral fractures and an osteotomy. Conventional X-ray as seen in Fig. 8.9(a) showed a misalignment in both A–P and M–L views and there was an uncertainty on the cross section geometry of the femoral shaft at the level below the lesser trochanter. In order to design a custom implant and determine the form of surgery, CT scans and a medical model were used. The model (Fig. 8.9(b)) was made of a transparent resin material so that the internal anatomical geometry of the femoral cavity was clearly visible. A subtrochanteric osteotomy was planned and a custom implant (Fig. 8.9(c)) was designed with middle cutting flutes at the region of osteotomy to provide torsional stability. Post-operative

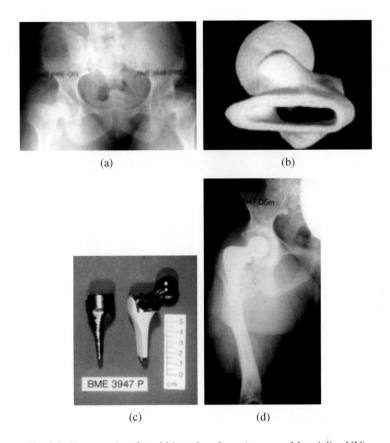

(a)   (b)

(c)   (d)

Fig. 8.8. Photographs of total hip arthroplasty (courtesy Materialise NV).

X-rays, as seen in Fig. 8.9(d), showed that the implant produced a good fit and stability.

It has been proven that RP models are very useful in terms of understanding bone deformity, designing implant and planning surgery.

8.2.3. Knee Implants

Engineers at DePuy Inc., a supplier of orthopedic implants, have integrated CAD and RP into their design environment, using it to analyze the potential fit of implants in a specific patient and then modify the implant design appropriately.[8] At DePuy, SLA plays a major role in the

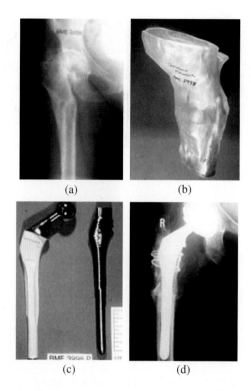

(a) (b)

(c) (d)

Fig. 8.9. Total hip arthroplasty for the case of femoral malunion (courtesy Materialise NV).

production process of all the company's products, standard and custom. The prototypes are also used as masters for casting patterns to launch a product or to do clinical releases of a product. For this application, there are several advantages over traditional casting tooling in that the lead times to manufacture the customized implants and the costs associated with these are significantly reduced.

8.2.4. Intervertebral Spacers

Human spinal vertebras can disintegrate due to conditions such as osteoporosis or extreme forces acting on the spine. In the management of such situations, a spacer is usually required as part of the spinal fixation process. RP has been investigated for the production of such spacers as

Fig. 8.10. Intervertebral spacers produced by the Nanyang Technological University (NTU) RT system.

it is an ideal process to fabricate 3D structures with good interconnecting pores for the promotion of tissue in-growth. Other considerations for producing such an implant are that the material is biocompatible and that mechanical compressive strength of the spacer is able to withstand spinal loads.

A process developed at the Nanyang Technological University for such a purpose uses a solid RP master pattern of the spacer to produce a soft mold. Stainless steel bearings coated with a formulated binder system are then cast into the soft mold under vacuum. Upon curing, the part is ejected from the mold. The part then undergoes debinding and sintering processes to produce the final part. The primary advantage of this process is its ability to use a solid RP pattern to produce from this a porous structure with controllable pore sizes and mechanical strengths.[15,16] Figure 8.10 shows the intervertebral spacers produced using the RT system.

8.2.5.   Buccopharyngeal Stent

A male child was diagnosed at birth with a persistent buccopharyngeal membrane.[17] The buccopharyngeal membrane forms a septum between the primitive's mouth and pharynx. Normally, it completely ruptures during embryo development, but was not the case for this child due to a genetic defect. Persistence of the buccopharyngeal membrane would have resulted in the partial fusion of his jaws, the inability in opening and thus speaking as he grows. Another problem was that the child would be rapidly growing and major anatomical changes were expected every six weeks up to the age of about four to five. There was also no readily available buccopharyngeal stent commercially as it was a rare disease and customization of the stent was essential to "morph" with the changing anatomy of growth and surgical procedures.

To solve this problem, the Kandang Kerbau Women's and Children's Hospital in Singapore worked collaboratively with the Biomedical Engineering Research Center (BMERC) and the Rapid Prototyping Research Laboratory (RPRL) at Nanyang Technological University to create a newly designed stent made with biocompatible materials. This material was soft, comfortable, yet rigid enough. The stent was designed to have excellent anti-migration properties when deployed at the pharynx region, yet easy to deploy and extract without causing any trauma to the patient (see Fig. 8.11).

To produce the stent, a master pattern was first fabricated using RP based on the patient's airway morphology. The stent master pattern

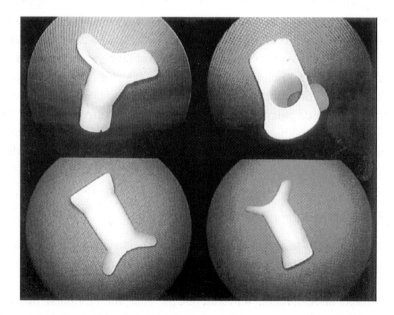

Fig. 8.11. Different views the biocompatible buccopharyngeal stent.

was then used to create a solid silicone mold. The silicone mold was parted to remove the master pattern, thereby leaving a negative cavity of the implant. The silicone mold was then sprayed with a safety release agent and reassembled. The polyurethane-based resin was then prepared and vacuum cast into the silicone mold under 10^4 Torr and in a controlled gas environment. The part was then ejected from the mold and post-processing carried out. Every six weeks when it is time to produce a new stent, the CAD model for the previous stent was modified using a growth surgical model semi-empirically developed with the surgeons. The process was repeated until the child reached the age of 5 years old when he was stable enough. With the newly developed stent, the patient was able to breathe, eat and initiate vocalized sounds as his anatomical structure was more stabilized, thereby learning to talk.

8.2.6. Customized Tracheobronchial Stents

Stents for maintaining the patency of the respiratory channel have been investigated for production using RP techniques.[18] Customization of these stents can be carried out to take into account compressive resistance with respect to stent wall thickness, as well as unique anatomical considerations. Measurements are taken of the actual forces required to open the airway channel to its original dimensions. The data is fed to the CAD system where modification of the stent design is carried out. Upon confirmation, the 3D data is fed to an RP system where the master pattern of the model is built. The master pattern then undergoes the silicone molding vacuum casting process, reproducing the stent master pattern with a biocompatible material with all its strength, spring-back and anti-migration properties in place. Figure 8.12 shows four vacuum cast tracheobronchial stents in slightly differing sizes for an ideal intrasurgery custom fit. The stent is sterilized, packaged and delivered to the operating theatre.

8.2.7. Obturator Prosthesis for Oncologic Patients

Figure 8.13 shows the cavity in the mouth of a patient after resection of a tumor.[19] In order to protect the tissue weakened by irradiation and for

Fig. 8.12. Production of the customized stents in slightly differing sizes for an ideal custom fit.

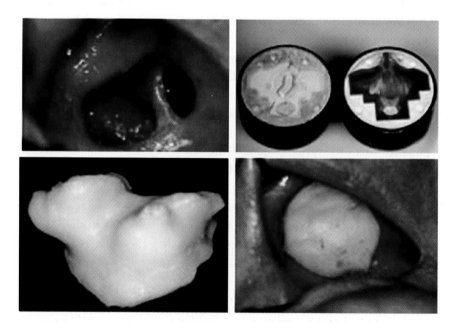

Fig. 8.13. Obturator prosthesis fits the cavity.

the patient to be able to breathe and eat normally, this hole needs to be filled by an implant. A CT-scan of the patient was made. The soft tissue around the cavity, clearly visible on the scans, was modeled using RP technology. This model served as a direct mold for the implant. The implant, called an obturator prosthesis, was cast from the mold in biocompatible silicone. No surgery was needed to implant the obturator prosthesis. As the silicone prosthesis is plastic deformable, it can be implanted very easily.

The prosthesis fitted the cavity much better than that achieved by using impression techniques. These traditional techniques produce a master of the obturator prosthesis by making an impression of the cavity in a plastic deformable material. The prostheses cast from such masters are always less accurate because of the presence of undercuts (the impression technique is not sensitive to local internal broadening of the cavity) and can severely damage the sensitive and vulnerable surrounding tissue. The soft prosthesis is fixed by means of magnets on a hard dental implant. This makes it possible to take it out for inspection and to replace it afterwards.

8.2.8. Tissue Engineering Scaffolds

Tissue engineering involves a combination of cells, scaffolds and suitable biochemical factors to improve or replace damaged or malfunctioning organs such as skin, liver, pancreas, heart valve leaflet, ligaments, cartilage and bone. A scaffold is a polymeric porous structure made of biodegradable materials such as PLA and PGA. They serve as supports to hold cells.

A typical process in tissue engineering is illustrated schematically in Fig. 8.14 and is as follows:

(1) Examine the defects and determine if a scaffold is necessary.
(2) Isolate functional (undamaged) cells from donor tissue to be cultured.
(3) Select suitable materials to be prototyped.
(4) The patient is introduced to CT or MRI scanning to obtain the geometric data of the defects.

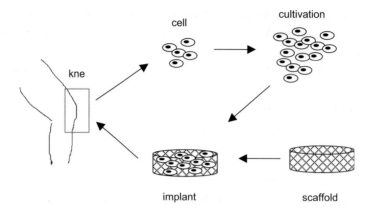

Fig. 8.14. An illustration of the tissue engineering process incorporating RP-made scaffolds.

(5) Reconstruct CT data into a virtual model.
(6) Design a scaffold with suitable porous networks to fit the virtual model.
(7) Convert CAD model of scaffold to STL file.
(8) Create the scaffold using a RP machine.
(9) Transplant cells to the scaffold.
(10) Implant the scaffold to the defected receiver site with growth factors.
(11) Scaffold degrades gradually while cells grow and multiply.

One challenge in tissue engineering is to improve the vitality of cells during transplantation. When transplanted to the scaffold, a high cell density means that the inner cells lack nutritional input from the exterior environment, causing some of them to die of malnutrition. Thus, the microstructures of the scaffold are very important to the normal functions of cells.

The conventional method to fabricate a scaffold is to use organic solvent casting or particulate leaching. However, it has three drawbacks: (1) the thickness is limited; (2) the sizes of the pores are not uniform; and (3) the interconnectivity or distribution of the pores is irregular. This certainly will affect the number and quality of the cells that are seeding on the scaffold. To solve these problems, RP was explored. Due to its advantage of fabricating intricate objects accurately, quickly and consistently, RP has become more and more popular for fabricating tissue engineering scaffolds.

Some researchers use RP techniques to study the optimal microstructure of the scaffolds. Based on cell type, interconnected networks of the scaffold can be pre-determined in a computer CAD model. After conversion to a STL file, a fine structure can be built layer by layer using RP machines. Some researchers apply RP techniques to study scaffold materials, for example, PCL/HA biocomposite, which has been shown to have favorable potential.[20] There are also researchers who study the ability of RP for high-throughput production of artificial human tissues or organs. In that study, more than 30 layers of a hepatocyte/gelatin mixture were laminated into a high spatial structure.[21]

8.3. DESIGN AND PRODUCTION OF MEDICAL DEVICES

RP has impacted the biomedical field in several important ways. Besides biomodeling for surgical planning and medical implants, another obvious application is to design, develop and manufacture medical devices and instrumentation. RP technology is also being used to fabricate drug dosage forms with precise and complex time release characteristics. In addition, the market value of new designs of medical devices or instrument can be proved with the help of RP.

8.3.1. Biopsy Needle Housing

Biomedical applications are extended beyond design and planning purposes. The prototypes can serve as a master for tooling such as a urethane mold. At Baxter Healthcare, a disposable-medical-products company, designers rely on two RP processes, SLA and SGC, to create master models from which they develop metal castings.[8] The masters also serve as a basis for multiple subtooling processes. For example, after a master model has been generated via one of the RP machines, the engineers might build a urethane mold around it, cut open the mold, pull out the master, then inject thermoset material into the mold to make prototype parts.

This process is useful in situations where multiple prototypes are necessary because the engineers can either reuse the rubber molds or make many molds using the same master. The prototypes are then delivered to customer focus groups and medical conferences for professional feedback. Design changes are then incorporated into the master CAD database. Once the design is finalized, the master database is used to drive the machining of the part. Using this method, Baxter Healthcare has made models of biopsy needle housing and many other medical products.

8.3.2. Drug Delivery Device

Drug delivery refers to the delivery of a pharmaceutical compound to humans or animals. The method of delivery can be classified into two ways, invasive and noninvasive. Most of the drugs adopt noninvasive ways by oral administration. Other medications, however, cannot be delivered by oral administration due to its ease of degradation, for example, proteins and peptide drugs. Generally, they are delivered by the invasive method of injection.

Currently, many research efforts are focusing on target delivery and sustained release formulation. Target delivery refers to delivery drugs at the target site only, for example, cancerous tissues and not elsewhere along the route. Sustained release formulation refers to releasing the drug over a period in a controlled manner, thus achieving optimally therapeutic concentration.

Polymeric drug delivery devices play an important role in sustained release formulation. However, the current fabrication methods lack precision, which impairs the quality of the device, resulting in a decrease in the efficiency and effectiveness of drug delivery. In one study, the SLS process is explored to fabricate a polymeric matrix of drugs layer by layer[22,23] using powdered materials. The capability to build controlled released drug delivery device is also demonstrated in a study conducted by the Nanyang Technological University.[24] In this study, a varying porosity circular disc with the outer region being denser acting as a diffusion barrier region and the inner more porous region acting as a drug encapsulation region is fabricated using the SLS process (see Fig. 8.15).

Fig. 8.15. Polymeric cylindrical DDD built with the SLS.

Biodegradable powder materials are used. This study also concludes that control of SLS process parameters, such as laser power, laser scan speed and bed temperature will influence the porous microstructure of the polymeric matrix.

Another study used three-dimensional printing (3DP) technology to create resorbable devices with complex concentration profiles within the device.[25] They showed that both the macro- and micro-structure of the device could be controlled. This study demonstrates several simple examples of such devices and several construction methods that can be used to control the release of the drugs.

8.3.3. Masks for Burnt Victims

Burn masks are plastic shields worn by patients with severe facial burns. They are used as a treatment device to prevent scar tissue from forming. Burn patients are required to wear the mask for at least 23 h a day between tissue reconstructive surgery sessions.

The traditional methods for producing burn masks vary and are never simple. The most conventional way is to form a mold by applying a plaster material to the patient's face. The process can be extremely anxiety-provoking for the patient since their eyes and mouth are completely covered throughout the drying process. The weight of the plaster can shift the tissue on the face, making it virtually impossible to get an exact replication.

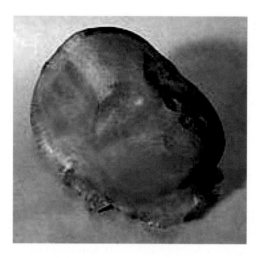

Fig. 8.16. A burn mask of a child developed by Total Contact (courtesy Total Contact).

RP can be used to produce custom-fit masks for burn victims. The process begins by digitizing the patient using noncontact optical scanning. The scan data is used to produce an RP model of a mask that is a very precise representation of the patient's facial contours. The mask applies pressure to the face to slow the flow of blood to the healing skin. This, in turn, reduces the formation of scar tissue.

A five-year-old child and his family flew from their Colorado home to have a burn mask developed.[26] Prior to their visit, physicians and occupational therapists had developed two burn masks using the conventional plaster method for the child, but encountered unsatisfactory results. The RP method enabled the child and his family to return to Colorado in less than 48 h with a complete, more accurately fitting mask (see Fig. 8.16). Today the child's soft tissue is healing better due to the accuracy of the burn mask, creating smoother skin surfaces and causing less abnormal scarring.

8.3.4.   Functional Prototypes Help Prove Design Value

Honeywell is one of the world's leading companies in militarized head-mounted display technology.[27] The display technology has also made minimally invasive surgeries easier. In actual surgeries, the functional prototype headset is worn by military surgeons, replacing the need for the conventional cathode ray tube monitors.

At the initial stage of developing the new miniature color-display technology, a prototype was not used. However, the research scientists found that without a prototype they could not get meaningful feedback on their design. On the other hand, they could not justify the tooling costs for making a real product. They then turned to a design firm to rapid prototype such a model. To let the surgeons provide truthful feedback, the prototype had to look and feel like the real product. A functional and durable material was required. After careful consideration, ABS was selected because the continuous copolymer phase gave the materials rigidity, hardness and heat resistance. In addition, it was easy to finish and approximated the injection-molded plastics well when built by RP. For the RP machine, FDM was chosen for the electronic covers and case housings because the operating temperatures needed to be high. For example, the electronics pack that supplies power to the displays was very small and can generate a great deal of heat in its prototype form. It contained a very high-density printed circuit board with the electronic equivalent of two television sets, plus video converters.

The final result was very pleasing. With the prototype, surgeons saw the benefits immediately and were quite sure that the newly developed display technology had proved its market value. RP had made the difference between being able to sell the concept to medical market leaders. The alternative could have resulted in a loss of funding for the program due to communication gaps and a lack of understanding in the technology.

8.4. FORENSIC SCIENCE AND ANTHROPOLOGY

Surgeons are not the only people interested in bones. Anthropologists and forensic specialists share the same interest too. RP and the related technologies have enabled these scientists to put a recognizable face on skeletons and share precious replicas of rare finds.

8.4.1. The Mummy Returns

To imagine how ancient people look different from the humans of today can be made viable to us by RP. Viewing the faces of people some

3,000 years ago through their skeletal finds is basically restoring history in real time.

In one study, an ancient Egyptian mummy from the collection of the Egyptian Museum (in Torino, Italy) was selected as a research example. This mummy was well-preserved and completely wrapped.[28] Without having to unravel the mummified wrappings, CT scans were obtained to establish the geometric data. CT is the most important noninvasive imaging technique for obtaining fundamental data for 3D-reconstructions of the skull and the body, especially with wrapped mummies. Based on the 3D CT scan data, a nylon model was built using the Selective Laser Sintering technology, which provided the frame for facial reconstruction of the mummy.

Anthropometric data, the conditions of the remaining dehydrated tissues and the most accepted scientific and anthropological criteria were used to restore the physical appearance of the mummy. This was done by progressive layering of plasticine on the nylon model.

8.4.2. "Officer Blue"

One quiet day in Central Park, New York City, two horse policemen were making their rounds on horseback. Suddenly, a scuffle broke out and one cop rushed over to the scene. However, he was shot in the back by a sniper. The bullet had gone right through his protective vest, passed his body and lodged in his horse, Officer Blue. The policeman was dead, but the horse was still alive, with the bullet inside its spine. To determine the type and origin of the bullet, investigators had to examine it. However, without having to "put down" the horse, it was impossible for the bullet to be removed.

In this case, one of the RP technologies, the 3D printer, provided the solution. The horse was first brought to a CT scan to obtain the digital data of the bullet. Then, the CT scanned data were reconstructed and converted to STL file, which was sent to the 3D printer to recreate an exact replica of the bullet. The 3D printer fabricated the bullet model layer by layer until a blue plaster prototype emerged. This bullet model was used as evidence in the court. In the end, the horse's life was spared.

Fig. 8.17. (Left) LOM model before decubing and (right) finished model.

8.4.3. The Woman in the River

In another criminal case, the laminated object manufacturing (LOM) was used[29] to perform forensic reconstruction of a murdered body. A dismembered body of a woman was uncovered in rural Wisconsin. However, it was impossible for the police to identify who it was, because the skin of the face had been removed. Fortunately, the skull of the victim was preserved whole. Using CT scanning, a virtual model of the skull was constructed. After conversion to STL file, a model of the victim was created layer by layer using the LOM machine (Fig. 8.17). Forensic anthropologists performed a facial reconstruction directly on the LOM model and photographs of the reconstructed face were distributed for identification (see Fig. 8.18). From the flyers, someone identified one of the images. Within a day, this led to the successful arrest of the murderer who is now serving a life sentence.

8.5. VISUALIZATION OF BIOMOLECULES

"If a picture is worth a thousand words, would a life model be worth a thousand pictures?"

— Lim C S, Daniel

8.5.1. Biomolecular Models for Educational Purposes

The stereo-structure of a molecule determines its physical and chemical properties. In biological sciences, the structures of proteins are intensively

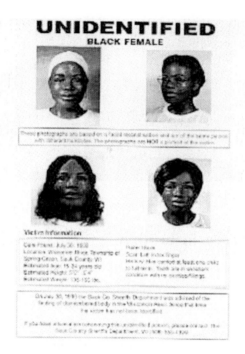

Fig. 8.18. The completed RP-assisted reconstructed face photographed with a variety of hairstyles and flyers were distributed for identification.

studied in order to understand biological activities. Researchers generally have a good spatial sense to imagine the structure by the virtual construction in the computer. However, to explain it to students clearly is very challenging, because words are always vague to some degree and each student has his or her own perception and understanding.

RP technology can be used to produce accurate, 3D physical models of proteins and other molecular structures. From protein data banks where proteins of known structures are stored, the structure file of the protein of interest can be found. Based on *xyz* coordinates obtained from the structure file, a CAD model could be constructed. After conversion to STL file, a 3D protein model can be rapidly prototyped (see Fig. 8.19).

Students are able to interact with the model and explore the shape and chemical properties. Color schemes can be used to identify the helixes and sheets.

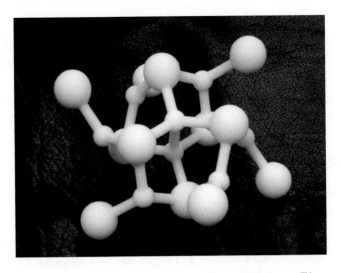

Fig. 8.19. RP model of a protein molecule made by Objet PolyJet™ system.

8.5.2. 3DP for Modeling Protein–Protein Interactions Study

Protein–protein interactions refer to the association and disassociation of protein molecules. They play a vital role for nearly every biological process in a living cell. A good understanding of the mechanisms of these interactions can help clarify the cause and development of disease and provide new therapeutic approaches.

One issue in protein–protein interactions study is to deduce the spatial arrangement of the complex (a protein carrying another one for a long time or a brief interaction just for modification); in other words, to predict the molecular structure of the protein complex using the known protein structures.[30] In principle, especially for a large complex, the interface of interaction is unique due to surface complementarity. This is where physical models come in and work.

Structures of the known proteins can be found in Protein Data Bank, thus *xyz* coordinates can be obtained to build CAD models. After conversion to STL file, accurate physical models are made using RP machines. Then these models, the components of the complex, are maneuvered by hand using translational and rotational orientations to quickly search for the most suitable and probable configuration. Based on the configuration identified by the models, a virtual complex is reconstructed in molecular graphics programs. After modification, the structure of the complex is considered to be finally determined.

In 2002, an investigative study proved that RP is in general suitable for the modeling of protein–protein interaction.[31]

REFERENCES

1. Materialise NV [On-line], available at http://www.materialise.be/medical-rpmodels/case3_ENG.html (June, 2007).

2. T. Kaneko, M. Kobayashi, Y. Tsuchiya, T. Fujino, M. Itoh, M. Inomata, M. Uesugi, K. Kawashima, T. Tanijiri and N. Hasegawa, Free surface 3-dimensional shape measurement system and its application to Mictotia ear reconstruction, *The Inaugural Congress of the International Society for Simulation Surgery* (1992).

3. C. K. Chua, S. M. Chou, W. S. Ng, K. Y. Chow, S. T. Lee, S. C. Ang and C. S. Seah, An integrated experimental approach to link laser digitizer, CAD/CAM system and rapid prototyping system for biomedical applications, *Int. J. Adv. Manuf. Technol.* **14**(2), 110–115 (1998).

4. J. Adachi, T. Hara, N. Kusu and H. Chiyokura, Surgical simulation using rapid prototyping, *Proc. Fourth Int. Conf. on Rapid Prototyping*, 14–17 June (1993), pp. 135–142.

5. M. Koyayashi, T. Fujino, H. Chiyokura and T. Kurihara, Preoperative preparation of a hydroxyapatite prosthesis for bone defects using a laser-curable resin model, *The Inaugural Congress of the International Society for Simulation Surgery* (1992).

6. Cyon Research Corporation, Rapid prototyping helps separate conjoined twins [On-line], available at http://www.newslettersonline. com/user/user.fas/s=63/fp=3/tp=47?T=open_article,484858&P= article (April 2006).

7. Materialise [On-line], available at http://www.materialise.be/medical-rpmodels/case22_ENG.html (2007).

8. D. P. Mahoney, Rapid prototyping in medicine, *Comput. Graph. World* **18**(2), 42–48 (1995).

9. B. Swaelens and J. P. Kruth, Medical applications in rapid prototyping techniques, *Proc. Fourth Int. Conf. on Rapid Prototyping*, 14–17 June (1993), pp. 107–120.

10. A. Jacobs, B. Hammer, G. Niegel, T. Lambrecht, H. Schiel, M. Hunziker and W. Steinbrich, First experience in the use of stereolithography in medicine, *Proc. Fourth Int. Conf. on Rapid Prototyping* 14–17 June (1993), pp. 121–134.

11. J. Faber, P. M. Berto and M. Quaresma, Rapid prototyping as a tool for diagnosis and treatment planning for maxillary canine impaction, *Am. J. Orthod. Dentofacial Orthop.* **129**(4), 583–589 (2006).

12. P. S. D'Urso, O. D. Williamson and R. G. Thompson, Biomodeling as an aid to spinal instrumentation, *Spine* **30**(24), 2841–2845 (2005).

13. Materialise NV [On-line], available at http://www.materialise.be/medical-rpmodels/case6_ENG.html (June 2007).

14. Materialise NV [On-line], available at http://www.materialise.be/medical-rpmodels/case7_ENG.html (June 2007).

15. C. S. Lim, M. Chandrasekeran and Y. K. Tan, Rapid tooling of powdered metal parts, *Int. J. Powder Metall.* **37**(2), 63–66 (2001).

16. C. S. Lim and M. Chanrdrasekeran, A process to rapid tool porous metal implants, Internal Report, Nanyang Technological University, Singapore, 1–6 February (1999).

17. S. S. Tan, H. K. Tan, C. S. Lim and W. M. Chiang, A novel stent for the treatment of persistent buccopharyngeal membrane, *Int. J. Pediatr. Otorhinolaryngol.* **70**(9), 1645–1649 (2006).

18. C. S. Lim, P. Eng, S. C. Lin, C. K. Chua and Y. T. Lee, Rapid prototyping and tooling of custom-made tracheobronchial stents, *Int. J. Adv. Manuf. Technol.* **20**(1), 44–49 (2002).

19. Materialise NV [On-line], available at http://www.materialise.com/medical-rpmodels/case2_ENG.html (June 2007).

20. F. E. Wiria, K. F. Leong, C. K. Chua and Y. Liu, Poly-epsilon-caprolactone/hydroxyapatite for tissue engineering scaffold fabrication via selective laser sintering, *Acta Biomater.* **3**(1), 1–12 (2007).

21. X. Wang, Y. Yan, Y. Pan, Z. Xiong, H. Liu, J. Cheng, F. Liu, F. Lin, R. Wu, R. Zhang and Q. Lu, Generation of three-dimensional hepatocyte/gelatin structures with rapid prototyping system, *Tissue Eng.* **12**(1), 83–90 (2006).

22. C. L. Liew, K. F. Leong, C. K. Chua and Z. H. Du, Dual material rapid prototyping techniques for the development of biomedical devices. Part I: Space creation, *Int. J. Adv. Manuf. Technol. United Kingdom* **18**, 717–723 (2001).

23. C. L. Liew, K. F. Leong, C. K. Chua and Z. H. Du, Dual material rapid prototyping techniques for the development of biomedical devices. Part II: Secondary powder deposition, *Int. J. Adv. Manuf. Technol.* **19**, 679–687 (2002).

24. K. F. Leong, C. K. Chua and W. S. Gui and Verani, Building porous biopolymeric microstructures for controlled drug delivery devices using selective laser sintering, *Int. J. Adv. Manuf. Technol.* **31**(5–6), 483–489 (2006).

25. B. M. Wu, S. W. Borland, R. A. Giordano, L. G. Cima, E. M. Sachs and M. J. Cima, Solid free-form fabrication of drug delivery devices, *J. Controlled Release* **40**(1), 77–87 (1996).

26. Total Contact [On-line], available at http://www.totalcontact.com (June 2007).

27. Materialise NV [On-line], available at http://www.cadinfo.net/editorial/honeywell.htm (June 2007).

28. F. Cesarani, M. C. Martina, R. Grilletto, R. Boano, A. M. Roveri, V. Capussotto, A. Giuliano, M. Celia and G. Gandini, Facial reconstruction of a wrapped Egyptian mummy using MDCT, *Am. J. Roentgenol.* **183**(3), 755–758 (2004).

29. R. S. Crockett and R. Zick, Forensic applications of solid freeform fabrication, *Proc. Solid Freeform Fabrication Symp.* 549–554 (2000).

30. T. S. Shimizu, N. Le Novère, M. D. Levin, A. J. Beavil, B. J. Sutton and D. Bray, Molecular model of a lattice of signalling proteins involved in bacterial chemotaxis, *Nature Cell Biol.* **2**(11), 792–796 (2000).

31. M. Laub, M. Chatzinikolaidou, H. Rumpf and H. P. Jennissen, Modelling of protein-protein interactions of bone morphogenetic protein-2 (BMP-2) by 3D-rapid prototyping, *Materialwissenschaft und Werkstofftechnik* **33**, 729–737 (2002).

PROBLEMS

1. List several possible applications for RP in medical and biomedical engineering.
2. Name several advantages and disadvantages of applying RP to the field of medicine and biomedical engineering.
3. List some possible materials for use with RP in relation to *in-vivo* applications.
4. Discuss how RP can create value for surgical procedures relating to the separation of conjoined twins joined at the head.
5. How can RP models be useful to the surgeons before and during the operating procedure to remove a tumor from the cranium?
6. Why and how is RP important when producing hip implants for non-standard sized patients requiring hip replacement?
7. Respiratory stents, such as the buccopharyngeal stent, has been used on babies with congenital buccopharyngeal defects. Explain how RP can be used to support the child with such a stent until he is old enough for major reconstructive surgery.
8. Discuss how RP can be used to support organ replacement by tissue engineering. Discuss what materials should be considered and why.
9. Explain the challenges of building a RP system for tissue engineering applications.
10. In what way can a RP model assist in the design and laboratory testing of a substitution replacement mechanical heart valve? Discuss the RP selection considerations for such an application.

11. How can RP prove useful in forensic science applications? Compare current techniques with RP for such applications.
12. Being able to visualize scientific concepts in three-dimensions can be quite a challenge when it comes to protein–protein interaction research. How can RP create value in the research or teaching laboratories for such an application?

Chapter 9
EVALUATION AND BENCHMARKING

9.1. USING BUREAU SERVICES

The best way to experiment with Rapid Prototyping (RP) is with a service bureau that owns and operates one or several systems. RP equipment is still fairly expensive (US$50,000 or more, though some concept modelers costs just above US$20,000) and the cost of operator training, materials and installation of the equipment can easily double this cost. The volume of prototyping work in your company will probably not justify the acquisition of such a system. Thus, it may be more economical to engage a service bureau. Some factors to consider when engaging a service bureau include:

(1) type of material needed for the prototype,
(2) size of the prototype,
(3) accuracy,
(4) surface finish,
(5) experience of service bureau,
(6) location of service bureau for communication and coordination of jobs,
(7) provision of secondary processes such as machining or sandblasting,
(8) cost.

Typically, service bureaus base their charges on the following items:

(1) modeling and preparation of model for building (if necessary),
(2) execution of program to convert computer model into STL file,
(3) "slicing" time in converting STL file into cross-sectional data,
(4) the amount of machining time required to make the part,
(5) the setup time in preparing the apparatus,

(6) the actual amount of material used to make the part,
(7) post-fabrication assembly such as gluing together several smaller pieces,
(8) secondary processes such as machining or sandblasting.

Needless to say, the company can save cost on the first two items if it has its own CAD facility and CAD–RP interface. Fees from service bureaus can range from a few hundred dollars for a simple part to thousands of dollars for large and complex prototypes. The Wohler's Report,[1] published annually, lists service bureaus worldwide.

9.2. SETTING UP A SERVICE BUREAU

There are three major considerations in assessing the economic feasibility of the proposed RP service bureau. These are given as follows:

(1) Is there sufficient demand for rapid prototypes in the country (or region) to justify the establishment of an additional service bureau?
(2) Can the proposed facility operate at a profit with the current market price for RP parts, proposed level of output and the investment necessary to establish the proposed facility?
(3) Will the profit realized from the operation of the proposed facility justify the investment?

To illustrate how a study on the setting up of a service bureau can be carried out, a case will be presented. The figures quoted in the case study are fictitious due to the confidentiality of such an actual study. However, they are sufficient to illustrate the working of such a study.

9.2.1. Preliminary Assumptions

Several basic assumptions are required in developing the substantive materials which will be analyzed to reach the conclusions. Firstly, the size of operation must be assumed. If it is assumed that the initial size of the operation will be small, this could mean that only one RP system can be considered for acquisition. Accordingly, the skilled manpower to be employed (or redeployed) to provide the service, both technically and administratively, will only be sufficient for that operation.

Secondly, it is assumed that the proposed service bureau would concentrate on the industries represented in the country (or region). The market potential existing in the country or region and the anticipated size of the operation warrant such an assumption.

The third assumption deals with the exact nature of the operation. For example, if it is assumed that the service bureau would not be an integrated operation, then there will be no additional processing of parts such as secondary milling, grinding processes, etc. Thus, these accessories and equipment need not be computed in the study.

A final assumption is that the service bureau would operate at a predetermined level of output. Certain basic facilities are required to start such an operation. However, the utilization of such equipment may vary, though not considerably, due to seasonal variations in the availability of big projects from major clients. This specified level of operation would entail an average-sized project (one or more prototypes, may be identical or different parts) lasting for three working days. Taking 262 working days per year, about 87 projects can be undertaken annually. For the purpose of calculation, it is assumed that each project handles two parts as all RP systems can build numerous parts simultaneously. This production (or sales) level is vital to the analysis that follows. Operating at some other level of output than the one specified would substantially influence cost and revenue and would not be reflected in the analysis of the study.

It should be pointed out that no attempt is made to analyze the managerial abilities of administrative personnel of the service bureau. The profitability of any business operation is, however, dependent on the possession of adequate managerial abilities by those personnel responsible for the decision making within the organization.

9.2.2. Market Potential for Prototyping

9.2.2.1. *Consumer Demand*

The demand of prototypes in the country may or may not be available directly. Sometimes, even without the official figures of the actual demand of prototypes, other forms of evidences can point to or indicate the state of demand, such as:

(1) Many multi-national companies relocating their design activities to the country.
(2) Shorter product life cycle for many products (e.g., cellular phones, MP3 players, etc.).
(3) More designed-in-country products.
(4) Government's or local agencies' substantial increase in R&D expenditure so as to promote research, which in turn will increase the demand for the services of prototypes.

A good source of statistics is from census or trade reports. For instance, in Singapore, from the Report on Census of Industrial Production, 2004,[2] it is possible to establish figures relating to the sales of various products and services.

From numerous RP reports, case studies and proceedings published, it can be inferred that many products and services can use the service bureau. These products and services are shown in Table 9.1, along with the number of establishments and total sales figures.

However, the sales figure is a total figure and will not reflect accurately the same figure that can be replaced by RP. Nevertheless, the figures provide a useful gauge of the total number of establishments and sales volume.

To be a step closer to realize the dollar value of prototype demand, a set of data is collected from the industries most active in RP. The collection of data is carried out while maintaining the confidentiality of the respondents. Table 9.2 shows a breakdown of demand by major industry groups which are assumed to be most active in RP. An average budget of US$1,000 is assumed for the building of a RP part. Therefore, column 5 is obtained by the multiplication of columns 1, 2 and 3 and US$1,000. This estimate of US$1,000 is an average as it is understandable that some more complex and bigger parts will cost more. On the other hand, simpler parts will cost slightly lesser.

The above estimated demand for RP is 5,572 parts and US$5,572,000 in dollar value. Suppose there are six service bureaus in the country

Table 9.1. Sales volume of products and services.[2]

Major industry group	Establishments	Sales (US$ '000)	Application area of RP
Dies, molds, tools, jigs and fixtures	130	299,274	Dies, molds and tools
Connectors	10	279,049	Product design and functional testing
Electrical household appliances	7	517,029	Product design and fitting
Electronic products and components	247	32,378,744	Product design and mock-up
Computer peripheral equipment	20	3,590,355	Product design and functional testing
Communication equipment	9	1,422,222	Product design and fitting
Television sets and subassemblies	5	1,587,638	Product design and fitting
Audio and video combination equipment	11	3,089,532	Product design and fitting
Motor vehicle parts and accessories	14	84,491	Product design and functional testing
Surgical and medical instruments	12	364,482	Product design
Watches and clocks	8	191,324	Product design
Jeweler	57	338,653	Product design and tooling
Toys	13	218,932	Product design and tooling
Perfumes, cosmetics and toilet preparations	11	31,778	Product design
Plastic household and kitchen ware	7	11,139	Product design and fitting
Plastic bottles, boxes and containers	29	145,822	Product design
Plastic precision parts	116	815,743	Product design and fitting
Pottery, earthenware and glass products	12	173,617	Product design and tooling
Footwear	27	47,632	Product design and tooling

Table 9.2. Estimated demand for RP in dollar values.

Major industry group	Establishments	Number of models per year	Number of parts per model	Sales (US$ '000)
Connectors	10	6	2	120
Computer peripheral equipment	20	8	10	1,600
Electrical household appliances	7	12	8	672
Communication equipment	9	16	4	576
Television sets and subassemblies	5	6	18	540
Audio and video combination equipment	11	8	12	1,056
Motor vehicle parts and accessories	14	2	36	1,008
Total				5,572

producing at the rate of two RP parts per week, the total output of present service bureaus is:

$$\text{Total output of present service bureaus}$$
$$= 6 \text{ service bureaus} \times 2 \text{ RP parts/week/}$$
$$\text{service bureau} \times 52 \text{ weeks/year}$$
$$= 624 \text{ parts per year.}$$

This number is far from meeting the RP demands of the industries. Therefore, the analysis of the prototyping business in the country compared to the total output of present service bureaus in the country shows a net excess of consumption when stated in prototype number $(5,572 - 624) = 4,948$ parts. In terms of equivalence in dollars, this net excess of demand is just under US$5.0 million. This figure represents the potential for an additional service bureau.

The growth in consumer demand for rapid prototypes is extremely promising based on a study by Terry Wohlers.[3] Revenue collected from both sales and service bureaus have increased over the years. However, the increase in revenue from service bureaus is more spectacular. Revenue

jumped exponentially in 2004 and continued to repeat similar performances in subsequent years.

Such a phenomenon can only be explained by two reasons: First, there
is greater industrial awareness now than before. The number of annual
conferences devoted to RP in USA and Europe, which started since 1990,
has grown to more than six. Many more technical papers, journals and
other articles have been written on RP.

Secondly, the capabilities of RP systems have improved tremendously
since its early beginning in 1988 — in terms of material range, accuracy,
mechanical properties, industrial applications, etc. These improvements
have won over many skeptics.

While there is no formal study done which indicates that the trend
experienced in the United States is applicable worldwide, it is believed
that other countries will experience more or less a similar growth.

9.2.2.2. Application Areas of RP Systems

This is important because there may be a strong correlation between the
type of RP system and the industry in which the prototypes are used. For
instance, the connector industries use prototypes which must meet certain
requirements (testability, form and fit, etc.) and these requirements may
be fulfilled easily by, say RP system A and not so by RP system B. It
should, however, be noted that with the trend in RP systems' advancements, the gap is rapidly narrowing.

9.2.2.3. Competitors

The competitors in the country or region are those which currently operate a service bureau. The six service bureaus are shown in Table 9.3.
Note that the tabulated column-wise by type of RP systems used in serv-
icing clients. This arrangement is important, it will be shown later that
the type of RP system is pivotal to the area of applications and indirectly
responsible for the potential amount of clients whom the service bureau
can serve.

In their years (or months) of operation, it is not possible to obtain their
output as these are confidential information. However, one can safely

Table 9.3. RP service bureaus.

RP systems used in service bureau	3D Systems' SLA	CMET's Rapid Meister	Sony's SCS	Cubic techno-logies' LOM	Stratasys' FDM	3D Systems SLA and SLS
Name of service bureau	AAA	BBB	CCC	DDD	EEE	FFF
Service available since	1990	1994	1992	1993	1995	1995

estimate that the current usage exceeds an average order of two parts per week.

An analysis of the level of expertise and experience and the number of operational staff available in the service can also be carried out. This relates directly to their capabilities and limitations in handling projects. Table 9.4 shows the number of people involved, the level of experience and the clients served by the service bureau.

9.2.2.4. *Competitive Analysis*

The strengths and weaknesses of the competitors are next analyzed. Considerations include the nature of the service bureau such whether it is primarily a vendor or distributor, or an academic or research entity. In such instances, they may be weak in providing a total integrated solution. The priority of a vendor/distributor is focused on sales whereas the academic or research entity's primary concern is education, teaching and research. Another consideration is the availability of additional facilities and equipment in supporting basic engineering works, which are important in providing a holistic approach to engineering solutions.

9.2.2.5. *Marketing Strategies*

Potential clients do not know who you are and what you offer when your services are first launched. Promotional activities must

Table 9.4. Service bureau staff, experience and clients.

Service bureau	RP systems	Number of staff	Clients	Experience
AAA	SLA	Five engineering staff	Jeweler, souvenir, medical, products, audio, video, computer and telecommunications industry	Most experienced of all in RP; goal is to transfer technology
BBB	Rapid Meister	Not known (20 people in company)	Telecommunications, computer, audio, video, industrial and consumer products industry	New comer in RP, but has been in tool and mold business
CCC	SCS	Two engineering staff	Consumer electronics, consumer products, automobile accessories, audio and video industry	Second most experienced; meets own needs but this is also its main disadvantage
DDD	LOM	Three engineering staff	Telecommunications, computer, audio, video, industrial and consumer products industry	Not main business
EEE	FDM	Not known (six people in company)	Telecommunications, computer, audio, video, industrial and consumer products industry	Machine not delivered yet, but has been in mold and electrode making business
FFF	SLA & SLS	Not started yet	NA	Machine not delivered yet

accomplish this for you. There are two basic forms of promotional activities:

(1) *Publications.* One of the most effective ways is to disseminate information through publications. There are five types of publications and they are described as follows:

  (a) *Service bureau brochure.* Firstly, a service bureau brochure will have to be created. No efforts should be spared in making IT a top-class-colorful, gloss-finished and graphics-illustrated one. Important details like total engineering solutions, lead-time and one-stop service should be emphasized. The brochure should not be clouded with technical details, but should be designed professionally and should be more business-oriented. A single folded sheet brochure is preferred.

  (b) *Technical brief.* This is where the technical details — process description, material properties and composition, etc. — can be included. This brief is especially useful for the technically minded clients. However, this brief should preferably be a single sheet. Alternatively, the service bureau brochure and Technical Brief can be combined into an A3-sized paper folded in the middle.

  (c) *Technical article.* In times where participation at a conference, seminar, or workshop requires the submission of an article on the service or its experience with special application projects, this will come in useful. The length of the article can be about six to ten pages.

  (d) *Technical video and/or disk (or CD).* Most RP systems come with a video showing technical details and possibly case studies and interviews with clients. It can be used directly for your purpose. Alternatively, a CD using multi-media technology can be created. This, however, will have to be carried out by the company, as many RP systems do not come with this.

  (e) *Internet website with easily identifiable URL.* Many potential clients will have access to the Internet and the information of the service bureau must be made available readily. An easily identifiable URL or website address will be necessary. Professional assistance in setting up the website may be necessary. Complete information must be made available, including details like total engineering solutions, lead-time and one-stop services which must be emphasized. Ease of navigation through the pages will be essential. Multi-media design created on the video CDs can also be edited and included here.

(2) *Publicity activities.* Publicity activities are useful in that you can publicize through an attentive audience. It may be organized solely by your company or in conjunction with interested parties. While it is organized by some other organization, it is worthwhile to keep in contact with the relevant active organizations so that you are kept informed and where possible, invited for participation. There are three types of publicity activities:

(a) *Technical talk.* This is a very common activity organized by many engineering societies. However, some societies discourage sales talk and therefore, the subject contents will have to be technical. In many instances, the talk is free for members and thus, the company will have to sponsor the event. This event can be held in tertiary institutions or at hotels, depending on your budget. It is typically one to one-and-a-half hours in the evening and should include refreshments.

(b) *Exhibition.* Participation in relevant, major exhibitions held nationally or regionally will be useful too. It is cautioned that substantial costs will be incurred (mainly transportation costs and rental of booth). An evaluation should be carried out to decide if it is worthwhile. Exhibitions are typically several days' affairs and would mean disruption of work for at least two staff to man the booth.

(c) *Seminars/conferences.* Like technical talk, speaking at a conference or seminar requires a fair amount of experience and knowledge of your product. Unlike the technical talk, submission of an article is required. Typically, a notification of such an event comes through a "Call for Papers", followed by submission of an abstract and if accepted, the full manuscript. Every two years, there are several relevant conferences devoted to "Manufacturing", "Automation" and "Computer Integrated Manufacturing (CIM)". Since 1990, many dedicated international conferences and seminars on "Rapid Prototyping" have also been held.

9.2.2.6. *Competency and Preparation*

Potential clients do not know how good you are and therefore whatever contact you make with them must demonstrate your competency and reliability. Some of the possible contacts with potential clients may be through telephone calls, visits, demonstrations, benchmarking tests, etc. Preparatory activities include really knowing your products, putting up big posters (create your own), or setting up a good website describing the process, using impressive showcases to display industrial parts made by the RP systems, as well as other secondary value-added processes put in by your company.

9.2.3. Cost of the Service Bureau Set-Up

9.2.3.1. *Land and Building Costs*

Land and building costs are important in any operation. Very likely, land and building are to be leased. Location will be one very significant consideration as it should be as close to the client as possible. In this case study, the lease on land and building of an appropriate size is taken to be US$20,000 per annum.

9.2.3.2. *Main Equipment Costs*

The list and current prices of equipment necessary to outfit the proposed facility is shown in Table 9.5. Total equipment costs include machine, materials, accessories and post-cure equipment (if applicable). Typically, pricing also includes installation, commissioning and training.

9.2.3.3. *Optional Equipment Costs*

The optional equipment includes a workstation and a CAD/CAM package such as Pro-Engineer. Table 9.6 shows an example of the likely cost of

Table 9.5. Cost of RP equipment.

RP systems	Estimated price*
SLA	US$300,000
SGC	US$295,000
SLS	US$366,000
LOM	US$90,000
FDM	US$120,000
SOUP	US$600,000
SCS	US$380,000

Table 9.6. Cost of optional equipment.

CAD/CAM	Estimated price
Software	Pro-Engineer
	Basic Software: US$25,000
Training	Basic: US$4,000 (5 days)
	Surface: US$2,000 (3 days)
Hardware	Pentium: US$2,000
	SGI: US$30,000

such equipment. Total optional equipment cost with an IBM compatible personal computer is US$33,000.

9.2.4. Comparison and Selection of RP Systems

A point-by-point comparison of the various RP systems under consideration is carried out. It is worthwhile to note that there are strengths and weaknesses for every system and the weaknesses of the systems are constantly improved upon by the vendors.

From the analysis of various factors, suppose that laminated object manufacturing (LOM) is recommended for the service bureau for the following supporting reasons:

(1) the machine cost is the lowest amongst all its competitors,
(2) its applications are substantially wide, based on sales figures,

(3) it also has the lowest material costs,
(4) the process is clean and can operate in an office environment,
(5) the material need not be shielded from sunlight and does not have a foul smell,
(6) supports are not required in the process.

9.2.5. Capital Investment Requirements

The proposed venture can now be stated in terms of its initial investment requirements. The initial capital required will include the RP machine and the CAD/CAM system, inclusive of all training needs.

(a) LOM   : US$90,000
(b) CAD/CAM   : US$33,000
     Total   : US$123,000

9.2.6. Revenues, Expenses and Return on Investment

9.2.6.1. *Revenues*

The revenue derived from operations will come from the revenues derived from the production of prototypes. In the future plans for the service bureau, other sources of revenue can come from the production of molds, tools and dies from the master pattern of the prototypes. This will complement the company's mission to provide a total solution to clients in meeting a diverse range of manufacturing applications.

The level of output has already been pre-determined to be 87 projects per annum. This project is assumed to be of average size and will take three days to complete. It is also assumed that the project will involve two prototypes to be produced. The stages of production will include model checking, model building and model post-processing. Other activities such as CAD modeling, file transfer and secondary processes will be excluded. From industrial data taken from competitors, it is fair to assume that an average project will cost US$1,000. Therefore, annual revenue will amount to 87 × 2 × US$1,000 = $174,000.

9.2.6.2. *Operating Expenses*

Three major categories of expenses are discussed in this section: cost of expendable, salaries and depreciation and maintenance.

(1) *Cost of expendables.* One of the major operating costs is the cost of raw materials. Raw materials may be liquid resin for the case of SLA, a roll of paper for LOM, or a roll of filament for fused deposition modeling (FDM). The price for each case varies. The recommended LOM uses paper, the cost of paper is US$100 per roll. This roll has a width of 210 mm and length of 150 m. For big parts, this roll of paper can be used to build two parts. It can build three to four parts for smaller components. Assume, therefore, that each roll of paper can be used for every two projects. For 87 projects in a year, paper cost per annum works out to be (US$100/roll × 87 projects/2 projects/roll = US$4,350).

The other major expendable is the laser, if used. The laser is used in RP systems such as SLA, LOM and selective laser sintering (SLS). The price for each system varies. As before, the laser used in LOM is used for the calculations. Assume that the laser costs US$20,000 and can last 10 years. Laser cost per annum works out to be (US$20,000/ 10 years = US$2,000). Other costs, if any, may be assumed to be negligible. Therefore, the summary of expendable per year = US$4,350 + US$2,000 = US$6,350.

(2) *Salaries.* Assume that the number of personnel required for the operation is two. Administrative and marketing functions are presumed to be already in existence. The functions of these two personnel are predominantly technical and both are equally proficient in the entire process of RP so that each can cover the other's duty without any problem or delay in the project.

However, for proper management and control, one should be an engineer and the other an engineering assistant. In this way, the engineer can focus on scheduling and prioritizing projects and the assistant on the project proper. The following figures show the yearly salaries:

(a) Engineer   : US$20,000
(b) Engineering assistant : US$12,000
    Total   : US$32,000

Total annual salaries amount to US$32,000. Other employee expenses are also included in the above salaries. These could be insurance, hospitalization and other fringe benefits and the total wage per annum is US$35,000.

(3) *Depreciation and maintenance.* Assume that several other categories of expenses (such as utilities, accounting charges, bad debt expenses) amount to 10% of depreciation of the equipment. Total annual capital expense is, therefore, depreciation to 110%. If the equipment is be depreciated over a 15-year period, the annual depreciation expense would total US$173,000/15 years = US$11,533. The total annual capital expense is US$11,533 × 1.1 = US$12,686. The annual maintenance of LOM and Pro-Engineer software is 15%, i.e., 0.15 × US$25,000 = US$3,750.

9.2.6.3. *Pro Forma Income Statement*

The results of the analysis of anticipated revenues and expenses are next used to develop the pro forma income statement shown in Table 9.7. Expenses total US$77,786. This amount is subtracted from total revenues to determine profit before taxes. Company income taxes are then subtracted to determine the net profit after taxes, which is US$70,236.

Table 9.7. Pro forma income statement.

Sales (total revenues)		S$$174,000
Lease	(US$20,000)	
Cost of expendable	(US$6,350)	
Salaries	(US$35,000)	
Depreciation	(US$9,020)	
Maintenance	(US$3,750)	
Total expenses		(US$74,120)
Profit before income taxes		US$99,880
Corporate taxes (27%)		US$26,968
Net profit after taxes		US$72,912

9.2.6.4.   *Return on Investment Analysis*

The rate of return on investment is computed by dividing net profit after taxes of US$70,236 by the total investment (including capital investment and annual operating expense less depreciation) of US$38,100. The rate of return = (US$70,236/US$238,100) × 100% = 29.5%.

Assuming a 15% minimum acceptable rate of return and since the calculated value of the rate of return of 29.5% is greater than 15%, the proposed venture is deemed economically viable.

9.3.   TECHNICAL EVALUATION THROUGH BENCHMARKING

The execution of a benchmark test is a traditional practice and necessary for all kinds of highly productive and expensive equipment such as CAD/CAM workstation, CNC machining center, etc. Wherever a relatively broad spread of possibilities is offered for specific users' requirements, the execution of a benchmark test is absolutely necessary. The dynamic development and increased range of commercially available RP systems on offer (currently close to 40 different types of equipment worldwide, partly in different types of equipment, partly in different sizes), mean objective decision making is essential. In analyzing the benchmark test piece, some tests have to be conducted including visual inspection and dimension measurement.

9.3.1.   Reasons for Benchmarking

Generally, benchmarking serves the following purposes:

(1) It is a valuable tool for evaluating strengths and weaknesses of the systems tested. Vendors have to produce the benchmark models in response to requests from potential buyers. In so doing, they will not have the choice to demonstrate what they want, but have to show what is requested. Consequently, the vendors cannot hide the limitations of the system. It is also a rigorous and, therefore, more revealing

means of testing so that the potential buyer can verify the claims of the vendor.

(2) Since the benchmark model is specifically designed by the potential buyer, it can be custom-made to its own requirements and needs. For example, in the case of a company that makes parts which frequently have very thin walls, then the ability of the system to produce accurately built thin walls can be tested, measured and verified.

(3) Benchmarking has also become a means for helping various departments within a company to comprehend what the RP system can do for them. This is vitally important in the context of concurrent engineering whereby designers, analysts, manufacturing engineers work on the product concurrently. Today, a RP system's application areas extend beyond design models, to functional models and manufacturing models.

(4) Sometimes, a benchmark test may also help to identify applications for a RP system, which had not previously been considered. Though this is not really a primary motivational factor for benchmarking, it is nevertheless a side benefit.

9.3.2.   Benchmarking Methodology

There are four steps in the proposed benchmarking methodology:

(1) Deciding on the benchmarking model type.
(2) Deciding on the measurements.
(3) Recording time and measurements, tabulating and plotting the results.
(4) Analyzing and comparing results.

9.3.2.1.   *Deciding on the Benchmarking Model Type*

In general, RP benchmarking models can be categorized according to the following types:

(1) *Typical company's products.* This is probably the most common type since the company needs to confirm how well its products can be

prototyped and whether its requirements can be fulfilled. The company is also in the best position to comment on the results since it has intimate product knowledge. Examples are turbine blades, jeweler, cellular phones, etc.

(2) *Part classification.* According to Wall,[4] a classification scheme based on general part structures is applicable for RP parts. This classification scheme based on part structure has ten part classes as seen in Table 9.8.

Figure 9.1 shows some examples of such a classification scheme. The part sizes and part structures are most related to shrinkage, distortion and curling effects.

(3) *Complex and hybrid parts.* Alternatively, determined complex parts can be designed, aiming to test the performance of available systems in specific aspects. Furthermore, this category can include a hybrid combining types 1 and 2 above.

9.3.2.2. *Deciding on the Measurements*

In deciding on the measurements, it must be stressed that, as far as possible, the benchmark model should:

(1) be relatively simple and designed with low expenditure,

Table 9.8. Part structure classification scheme of 10 part classes.

Part class number	Part structure
Part Class 1	Compact parts
Part Class 2	Hollow parts
Part Class 3	Box type
Part Class 4	Tubes
Part Class 5	Blades
Part Class 6	Ribs, profiles
Part Class 7	Cover type
Part Class 8	Flat parts
Part Class 9	Irregular parts
Part Class 10	Mechanisms

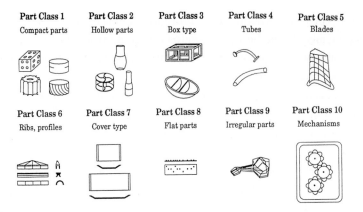

Fig. 9.1. Part structure classification scheme for RP systems.

(2) not utilize too much material and

(3) allow simple measuring devices to determine the measurements.

In general, two types of measurements can be taken namely, main (large) measurements and detailed (small) measurements.

9.3.2.3. *Recording Time and Measurements, Tabulating and Plotting the Results*

The measurement results are based on the deviations of the built part from the CAD model. These deviations of both the main and detailed measurements are tabulated for each of the RP systems. Subsequently, the results can be plotted to show graphically the performance of the systems in a single diagram. Thus, for the main measurements, one can visually compare the systems' performance from the main measurement diagram and similarly for the detailed measurement diagram.

The time results are based on three components — data preparation, building time and post-processing. The total time is based on the addition of the three components. A table of all four-time results can be tabulated with each RP system alongside another.

9.3.2.4. *Analyzing and Comparing Results*

In a general analysis, the deviations of all systems may or may not be in an acceptable range. The evaluator of the systems is usually decided by this acceptable range. The time component by itself gives one an idea of the length required for a task and directly affects the cost factor. Therefore, the time data can become useful for a full economic justification and cost analysis.

In comparison, one can determine, based on the time and measurement results, the strengths and weaknesses of the systems. In arriving at a conclusion, the evaluator must only consider the benchmarking results as *a component* of the overall evaluation study. The benchmarking results should never be taken as the only deciding factor of an evaluation study.

Finally, the above approach ignores the human aspects. For example, the skills, expertise and experience of the vendors' operators are not accounted for in the benchmark.

9.3.3. Case Study

9.3.3.1. *The Button Tree Display*

With the compliments of Thomson Multi-media in Singapore, the results of a benchmarking study involving five machines are made available here. The test piece is a button tree display that is mounted in between the front cabinet of a hi-fi set and printed circuit board (PCB) as shown in Fig. 9.2. The button tree display has a frame of length 128 mm and width 27 mm. It consists of three round buttons joined to the frame by 0.6 mm hinges. The buttons have "legs" that contact the tact switch on the PCB when it is depressed. There are also locating pins for location and light emitting

Fig. 9.2. Button tree display is mounted on front cabinet of a hi-fi.

diode (LED) holder on the frame. Catches are made from the side of the frame so that the button tree display can hold down firmly to the PCB. The five different test pieces are made from the principals of solid ground curing (SGC), LOM, solid object ultraviolet plotter (SOUP), solid creation system (SCS) and FDM. A photograph of each test piece is shown in Figs. 9.3, 9.4, 9.5, 9.6 and 9.7, respectively.

The measurements taken are linear, radial and angular dimensions. Coordinate measuring machine (CMM), profile projector and vernier calliper are used for the measurements. When choosing the type of dimensions to measure, a few criteria are considered:

(1) the overall dimensions are to be included,
(2) the important dimensions that will affect the operation of the button tree,

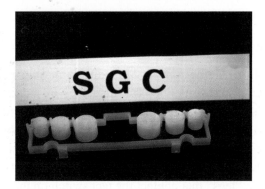

Fig. 9.3. Benchmark test piece made from Solid Ground Curing (SGC).

Fig. 9.4. Benchmark test piece made from Laminated Object Manufacturing (LOM).

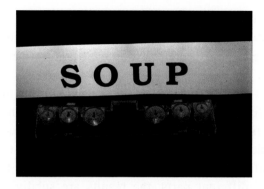

Fig. 9.5. Benchmark test piece made from Solid Object Ultraviolet Plotter (SOUP).

Fig. 9.6. Benchmark test piece made from Solid Creation Systems (SCS).

Fig. 9.7. Benchmark test piece made from Fused Deposition Modeling (FDM).

(3) there must be a variety of dimensions,
(4) there must be sufficient main and detailed dimensions to plot the graph and give an indication of which method is superior.

Figures 9.8–9.10 show the technical drawings for the button tree display. The dimensions of the test piece are divided into three parts. The first part includes dimensions taken from the frame, locating pins and LED holder. The second part has the dimensions taken from the button set and the dimensions of the catch fall into the third part. Related to the different parts, the results are subdivided into main measurements (>10 mm) and detailed measurement (=10 mm). For both the readings, the deviations from the actual reading are computed and tabulated. The deviations for both the measurements are plotted and compared.

9.3.3.2. *Results of the Measurements*

The graphs plotted facilitate a detailed evaluation concerning the five different techniques namely the SCS, LOM, SOUP, SCS and FDM, in terms of main and detailed measurements and their deviations from the designed dimensions.

From visual inspections, none of the thin rib or wall is missing. This shows that all the five processes are capable of making wall thickness as thin as 0.5 mm.

However, one of the locating pins of the LOM test pieces is tilted and the catches of the SGC and FDM test pieces are missing. This is due to the catches being too weak and coming off when the supports were removed. SLA objects are built on supports rather than directly on the elevator platform. Supports are used to anchor the part firmly to the platform and prevent distortion during the building process. The SOUP, FDM and SCS test pieces came with supports, thus they had to be removed before any measurements were taken. The button sets of the five test pieces are slanted due to its weight and the weak hinges. Tables 9.9 and 9.10 list the measurements taken and their deviations from the nominal values.

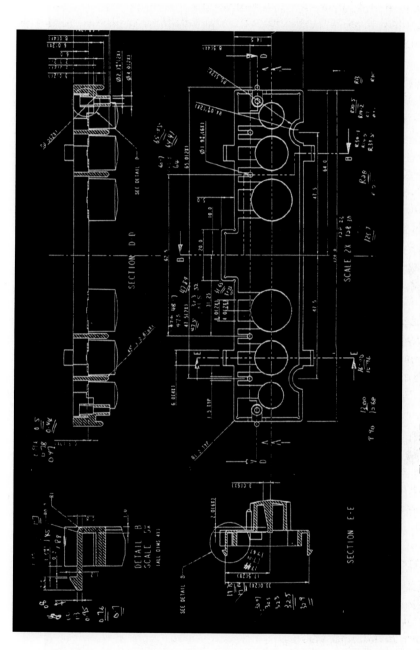

Fig. 9.8. Front and plan views of the Button Tree Display.

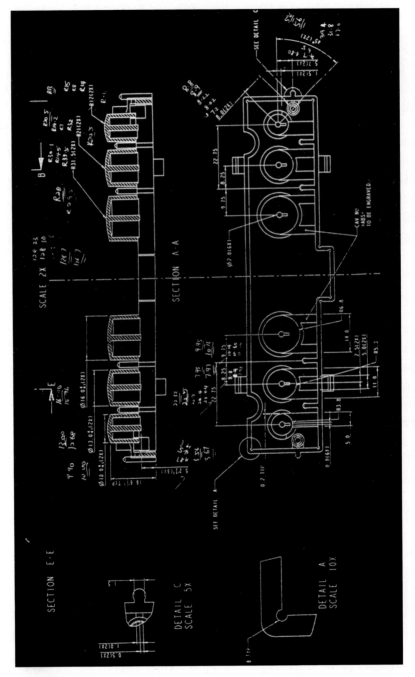

Fig. 9.9. Sectional view of the Button Tree Display.

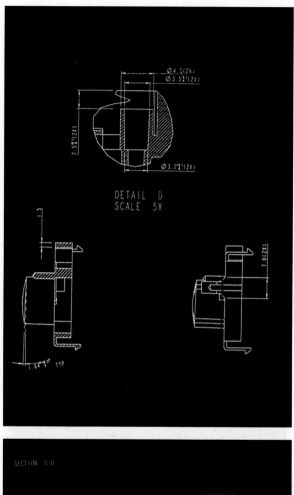

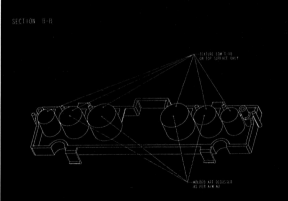

Fig. 9.10. (Top) Sectional and (bottom) isometric views of the Button Tree Display.

Table 9.9. Main measurements (>10 mm) of the five benchmark test pieces.

		Main measurements				
Drawing dimensions (mm)		Measured dimensions (mm) and deviations (mm) in italics				
		SGC	SOUP	LOM	SCS	FDM
Frame, location	128.0	128.3	129.1	128.2	128.7	128.7
pins and		*+0.3*	*+1.1*	*+0.2*	*+0.7*	*+0.7*
LED hold	27.0	27.1	27.3	27.1	27.1	26.8
		+0.1	*+0.3*	*+0.1*	*+0.1*	*−0.2*
	65.0	65.7	65.5	64	64.4	65.6
		+0.7	*+0.5*	*−1.0*	*−0.6*	*+0.6*
	47.5	47.6	47.5	48.7	47.9	48.7
		+0.1	*0.0*	*+1.2*	*+0.4*	*+1.2*
	31.25	31.3	31.5	32	31.7	31.2
		+0.05	*+0.25*	*+0.75*	*+0.4*	*−0.05*
Button tree	R31.5	30.1	14.5	33.5	28.0	29.5
		−1.4	*−17.0*	*+2.0*	*−3.5*	*−2.0*
	R21.0	14.2	7.0	32.0	20.5	20.5
		−6.8	*−14.0*	*+11.0*	*−0.5*	*−0.5*
	R12.0	15.0	8.0	14.0	13.0	11.0
		+3.0	*−4.0*	*+2.0*	*+1.0*	*−1.0*
	16.0	16.1	16.1	15.9	16.0	16.0
		+0.1	*+0.1*	*−0.1*	*0.0*	*+0.0*
	13.0	13.0	13.1	12.9	13.0	12.7
		+0.0	*+0.1*	*−0.1*	*0.0*	*−0.3*
	22.75	23.0	23.8	24.4	22.45	22.3
		+0.25	*+1.05*	*+1.65*	*−0.3*	*−0.45*
Catch	33.0	32.3	33.1	32.3	32.5	32.9
		−0.7	*+0.1*	*−0.7*	*−0.5*	*−0.1*
	17.5	17.4	17.0	17.6	17.8	17.4
		−0.1	*−0.5*	*+0.1*	*+0.3*	*−0.1*

The plots illustrated in Figs. 9.11 and 9.12 show that SGC achieved better measurements or fewer deviations as compared to other processes in detailed measurements. For the main measurements, FDM attained lesser deviations. On the other hand, SOUP produced the higher

Table 9.10. Detailed measurements (=10 mm) of the five benchmark test pieces.

		Detailed measurements				
		Measured dimensions (mm) and deviations (mm) in italics				
Drawing dimensions (mm)		SGC	SOUP	LOM	SCS	FDM
Location pins	8.0	7.2	8.2	8.0	8.1	8.8
		−0.8	*+0.2*	*+0.0*	*+0.1*	*+0.8*
	5.7	4.7	6.8	6.5	7.6	7.4
		−1.0	*−1.1*	*−0.8*	*−1.9*	*−1.7*
	45°	39.4	31.8	37	56	54
		−5.6	*−13.2*	*−8.0*	*+11.0*	*+9.0*
	8.25	8.8	8.0	8.4	8.0	7.9
		+0.55	*−0.25*	*+0.15*	*−0.25*	*−0.35*
Button tree	9.75	10.1	10.2	10.7	10.1	10.0
		+0.35	*+0.45*	*+0.95*	*+0.35*	*+0.25*
	10.0	10.1	10.1	9.9	10.0	9.9
		+0.1	*+0.1*	*−0.1*	*0.0*	*−0.1*
	5.2	4.9	4.8	4.5	3.6	3.7
		−0.3	*−0.4*	*−0.7*	*−1.6*	*−1.5*
	0.5	0.5	1.0	0.8	0.5	0.4
		0.0	*+0.5*	*+0.3*	*0.0*	*−0.1*
Catch	1.6	1.5	1.1	1.4	1.7	1.9
		−0.1	*−0.5*	*−0.2*	*+0.1*	*+0.3*
	0.8	1.0	1.0	1.3	1.0	0.7
		+0.2	*+0.2*	*+0.5*	*+0.2*	*−0.1*

deviations for both the detailed and main measurements. From all the measurements taken, the greater deviations are observed for the radii of the button set and the angles of the LED holder. These indicate that RP systems have limitations in producing good curve surfaces.

The deviations could be due to the following errors in taking the measurements:

(1) The poor surface finishes of the test pieces result in inaccurate dimensions.

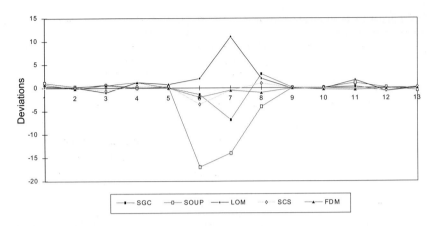

Fig. 9.11. Deviation of main measurements (>10 mm) from the nominal values for the five benchmark test pieces.

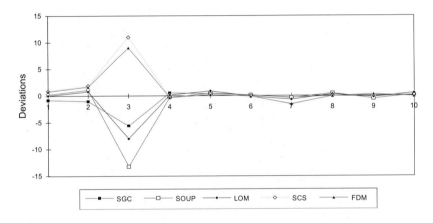

Fig. 9.12. Deviation of detailed measurements (=10 mm) from the nominal values for the five benchmark test pieces.

(2) The test pieces are not strong as they became deformed under pressure.

(3) The support of the SOUP test piece was not properly removed.

9.3.4.   Other Benchmarking Case Studies

Other than the benchmarking performed by Thompson Multi-media, there are also other benchmarking case studies done by various researchers and Tables 9.11(a)–9.11(e) list a few from the year 2000 to 2006.[5–16]

Table 9.11. Benchmarking studies published from (a) 2006, (b) 2005, (c) 2004, (d) 2003 and (e) 2000–2002.

	(a) Benchmarking studies published in 2006		
1	K. Abdel Ghany and S. F. Moustafa	Comparison between the products of four RPM systems for metals	*Rapid Prototyping Journal* **12**(2), 86–94 (2006)

This work evaluates and compares the quality of four identical benchmarks fabricated from different metallic powders by using four recently developed RPM systems for metals. The evaluation considers benchmark geometry, dimensional precision, material type, product strength and hardness, surface quality, building speed, materials, operation and running cost.

2	D. Dimitrov, W. van Wijck, K. Schreve and N. de Beer	Investigating the achievable accuracy of three dimensional printing	*Rapid Prototyping Journal* **12**(1), 42–52 (2006)

This paper deals with current research towards the building of a full capability profile — accuracy, surface roughness, strength, elongation, build time and cost — of this important process.

3	M. Mahesh, Y. S. Wong, J. Y. H. Fuh and H. T. Loh	A six-sigma approach for benchmarking of RP&M processes	*International Journal of Advanced Manufacturing Technology* **31**, 374–387 (2006)

This paper presents a methodology of using six-sigma quality tools for benchmarking of rapid prototyping manufacturing (RP&M) processes. It involves the fabrication of a geometric benchmark part and a methodology to control and identify the best performance of the process to reduce the variability in the fabricated parts.

	(b) Benchmarking studies published in 2005		
4	Todd Grimm	3D printer dimensional accuracy benchmark	T. A. Grimm and Associates, Inc.

This benchmark is to analyze and quantify the dimensional accuracy available from the Dimension® SST, InVision™ SR and ZPrinter® 310. This report illustrates the accuracy of each system with reverse engineering color maps and comparative charts.

(Continued)

Table 9.11. (*Continued*)

5	V. R. Gervasi, A. Schneider, and J. Rocholl	Geometry and procedure for benchmarking SFF and hybrid fabrication process resolution	*Rapid Prototyping Journal* **11**(1), 4–8 (2005)

This paper shares with the solid free-form fabrication community a new procedure and benchmark geometries for evaluating SFF process capabilities. The procedure evaluates the range capability of various SFF and SFF-based hybrid processes in producing rod and hole elements.

(c) Benchmarking studies published in 2004

6	M. Mahesh, Y. S. Wong, J. Y. H. Fuh and H. T. Loh	Benchmarking for comparative evaluation of RP systems and processes	*Rapid Prototyping Journal* **10**(2), 123–135 (2004)

This paper presents issues on RP benchmarking and aims to identify factors affecting the definition, fabrication, measurements and analysis of benchmark parts.

7	K. W. Dalgarno and R. D Goodridge	Compression testing of layer manufactured metal parts: The RAPTIA compression benchmark	*Rapid Prototyping Journal* **10**(4), 261–264 (2004)

This paper reports the results of a compression test benchmarking study carried out to investigate the mechanical properties of layer manufactured metal components in order to assess their suitability in load bearing application. Compression tests were carried out on the DTM LaserForm St-100 material, ARCAM processed H13 tool steel, EOS DirectSteel (50μm) and the ProMetal material.

(d) Benchmarking studies published in 2003

8	Todd Grimm	Rapid prototyping benchmark: 3D Printers	T. A. Grimm and Associates, Inc.

A detailed benchmark analysis of the fabrication process, part quality, cost and others from the tested products of Dimension from Stratasys, Z406 from Z Corporation, MDX-650 from Roland, Quadra Tempo from Objet Geometries, Viper si2 from 3D Systems, ThermoJet form 3D Systems and PatternMaster from Solidscape.

Table 9.11. (*Continued*)

9	A. Pfister, R. Landers, A. Laib, U. Hübner, R. Schmelzeisen and R. Mülhaupt	Biofunctional rapid prototyping for tissue-engineering applications: 3D bioplotting versus 3D printing	*Journal of Polymer Science: Part A: Polymer Chemistry* **42**, 624–638 (2004)

This paper presents two important rapid-prototyping technologies (3D Printing and 3D Bioplotting) where comparisons are made with respect to the computer-aided design and free-form fabrication of biodegradable polyurethane scaffolds meeting the demands of tissue engineering applications. For 3D Printing, Z-Corp system (Z402) is being used.

10	H.-S. Byun and K. H. Lee	Design of a new test part for benchmarking the accuracy and surface finish of rapid prototyping processes	*Lecture Notes in Computer Science* **2669**, 731–740 (2003)

This paper presents a new test part that can benchmark various RP techniques. The test part was made on several major RP machines and measured by a coordinate measuring machine (CMM). The evaluation of the accuracy, as well as the surface finish, is discussed for different RP processes. Systems used are SLA3500, FDM8000, SLS2500, Z402 and LOM1015.

(e) Benchmarking studies published in 2002

11	R. I. Campbell, M. Martorelli and H. S. Lee	Surface roughness visualisation for rapid prototyping models	*Computer Aided Design* **34**(10), 717–725 (2002)

This paper describes a methodology and software implementation that provides the designer with a computer graphic-based visualization of RP model surface roughness. The surface roughness values were obtained through an extensive empirical investigation with five different RP systems: SLA 350, Actua 2100, FDM 1650, LOM 1015 and Z402.

Benchmarking studies published in 2000

12	F. Xu, Y. S. Wong, and H. T. Loh	Toward generic model for comparative evaluation and process selection in rapid prototyping and manufacturing	*Journal of Manufacturing Systems*

Benchmarking in terms of surface roughness, building time and building cost on model part fabricated by SLA, SLS, FDM and LOM.

9.4. INDUSTRIAL GROWTH

The RP industry has enjoyed tremendous growth since the first system was introduced in 1988. The rate of growth has also been significant. Right up to 1999, the industry was enjoying two-digit growth rate annually. In 2001, the RP industry continued to expand, though not at the same rate anywhere near as before. More systems were installed, more materials for these systems used and more applications for the technology were uncovered.[17] The rate of growth has tapered off significantly since 2000. The events and economic conditions of 2001 did not help to improve the situation. Then in 2003, the RP industry took a turn and revenues returned to levels of the past.[18] The growth of the RP industry once again leaped upwards in 2004.[19] There has been no sign of slowing down in the RP industry growth since then.

Observation made on four leading RP companies in 2006 revealed an approximate of 15% sale increases compared to 2005.[20] 3D Systems still remains as the leading company in the RP industry with revenue of $135 million in 2006. But due to the fierce price competition it is facing from competitors like Stratasys; its revenue has slipped compared to the preceding year. Stratasys reported an approximate $104 million in revenue for year 2006 which was about a 26% increase from 2005.[21] With the introduction of the only high-definition color 3D printer in the market, Z Corporation has enjoyed a 50% revenue growth.[22] EOS GmbH (Germany), the leading company for sintering systems, reported an approximately $68 million in revenue during the fiscal year ending September 2006.[23]

9.5. RECENT AND FUTURE DEVELOPMENT TRENDS

As the whole RP industry moves forward, RP-driven activities will continue to grow. Compared to several years ago, more significant trends have appeared. Among them are the proliferation of low-cost 3D printers, the rise and fall of metal-based RP companies, RP in rapid manufacturing and mass customization and RP use in the biomedical engineering field. Continuous improvements in the area of speed and quality, as well as ease of use have also been observed.

9.5.1. Low-Cost Office and Desktop 3D Printers

Office and desktop 3D printers may have resulted from further development of concept modelers, the breed of RP system geared towards producing prototypes for design reviews instead of physical testing or fully functional parts. This class of RP system is characterized by its higher speed, lower cost and weaker accuracy and resolution compared to the higher-end class. It has been found that since the concept modeler is designed to operate in design offices, not at workshops, they feature clean and safe operations. These aspects apparently become very appealing to the market and now become one of the most important aspects in choosing a RP system.

Over time, the problems with accuracy, surface finish and material properties of 3D printers have been gradually addressed. The 3D printers now are still of a different class from the higher-end systems, but the gap is narrowing. With the quality of 3D printers increasing, the system may capture more and more market share from the high-end systems in the future.[4] An example of a high resolution desktop machine is the Solidscape's T76 Benchtop system. It offers layer thickness of only 0.013 mm and is even used for creating investment casting patterns.

The growth of office and desktop 3D printers is also accompanied by the lowering of cost to purchase and maintain such systems. While in 2003 such a system may have cost around $50,000, in 2006, the price was already reduced to around $20,000. Examples of machines of this price class are the Stratasys Dimension and 3D System's LD 3D Printer. The trend continues that recently in January 2007, 3D System introduced the V-Flash 3D modeler which costs only around $9,900, half the price of previously existing low-cost 3D printer. The V-Flash was launched towards the last quarter of 2007.[24]

3D printers encompass 70% of the total RP system sold in 2005.[25] Its sales previously doubled in 2004 with Stratasys taking the lead.[19] The increase of the number of office and desktop 3D printers produced and purchased is an indication that the market is enlarging to encompass smaller companies that do not have workshop-class facility, as well as funds to purchase and maintain high-end RP system. With the reduction in the price of 3D printers, it will become more affordable for educational institutions below university level such as polytechnics, colleges and schools to purchase 3D printers.

9.5.2. The Rise and Fall of Metal-Based RP System Companies

As its name implies, metal-based RP systems offer output material of metal, as such they have excellent potential to produce ready-to-use prototypes of metal parts. Also called direct metal technologies,[26] these systems possess absolute advantage in delivering prototypes of metal parts with closer material properties with the final product than resin-based or plastic-based RP systems, while keeping the speed above regular CNC machining. Once the technology is perfected, it is not impossible for the systems to become rapid manufacturing systems, replacing traditional metal manufacturing systems.

As if tempted by the potential of metal-based RP systems, at the time of writing, at least eleven companies[27] offer metal-based RP systems: 3D Systems, EOS, Concept Laser, F&S/MCP, Phenix Systems, Trumpf, Arcam, Optomec, POM, ProMetal and Solidica. The technologies used by these companies mostly revolve around laser sintering and laser melting of a layer of powder. Certain companies like ProMetal and Solidica use different techniques. In the case of Solidica, CNC milling becomes one of its main processes other than its layered aluminum lamination. The details of the companies and their technologies have been discussed in earlier chapters.

Most of current metal-based RP system still may not be able to produce part quality in terms of surface finish on par with CNC milling or investment casting, but it is much closer to the quality of sand casting.[27] Techniques that introduce a hybrid between layer deposition and CNC milling like Solidica's may solve the surface finish problem, though it adds complexity to the process. A key advantage is that metal-based RP systems can also be used to make tooling parts and casting molds.

The large number of companies producing metal-based RP systems may be an advantage for the consumers. As they compete with each other, the speed and quality of the machines, as well as the price will become competitive. However, as the survival of the fittest dictates, few will prevail in the end and the rest may disappear along the way. As such, it becomes quite risky for a customer to purchase one of the systems, as the company may cease to operate resulting in the loss of the company's support and service to the customer.

9.5.3. Rapid Manufacturing and Tooling

Rapid manufacturing and tooling (RM&T) has evolved from rapid prototyping and manufacturing (RP&M). It refers to a process that uses RP technology to produce finished manufactured parts or templates, which are then used in manufacturing processes like molding or casting, directly building a limit volume of small prototypes. However, it is highly unlikely that this RP-driven process will ever reach the kind of production capacity of processes such as injection molding.[28] It refers to a process that uses RP technology to produce finished manufactured parts or templates. With enhancement of RP technologies, there is an increasing number of companies from aerospace, motor sports, medical, dental and consumer products industries turning to RM&T to produce customize and short-run production. The prospect of RM&T to grow and become the largest application of additive fabricate is optimistic.[28]

Typically, the costs incurred using RM&T application is much lower than conventional tooling cost. The tooling time is also notably lesser than conventional tooling time. This is mainly due to the fact that RP processes are able to fabricate models directly and quickly without the need to go through various stages of the conventional manufacturing process (i.e., post-processing). Moreover, it is more economical to build small intricate parts in low quantities using RM&T as there is compelling evidence that rapid manufacturing may be less expensive than traditional manufacturing approaches.[29]

In addition, RM&T has the ability to make changes to design at low cost even after the product has reached the production stage. Such a feature is highly desirable as it is usually very expensive to make modifications to the molds (if molding process is the fabrication for the finished product) and the time needed to modify the mold is also likely to be lengthy. Thus, RM&T is much preferred over the conventional tooling method in the early stages of the manufacturing cycle as modifications can be done almost instantaneously by simply editing the STL files and the cost incurred is relatively low.

However, the downside to rapid manufacturing processes is its lack of an appropriate range of materials for this application. Not all materials available are suitable for all products but nevertheless, current materials such as sintered nylon, epoxy resin and composite materials are often good enough for most low volume applications. Moreover, since RM&T has been successfully implemented in the aerospace industry, which imposes stringent quality demand, it is entirely possible that such applications can also be effectively implemented in other industries.

9.5.4. RP in Mass Customization

Tightly related with the concept of rapid manufacturing, the rapid production technology may be one of the keys in putting mass customization of product towards reality.[30] Mass customization has been growing stronger and stronger in recent years and it has been seen as the future paradigm of manufacturing.[31] Multiple cases have been around that indicate the shifting of the market trend towards mass customization, one well-known case being Dell's mass customization.[32] Backed by the market trend, this will put RP in an even stronger status in the future.

The concept of mass customization with RP does not stop at just high-end products. At the time of writing, there is even an innovative business concept in the toy industry being developed. Z Corp and SolidWorks have been working together to bring the usage of RP to produce customized parts to kids. The project is named Cosmic Modelz, where kids can use SolidWorks' Cosmic Blobs software to create their own custom figures and toys. The data can then be sent to Z Corp to be produced for as low a price as around $25 to $50 per model.[33] With the constant decreasing cost of RP, it is not impossible to see more and more of this kind of services in the future operated by smaller companies or even by individuals.

9.5.5. Biomedical Engineering Application

When RP technologies were first introduced, they were mainly used in the automotive industry, where verification and evaluation of new design ideas are required before marketing new products. Today, the use of RP technologies has expanded beyond the automotive industry into a wide variety of industries and fields.

RP technologies have been used in a broad spectrum of applications in the field of biomedical engineering. A variety of RP systems have been used in the production of scale replicas of human bones and body organs to advance customized drug delivery devices and other areas of medical sciences including anthropology, paleontology and medical forensics.[34]

FDM, SLS, 3DP and SLA are the most common systems employed in fabrication of tissue engineering scaffolds.

Intensive studies conducted on RP techniques related to biomedical application have been observed in recent years. Studies in various disciplines are done to evaluate the RP techniques, compare RP techniques with conventional techniques and improve the scaffold structures fabricated with various RP techniques, to customize implants and to explore new techniques/materials.[34-43]

Although some success in studies for tissue engineering using RP technologies are observed, this technique has yet to deliver significant progress in the clinical use. The utilization of RP technologies will grow as there are more low-cost systems with improved features being introduced to the market thus encouraging the wider use of RP technologies.

9.5.6. Others

Besides the above prominent development trends discussed, there are also two other trends observed in the development of RP.

9.5.6.1. *Ease of Use*

One of the development trends that has been going on continuously since the first RP system is to make the RP systems easier to use than before.

All RP systems are supplied with user friendly software that provides good visual feedback as well as useful automatic or semi-automatic functions to use with the 3D model and the machine.

Making post-processing easier to carry out or attempts to eliminate it altogether is another notable development trend. As systems are able to produce parts with high-dimensional accuracy and better surface finish, the need for elaborate post-processing is also diminished. Companies introduced various technologies to simplify, if not eliminate, post-processing. Stratasys introduced water-soluble supports even on their Dimension low-cost system line, which is the Soluble Support Technology (SST), which requires users to simply immerse the prototype into a water-based solution and wash away the supports.

9.5.6.2. *Improvement of Building Speed and Prototype Quality*

One other development trend is the improvement of building speed and prototype quality of RP systems. As with the ease of use, this development has been continuously observed. Newer systems introduced are rapidly replacing older systems that are slow in building time and weak in accuracy. Some years ago Stratasys introduced the FDM Titan upgrade which offered an average of 54% increase in building speed.[44] Recently, Stratasys launched FDM 200MC to replace the predecessor system, Prodigy Plus. When used with ABSplus material, FDM 200MC systems can fabricate parts achieving up to 67% stronger mechanical properties. ABSplus material is engineered to work optimally with FDM200MC.[45] Kira produced Katana that demonstrates building of a camera mockup in just an hour.[46] Solidscape made a huge improvement in its T76 Benchtop printer with three times speed increase and material having twice the strength of the previous model.[47]

REFERENCES

1. Wohlers Report 2007, State of the Industry Annual Worldwide Progress Report, *Wohlers Association, Inc.* (2007).

2. Economic Development Board, Singapore, Report on census of industrial production, *Research and Statistics Unit, Economic Development Board, Singapore* (2004).

3. T. Wohlers, Rapid prototyping growth continues, *Rapid Prototyping Rep.* **4**(6), 6–8 (1994).

4. M. B. Wall, Making sense of prototyping technologies for product design, *MSc Thesis,* MIT, USA (1991).

5. K. Abdel Ghany and S. F. Moustafa, Comparison between the products of four RPM systems for metals, *Rapid Prototyping J.* **12**(2), 86–94 (2006).

6. D. Dimitrov, W. van Wijck, K. Schreve and N. de Beer, Investigating the achievable accuracy of three dimensional printing, *Rapid Prototyping J.* **12**(1), 42–52 (2006).

7. M. Mahesh, Y. S. Wong, J. Y. H. Fuh and H. T. Loh, A six-sigma approach for Benchmarking of RP&M processes, *Int. J. Adv. Manuf. Technol.* **31**(3–4), 374–387 (2006).

8 T. A. Grimm, 3D Printer Dimensional Accuracy Benchmark, *T. A. Grimm & Associates, Inc.* (2005).

9. V. R. Gervasi, A. Schneider and J. Rocholl, Geometry and Procedure for benchmarking SFF and hybrid fabrication process resolution, *Rapid Prototyping J.* **11**(1), 4–8 (2005).

10. M. Mahesh, Y. S. Wong, J. Y. H. Fuh and H. T. Loh, Benchmarking for comparative evaluation of RP systems and processes, *Rapid Prototyping J.* **10**(2), 123–135 (2004).

11. K. W. Dalgano and R. D. Goodridge, Compression testing of layer manufactured metal parts: The RAPTIA compression benchmark, *Rapid Prototyping J.* **10**(4), 261–264 (2004).

12. T. A. Grimm, Rapid prototyping benchmark: 3D printers, *T. A. Grimm & Associates, Inc.* (2003).

13. A. Pfister, R. Landers, A. Laib, U. Hübner, R. Schmelzeisen and R. Mülhaupt, Biofunctional rapid prototyping for tissue-engineering

applications: 3D bioplotting versus 3D printing, *J. Poly. Sci.: Part A: Polym. Chem.* **42**, 624–638 (2004).

14. H. S. Byun and K. H. Lee, Design of a new test part for benchmarking the accuracy and surface Finish of Rapid Prototyping Processes, *Lecture Notes in Comput. Sci.* **2669**, 731–740 (2003).

15. R. I. Campbell, M. Martorelli and H. S. Lee, Surface roughness visualisation for rapid prototyping models, *Comput. Aided Design* **34**(10), 717–725 (2002).

16. F. Xu, Y. S. Wong and H. T. Loh, Toward generic model for comparative evaluation and process selection in rapid prototyping and manufacturing, *J. Manuf. Sys.* **19**(5), 283–296 (2000).

17. T. Wohlers, Wohlers Report 2000: Rapid Prototyping and Tooling State of the Industry, *Wohlers Association, Inc.* (2000).

18. Wohlers Associates, Industry Briefing [On-line], available at http://www.wohlersassociates.com/brief05-04.htm (May 2004).

19. T. Wohlers, Wohlers Report 2005: RP, RT and RM State of the Industry, Time-Compression Technologies [On-line], available at http://www. timecompress.com/magazine/magazine_articles.cfm? article_id=344& issue_id=83&articles=344 (November 2005).

20. Castle Island Co., The Rapid Prototyping Industry — Overview [On-line], available at http://home.att.net/~castleisland/ind_10.htm (June 2007).

21. Castle Island Co., The Rapid Prototyping Industry — Major US-Based Vendors [On-line], available at http://home.att.net/~castleisland/ind_21.htm. (July 2007).

22. Z. Corporation, Z Corporation Announces 50-Percent Revenue Growth in 2005 [On-line], available at http://www.zcorp.com/news/newsdetail.asp? ID=412&TYPE=1 (January 2006).

23. Castle Island Co., The Rapid Prototyping Industry — Vendors Outside the US [On-line], available at http://home.att.net/~castleisland/ind_22.htm (July 2007).

24 3D Systems Corporation, V-Flash™ 3-D Desktop Modeling System [On-line], available at http://www.modelin3d.com/ (July 2005).

25. T. Wohlers, The Terminology Challenge, *Wohlers Associates Inc.* [On-line], available at http://wohlersassociates.com/blog/2006/11/the-terminology-challenge/ (November 2006).

26 G. Todd, Direct Metal Technologies Tackle the Impossible [On-line], *Time-Compression Technologies*, Available: http://www.time-compress.com/magazine/magazine_articles.cfm?article_id=323&issue_id=78&articles=323 (March 2005).

27. T. Wohlers, An explosion of metal-based RP systems, *Wohlers Associates Inc.* [On-line], available at http://wohlersassociates.com/blog/2003/12/an-explosion-of-metal-based-rp-systems/ (December 2003).

28. T. Wohlers, Wohlers Report 2001: Rapid Prototyping & Tooling State of the Industry, *Wohlers Association, Inc.* (2001).

29. R. Wuensche, Using direct croning to decrease the cost of small batch production, *Engine Technology International 2000 Annual Showcase Review for ACTech* GmbH (2000).

30. W. Terry and G. Todd, Rapid prototyping in 2006, *Time-Compression Technologies* [On-line], available at http://www. timecompress.com/magazine/magazine_articles.cfm?article_id=34&issue_id=45&articles=34 (2006).

31 D. R. Butcher, Mass customization: A leading paradigm in future manufacturing, *ThomasNet® IndustrialNewsRoom* [On-line], available at http://news.thomasnet.com/ IMT/archives/2006/02/innovative_manu.html (February 2006).

32. E. J. Fern, MS, PMP (Time-to-Profit, Inc.), Six steps to the future: How mass customization is changing our world [on-line], *American Society for the Advancement of Project Management*, available at http://www.asapm.org/asapmag/articles/SixSteps.pdf.

33. T. Wohlers, Cosmic Modelz, *Wohlers Association, Inc.* [On-line], available at http://wohlersassociates.com/blog/2006/08/cosmic-modelz/ (August 2006).

34. K. F. Leong, C. M. Cheah and C. K. Chua, Solid freeform fabrication of three-dimensional scaffolds for engineering replacement tissues and organs, *Biomaterials* **24**, 2363–2378 (2003).

35. W. Y. Yeong, C. K. Chua, K. F. Leong, M. Chandrasekaran and M. W. Lee, Comparison of drying methods in the fabrication of collagen scaffold via indirect rapid prototyping, *J. Biomed. Material Res. Part B: Appl. Biomaterials* **82B**(1), 260–266 (2007).

36. Y. N. Yan, R. Wu, R. Zhang, Z. Xiong and F. Lin, Biomaterial forming research using RP technology, *Rapid Prototyping J.* **9**(3), 142–149 (2003).

37. M. Wagner, N. Kiapur, M. Wiedmann-Al-Ahmad, U. Hübner, A. Al-Ahmad, R. Schön, R. Schmelzeisen, R. Mülhaupt and N.-C. Gellrich, Comparative in vitro study of the cell proliferation of ovine and human osteoblast-like cells on conventionally and rapid prototyping produced scaffolds tailored for application as potential bone replacement material, *J. Biomed. Materials Res. Part A* **83A**(4), 1154–1164 (2007).

38. Jiankang He, Dichen Li, Bingheng Lu, Zhen Wang and Tao Zhang, Custom fabrication of a composite hemi-knee joint based on rapid prototyping, *Rapid Prototyping J.* **12**(4), 198–205 (2006).

39. S. Singare, L. Dichen, L. Bingheng, G. Zhenyu and L. Yaxiong, Customized design and manufacturing of chin implant based on rapid prototyping, *Rapid Prototyping J.* **11**(2), 113–118 (2005).

40. X. Li, D. C. Li, B. Lu, Y. Tang, L. Wang and Z. Wang, Design and fabrication of CAP scaffolds by indirect solid free form fabrication, *Rapid Prototyping J.* **11**(5), 312–318 (2005).

41. L. L. Tan, FEATURE: Plugging Bone the Painless Way, *Innovation magazine* [On-line], available at http://www.innovationmagazine.com/innovation/volumes/v4n3/features1.shtml (2007).

42. D. W. Hutmacher, J. T. Schantz, C. X. F. Lam, K. C. Tan and T. C. Lim, State of the art and future directions of scaffold-based bone engineering from a biomaterials perspective, *J. Tissue Eng. Regenerative Medicine* **1**(4), 245–260 (2007).

43. B. R. Ringeisen, C. M. Othon, J. A. Barron, D. Young and B. J. Spargo, Jet-based methods to print living cells, *J. Biotechnol.* **1**, 930–948 (2006).

44. Stratasys Rapid Prototyping Upgrades Increase Speed and Capacity; FDM Titan gets 54% Speed-up; FDM Vantage gets 150% Build Volume Increase [On-line], available at http://www.theautochannel. com/news/2003/11/ 26/173492.html (November 2003).

45. Stratasys Inc, Stratasys Introduces FDM 200mc Rapid Prototyping and Manufacturing System [On-line], available at http://intl.stratasys. com/ media.aspx?id=873 (May 2007).

46. Kira Corporation, What is RapidMockup System? [On-line], available at http://www.rapidmockup.com/eg/menu1_5_e.htm (May 2007).

47. Solidscape®, Inc., Solidscape®, Inc. Introduces World's First Benchtop 3D Printer...including 3X speed increase and 200% stronger material [On-line], available at http://www.solid-scape.com/t6x_benchtop_ release.html (December 2004).

PROBLEMS

1. What are the considerations when choosing a service bureau?
2. What are the components that make up the total cost of a part built by a service bureau?
3. What are the major considerations in assessing the economic feasibility of a proposed RP service bureau?
4. Is there a correlation between the type of RP system and the industry in which the prototypes are used? Why?
5. Describe the marketing strategies a service bureau can employ.
6. Given that the total revenue is US$100,000, total expenses are US$70,000, corporate tax is 30% and total capital investment is

US$200,000, calculate the rate of return. Assuming a 15% minimum acceptable rate of return, would you proceed with the venture?

7. Name the reasons for benchmarking.
8. Describe in detail the RP benchmarking methodology.
9. How have the primary and secondary markets for RP performed over the years?
10. What are the likely trends in RP systems growth?

Appendix
LIST OF RP COMPANIES

3D-Micromac AG
Annaberger Str. 240
D-09125 Chemnitz
Germany

Tel.: +49(0) 371-400 43-0
Fax: +49(0) 371-400 43-40
E-mail: info@3d-micromac.com
URL: http://www.3d-micromac.com

3D Systems Inc.
333 Three D Systems Circle
Rock Hill, SC 29730, USA
US toll free Tel.: (800) 889-2964

Tel.: (803) 326-4080
URL: www.3dsystems.com

Arcam AB (publ.)
Krokslätts Fabriker 30
SE-431 37 Mölndal
Sweden

Tel.: +46-31-710 32 00
Fax: +46-31-710 32 01
E-mail: info@arcam.com
URL: www.arcam.com

Autostrade Co. Ltd.
13-54 Ueno-machi
Oita-City
Oita 087-0832, Japan

Tel.: +81-97-543-1491
Fax: +81-97-545-3910
URL: www.autostrade.co.jp

CAM_LEM Inc.
1768 E. 25th St.
Cleveland, OH 44114, USA

Tel.: 216-391-7750
Fax: 216-579-9225
E-mail: sales@camlem.com
URL: http://www.camlem.com

CMET Inc.
Sumitomo Fudosan Shin-Yokohama Bldg,
2-5-5, Shin-Yokohama
Kouhoko, Yokohama,
Kanagawa 222-0033, Japan

Tel.: +81-45-478-5561
Fax: +81-45-478-5569
URL: www.cmet.co.jp

CONCEPT Laser GmbH
An der Zeil 8
96215 Lichtenfels
Germany

Tel.: +49 (0) 9571/949-228
Fax: +49 (0) 9571/949-239
E-mail: info@concept-laser.de
URL: http://www.concept-laser.de/

Cubic Technologies
1000 E. Dominguez Street
Carson, CA, USA

Tel.: (310) 965-0129
Fax: (310) 965-0141
E-mail: meygin@cubictechnologies.com
URL: www.cubictechnologies.com

D-MEC Ltd.
Hamarikyu Park Side Place
5-6-10 Tsukiji, Chuo-ku
Tokyo 104-0045, Japan

Tel.: +81-3-5565-6661
Fax: +81-3-5565-6643
E-mail: Tokyo@d-mec.co.jp
URL: www.d-mec.co.jp

The Ennex™ Companies
Los Angeles, CA, USA

Tel.: (805) 451-4507
E-mail: contact@Ennex.com
URL: http://www.ennex.com/

EnvisionTec.
Brüsseler Straße 51
45968 Gladbeck
Germany

Tel.: 02043-98750
Fax: 02043-98751
URL: www.envisiontec.de

EOS GmbH Electro Optical Systems
Robert-Stirling-Ring 1
D-82152 Krailling/Munich
Germany

Tel.: +49 89 89336-0
Fax: +49 89 89336-285
E-mail: info@eos.info
URL: www.eos.info

The ExOne Company
P. O. Box 1111
Irwin, PA 15642, USA

Tel.: 724-863-9663
E-mail: info@exone.com
URL: http://www.exone.com

Fraunhofer Rapid Prototyping Alliance
Sandtorstraße 22
39106 Magdeburg, Germany
Spokesman of the Alliance:
Dr Rudolf Meyer

Tel.: +49 (0)391/4090-510
Fax: +49 (0)391/4090-512
URL: http://www.fraunhofer.de/fhg/EN/profile/
    alliances/Rapid_Prototyping.jsp

Kira Corporation
Tomiyoshi shinden, Kira-cho Hazu-gun
Aichi Pref, Japan

Tel.: 0563-32-1161
Fax: 0563-32-3241
URL: http://www.kiracorp.co.jp/

MCP HEK Tooling GmbH
MCP Tooling Technologies
Kaninchenborn 24-28
D-23560 Lübeck, Germany

Tel.: +49(0) 451 53004-10
Fax: +49(0) 451 53004-50
E-mail: info@mcp-group.de
URL: http://www.mcp-group.de

MicroFabrica Inc.
7911 Haskell Ave.
Van Nuys, CA 91406, USA

Tel.: 818-997-3322
Fax: 818-997-3322
URL: www.microfabrica.com

Objet Geometries Ltd.
Headquarters
2 Holtzman St.
Science Park
P. O. Box 2496
Rehovot 76124, Israel

Tel.: +972-8-931-4314
Fax: +972-8-931-4315
URL: www.2objet.com

Optomec Inc.
3911 Singer N.E.
Albuquerque, NM 87109, USA

Tel.: (505) 761-8250
URL: http://www.optomec.com

Phenix systems
29 rue Georges Besse
63100 Clermont Ferrand, France

Tel.: +33(0) 4 73 98 22 50
Fax: +33(0) 4 73 98 22 51
URL: http://www.phenix-systems.com

SDM
Lee E. Weiss (Principal Research Scientist)
The Robotics Institute, Carnegie Mellon University
217 Smith Hall, Pittsburgh, PA 15213, USA

Tel.: (412) 268-7657
URL: http://www.cs.cmu.edu/~sdm/

Solidimension Ltd.
Shraga Katz Bldg.
Be'erot Itzhak 60905
Israel

Tel.: (+972) 3 933-9733
Fax: (+972) 3 933-7880
E-mail: info@solidimension.com
URL: http://www.solidimension.com

Solidscape, Inc.
316 Daniel Webster Highway
Merrimack, NH 03054-4115, USA

Tel.: 603-429-9700
Fax: 603-424-1850
E-mail: precision@solid-scape.com
URL: http://www.solid-scape.com

SOLIGEN Inc.
Northridge Division
Tel.: (818) 718-1221
Fax: (818) 718-0760
URL: http://www.soligen.com/home.shtml

Speed Part RP AB
Krokslätts Fabriker 30, 2
431 37 Mölndal
Sweden
Tel.: +46-31-338 39 90
Fax: +46-31-338 39 91
E-mail: info@speedpart.se
URL: http://www.speedpart.se

Stratasys Headquarters
Stratasys, Inc.
7665 Commerce Way
Eden Prairie
MN 55344-2020, USA
Tel.: + 1 800 9373 010
Fax: +1 952 937 0070
E-mail: info@stratasys.com
URL: www.stratasys.com

Therics, Inc.
283 E. Waterloo
Akron, OH 44319, USA
Tel.: 330 773 7677
Fax: 330 773 7697
E-mail: info@therics.com
URL: http://www.therics.com/

Voxeljet Technology GmbH
Am Mittleren Moos 15
D-86167 Augsburg, Germany

Tel.: 0821-7483-0
Fax: 0821-7483-111
E-mail: info@voxeljet.de
URL: http://www.voxeljet.de

Z Corporation
32 Second Avenue
Burlington, MA 01803, USA

Tel.: +1 781-852-5005
Fax: +1 781-852-5100
URL: http://www.zcorp.com

CD-ROM ATTACHMENT

Multimedia is a very effective tool in enhancing the learning experience. At the very least, it enables self-paced, self-controlled interactive learning using the media of visual graphics, sound, animation and text. To better introduce and illustrate the subject of *Rapid Prototyping*, an executable multimedia program has been encoded in a CD-ROM attachment that comes with the book. It serves as an important supplement learning tool to understand better the principles and processes of RP.

More than 30 different commercial RP systems are described in Chaps. 3–5. However, only the six most matured techniques and videos on their processes are demonstrated in the CD-ROM. These six techniques and the length of their videos are listed in Table I.

Other than the videos for the six RP techniques, there is a virtual laboratory available for the SLS system. The virtual laboratory allows the user to try out the process of a real SLS system.

The working mechanisms of these six methods are interestingly different from one another. In addition, the CD-ROM also includes a basic introduction, the RP process chain, RP data formats, applications and benchmarking. While the book describes the principles and illustrates

Table I. The six RP techniques and their corresponding video lengths.

Technique	Movie length/min
Stereolithography Apparatus (SLA)	1:17
Polyjet	2:43
Laminated Object Manufacturing (LOM)	4:30
Fused Deposition Modeling (FDM)	2:07
Selective Laser Sintering (SLS)	1:17
Three-dimensional Printing (3DP)	3:40

with diagrams the working principle, the CD-ROM goes a step further and shows the working mechanism in motion. Animation techniques through the use of Macromedia Flash enhance understanding through graphical illustration. The integration of various media (e.g., graphics, sound, animation and text) is done using Macromedia's Director.

Additional information on each of the techniques, such as product information, application areas and key advantages, makes the CD-ROM a complete computer-aided self-learning software about RP systems. Together with the book, the DVD-ROM will also provide a directly useful aid to lecturers and trainers in the teaching of the subject on Rapid Prototyping.

CD-ROM USER GUIDE

This user guide will provide the information needed to use the program smoothly.

SYSTEM REQUIREMENTS

The system requirements define the minimum configurations that your system needs in order to run the multimedia package smoothly. In some cases, not meeting this minimum requirement may also allow the program to run. However, it is recommended that the minimum requirements be met in order to fully benefit from this multimedia course package.

The recommended system requirements are as follows:

* Intel Pentium PC 1.5 GHz or 100% compatible with 512 MB of RAM.
* Intel Integrated 3D Graphics card.
* Windows XP or higher.
* 4 GB of hard disk free space (if program is installed into the hard disk).
* Sound card and speakers.

The preferred system requirements are as follows:

* Intel Pentiun PC 1.8 GHz or 100% compatible with 1 GB of RAM.

INSTALLING RP MULTIMEDIA CD-ROM

The CD-ROM can be run on an Autorun mode without any installation simply by inserting it into the CD-ROM or DVD-ROM drive. The program will start automatically.

If the "Autorun" does not activate, you can run the program directly from the CD-ROM drive by following these steps below after you have inserted the CD-ROM into the CD-ROM or DVD-ROM drive:

(1) Start Windows XP or higher by turning on your computer.
(2) Insert the RP Multimedia CD into your CD-ROM drive.
(3) Click on the "Start" menu button and select "Run".
(4) In the "Run" dialog box, type "E:\start" and click on the "OK" button. If your CD-ROM drive is not "E", substitute its drive letter in place of "E".
(5) The program will start.

GETTING AROUND THE PROGRAM

The RP multimedia program is a courseware developed for the students, lecturers and any person who is interested in learning more about RP. The user interface has been designed to be very simple with a short animation introduction in order for first time users to understand how to navigate around the chapters and subtopics in the program.

Nevertheless, the next section will guide you through the various screens in the courseware and explain the functions of the icons on the screens in detail.

MAIN INTRODUCTION SCREEN

After the program has been started, a short introduction movie will be played. To skip this introduction movie, simply click on the "Skip" button. After playing the movie, you will be presented with the Main Introduction screen (see Fig. I). When you click on the "Main Area" button shown in Fig. I, it will direct you to the Main Menu of the courseware (see Fig. II) where you can learn more of the key contents of the RP techniques, processes and the applications by clicking on any of the eight individual chapters desired.

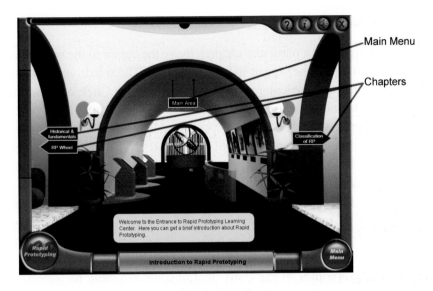

Fig. I. Main introduction page.

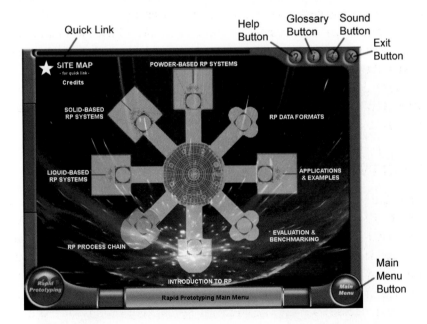

Fig. II. Main menu (after you have clicked on the "Main area" in Fig. I).

CHOOSING A CHAPTER

To learn more about a particular chapter, move the cursor to the word over the chapter in the homepage or in the main menu. The words will light up to indicate that the particular chapter can be selected. Click once on the left key on the mouse when the word is lighted up to go into that chapter.

The eight chapters are as follows: Introduction to RP; RP process chain; Liquid-based RP systems; Solid-based RP systems; Powder-based RP systems; RP data formats; Applications and examples; and Evaluation and benchmarking.

If you want to go to a specific subtopic of any of the chapters, you can click on the quick link identified as "SITE MAP" located on the top left-hand corner of the page. You will see the screen shown in Fig. III once you click on the words. By moving the cursor over the title of each chapter, the subtopics of the individual chapter available for selection will be shown on the right-hand side of the screen. If you click on any of them, you will be guided directly into the page of the requested subtopic.

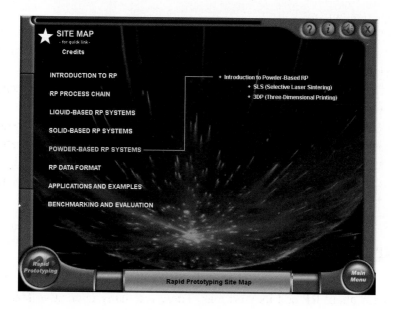

Fig. III. SITE MAP screenshot.

GUI LAYOUT DESIGN

The GUI Layout Design (the gray border in Figs. I and II) will be the same throughout the whole courseware. More advanced interactive features of this program will be covered in a later section "Exploring the Interactive Features."

COMPONENTS OF THE GUI LAYOUT DESIGN

The GUI Layout Design comprises of the main text area and various icons as shown in the screen shot (Figs. I and II) above. The main text area is where the text information of a particular chapter and subtopic will be displayed. It is also where the user will do most of the learning.

The icons are explained as follows:

(1) *Help button*: Click on the help button to assist you in navigating the specific chapter.
(2) *Glossary button*: Click on the glossary button icon and definitions of some terms used in the content of the chapter are provided for your understanding.
(3) *Sound button*: This is a toggle button. Click on the sound button icon to toggle the music on or off. If the music is playing, clicking on the icon will make the music stop playing. Clicking on the icon again and the music will start again. Note that the Sound button is not active when playing any of the movies.
(4) *Exit button*: Click on the exit button to end this program and return to Windows XP. When the exit button is clicked, a confirmation screen will appear. To confirm and exit the program, click on the "Yes" button. To cancel the exit program request, click on the "No" button.

EXPLORING THE INTERACTIVE FEATURES

The RP multimedia courseware is an interactive, fun and lively way to learn about RP and its workings. There are many interactive features that can be found in the program. Interactive features such as animations,

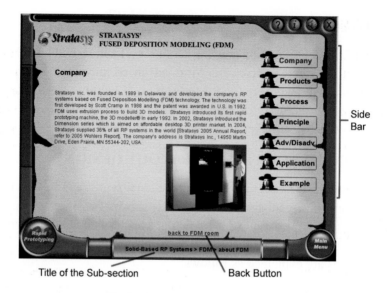

Fig. IV. Chapter on FDM screenshot.

graphics, tables and movies are "scattered" around the courseware to help enhance learning of those topics and concepts. Do enjoy and have fun exploring around!

Each of the chapters is designed with its own theme in order to provide you with a fresh look after you finish a chapter. The section of the chapter will highlight some of the interactive features (see Fig. IV) that can be found within the courseware and how to activate them.

MOUSE OVER A HOT SPOT

There are many "hot spots" within the main text area in the learning screen. By moving the cursor over such "hot spots", hidden text, graphics, or animations will appear (or move) on the screen.

For example, when the mouse cursor moves over "Data Preparation", the text explaining data preparation appears. There are many different "hot spots" in the entire courseware. Hence, do explore and discover the hidden "hot spots" to help in the understanding of the concepts of RP in a fun and quick manner.

BUTTONS

In each of the six RP techniques presented (two for each of liquid-based, solid-based and powder-based RP systems), there is a side bar for navigation within that section. Whenever the cursor is moved over any of the side bar buttons, they will be changed or highlighted in color. By clicking once on the side bar button, it will bring you to the respective page. Figure IV shows the side bar at the "Company" page of the FDM technique.

PLAYING A MOVIE

There are six movies in this courseware to help explain the concepts of the RP technique (again, two for each of liquid-based, solid-based and powder-based RP systems). Whenever a particular subtopic or page contains a related movie, the icon and words "movie clip" will appear when you move over an active icon as shown in Fig. V. Clicking on the movie icon will launch the movie player and play the movie automatically. You can return to the program during or after the movie is played.

The Movie player has the following controls:

(1) *Play/Pause button*: Play or pause the movie at the period of time.
(2) *Forward button*: Restart the movie by clicking on the button.

Fig. V. Screenshot with active movie icon.

(3) *Sound scroll bar*: User is able to adjust the loudness of the movie.
(4) *Movie scroll bar*: User is able to fast forward or scroll back to the part
    of the movie, which he/she had missed out.

SELF-CHECK QUIZ

There are self-check-quizzes in every part of the room for you to do self-
testing to check if you have understood the content. To start the quiz, the
user just has to click the "Start Quiz" button. After finishing the first round
of self-quiz, you can choose to review the answer, retry the questions again
or exit from the quiz (Fig. VI).

Fig. VI. Screenshot with active quiz icon.

INDEX

Symbols

3D model 11, 30, 49, 64, 70, 81, 156, 161, 168, 307, 389, 390, 407, 476

3D modeling 26, 27, 52, 151, 153

A

ABS (Acrylonitrile Butadiene Styrene) 121, 123, 138–142, 144, 145, 205, 427

ABS*plus* plastics 138

accuracy, 351, 359, 366, 369, 381, 392

adhesives, 8, 11, 162, 205

Adidas-Salomon AG 60

AeroMet Corporation 21, 291

air bubbles 159, 178

AlliedSignal Aerospace 381, 399

alumina 103, 109, 221, 264, 286–288

aluminum 8, 59, 73, 164, 206, 221, 229, 230, 233, 237, 250, 251, 253, 261, 274, 275, 367, 371, 373, 375, 383, 386, 472

ameloblastoma 409, 410

anthropology 427, 475

AOM. *See* Acoustic Optical Modulator

application–materials relationship 357

applications 357

Arcam AB 243

architecture 209, 278, 394

argon 119, 237, 239, 292

art 186, 391

ARTCAM 388, 389

artificial bone 406

Asahi Denak Kogyo K.K 80

ASCII 301, 334, 341, 345, 346, 348

Autostrade Co. Ltd 76

B

Battell Development Corporation 111

Baxter Healthcare 423, 424

Bell Helicopter 382

benchmark 453–455, 457–459, 464–469

benchmarking 437, 448, 453, 454, 457, 466–469

BFGoodrich company 199

binary 301, 334, 345

bio-modeling 412, 423
bio-molecular models 429
Boeing 164, 165
bone tumor 410, 411
boundary representation 336
button tree display 457, 458, 460–463

C

CAD interface 201, 224
CAD. *See* Computer-aided Design
CAM. *See* Computer-aided Manufacturing
CAM-LEM Inc. 182–184
CastForm™ 204, 206, 373
ceramic mold 221, 286, 371
ceramics 13, 101, 161, 191, 192, 208, 221, 233, 264, 267, 272, 289, 367, 368, 377, 393
Ciba-Geigy Ltd 52
CLI format 345, 346
CMET 19, 80, 81, 83, 444
CMM 12, 34, 147, 458, 469
CNC. *See* Computer Numerical Control
coincidental vertex 320–322, 325, 326, 355
COLAMM 130
collimated UV lamp 113, 225
Common Layer Interface. *See* CLI
complex geometric entities 306

Computer Modeling and Engineering Technology. *See* CMET
Computer Numerical Control 7, 25, 117
Computer-aided 7, 10, 27, 40, 84, 85, 182, 359, 387, 469
Computerized Tomography 338
Concept Laser GmbH 21, 248, 256
concept modeling 140, 172
congenital malformation 404
conjoined twins 407, 408, 435
cosmetic facial reconstruction 406
cranioplasty 403
CSG 336, 347–349
CT data 338, 407, 413, 422
CT. *See* Computerized Tomography
Cubic Technologies 20, 153, 154, 157, 158, 160
Cubital Ltd. 111

D

Daimler-Benz AG 385
DARPA 102
data 10, 26, 40, 156, 200, 301, 362, 403, 437
de facto standard 301, 350
decision table
Defense Advanced Research Program Agency. *See* DARPA
Delco Electronics 384, 385

Denken Engineering Co. Ltd
    130
dental   75, 128, 166, 171, 232,
    233, 411, 412, 421, 473
DePuy Inc.   50, 415
designer's intent   309
Desktop Manufacturing   10,
    108
DFE (Data Front End). *See*
    Cubital Ltd
Digital Light Processing   19, 69,
    73
dimension measurements   453
dimensional accuracy   112, 193,
    260, 392, 467
Direct CAD Manufacturing   18
Direct Hard Tooling   373, 377,
    400
Direct Shell Production Casting.
    See Soligen Inc.
distortion   83, 84, 145, 159, 161,
    162, 193, 207, 257, 261, 359,
    455, 460
DLP™. *See* Digital Light
    Procesing
D-MEC Ltd   63, 64, 106, 109
droplet   93, 192, 219
droplets   19, 90, 91, 93–95, 191,
    192, 219, 278, 280, 282, 367,
    407
drug delivery devices   424,
    475
DSPC. *See* Soligen Inc.
DTM Corporation   199, 374
Du Pont   118

E
economic feasibility   438, 481
economic justification   457
elastomer   67, 205, 215, 218,
    219
electrical charge   98
electrical generators   380
electric-discharge machining
    378
Electro Optical Systems. *See* EOS
    GmbH
Electrolux Rapid Prototyping   227
Electron Beam Melting.   *See*
    Arcam AB EBM
electronic mail   29
electronic products   362, 441
elevation control system   35
elevation mechanism   42, 43,
    45
elevator   41, 58, 66, 77, 82, 91,
    93, 112, 158, 229, 460
EnnexCorporation   186
envelope. *See* work volume
Envision Technologies GmbH
    68
EOS GmbH   35, 223, 224, 470
Ex One Company   278, 279
explorer   386

F
FAA. *See* Federal Aviation
    Authority
fabrication processes   25, 26
fabricators   25, 33
facet model   301, 303, 308, 348

facets   30, 159, 301, 303–307,
    309–320, 326, 330–333, 335,
    336, 339, 354, 355
fanjet engine   381
FDM. *See* Stratasys Inc.
FEA. *See* finite element analysis
Federal Aviation Authority   382
feedback mechanism   159
feedback system, closed-loop   161
finishing   32–34, 61, 138, 157,
    158, 163, 180, 203, 216, 228,
    239, 246, 359, 392, 400
finishing processes   359, 400
finishing techniques   158
finite element analysis   10
fixtures   216, 364, 441
flame retardant   230
flight-certified production castings
    382
Fockele und Schwarze GmbH.
    See LMS
footwear   53, 60, 205, 220, 221,
    441
Ford Motor Company   386
forensic science   427
Formigraphic Engine Co.   111
Fraunhofer-Gessellschaft   289
Fraunhofer Institute for Applied
    Materials Research. *See* IFAM
Fraunhofer Institute for
    Manufacturing Engineering
    and Automation. *See* IPA
Freed, Raymond S.   35, 199
French National Center for
    Scientific Research (CNRS)
    111

functional analysis   59, 116
functional evaluation   357
Fused Deposition Modeling   20,
    32, 137, 196, 290, 451, 459
    See also Stratasys Inc.

G
galvanometer mirror system   65,
    154
gaps   29, 42, 82, 303, 304,
    308–311, 313, 315–318, 320,
    321, 325–328, 330, 333, 343,
    344, 427
gates   122, 219
gating system   286, 373, 376
gearbox   383, 384
General Motors   384
Genisys   195
Geographic Information System
    (GIS)   393
geometric modeling   9, 22, 26,
    34
Gintic Institute of Manufacturing
    Technology   387, 388
Global Positioning System   384
golf club   235
GPS. *See* Global Positioning
    System
Gray Scale   389

H
Hekatron   211, 212
helicopter   382, 383
Helisys Inc.   153, 154
Helsinki University of
    Technology   348

Hewlett-Packard Graphics
     Language   337
historical development   7, 22
HPGL. *See* Hewlett-Packard
     Graphics Language
Hull, Charles W.   35, 199
humidity   56, 180, 216, 222, 273
hydraulic system   383
Hyundai Mobis   147, 148

I
IBM   449
IFAM   289
IGES. *See* Initial Graphics
     Exchange Specifications
IH process. *See* integrated hard-
     ened polymer
     stereolithography
Ikuta, Koji   98
implants   232, 233, 237, 240,
     247, 248, 261, 262, 278,
     338, 364, 413, 415, 416,
     423, 475
INCS Prototyping and
     Manufacturing Services   48
Indirect Hard Tooling   377,
     400
inert gas   191, 208, 209, 251,
     252
Initial Graphics Exchange
     Specifications   11
injection mold   121, 122, 146,
     212, 256, 257, 367, 401
injection molding   25, 34, 109,
     120–123, 163, 206, 211, 221,
     231, 235, 253, 256, 261, 267,

360, 365, 366, 371, 374, 375,
     377, 378, 386, 473
ink-jet   148–151, 168, 170, 219,
     220, 282–284, 287, 377
installation   39, 40, 49, 61, 62,
     79, 224, 383, 437, 448
Instant Manufacturing   18
instruments   73, 163, 262, 441
integrated hardened polymer
     stereolithography   98
interface   28, 29, 97, 108, 156,
     208, 224, 244, 285, 301, 337,
     338, 345, 350, 431, 438
interference studies   27
interlaminate strength   159
International Business Machine.
     See IBM
invalid models   303
invalid tessellated model   306,
     308, 309, 355
inventory   16, 221
investment casting   36, 47, 59,
     68, 75, 80, 83, 90, 94, 96, 97,
     116, 145, 152, 163, 164, 171,
     206, 209, 215, 216, 219, 222,
     235, 372–375, 381–383, 386,
     387, 401, 471, 472
ionographic process   113
IPA   289

J
Japan Synthetic Rubber   63, 64,
     106
jewelry   19, 60, 122, 124,
     126–128, 135, 136, 152, 153,
     171, 387, 388, 390, 401

jewelry industry   19, 60, 122,
124, 387, 388
jets   150, 168, 170, 215
JSR. *See* Japan Synthetic Rubber

K
Keio University Hospital   406
kiln   288
Kira Corporation Ltd   176, 177
kitchen ware   441
Kochan   14
Kraft paper   160
Kyushu Institute of Technology
98

L
Laminated Object Manufacturing.
See Helisys Inc.
laminates   13, 18, 20, 157
laser   12, 25, 35, 154, 199, 307,
405, 451, 472
Laser Engineered Net Shaping.
See Optomec Inc. LENS®
LaserCAMM. *See* Scale Models
Unlimited
Layer Exchange ASCII Format.
See LEAF
Layer Manufacturing   10, 18, 64,
143, 189, 345
layer thickness   31, 36, 41, 45,
57, 65, 77, 93, 104, 112,
130, 139–142, 149, 173, 176,
181, 182, 191, 207, 231, 253,
288, 291, 340, 342, 345, 387,
471
layering technology   45

LEAF (Layered Exchange ASCII
Format)   348–350, 356, 421
LED. *See* light emitting diode
leveling wiper   41
LIGA   98, 99
light emitting diode   457, 458
Light Sculpting   130
liquefied alloys   291
LMS (Layer Modeling System)
130
LMT-file. *See* LEAF
LOM. *See* Helisys Inc.
LOMSlice™. *See* Helisys Inc.
lost wax casting. *See* applications

M
Maestro   41, 384, 385
MAGICS. *See* Materialise, N.V.
manpower   130, 438
market potential   439
mask generation   112, 113
mask plate   113, 114
Maskless Mesoscale Material
Deposition. *See* Optomec
Inc. M³D
mass customization   470, 474
Massachusetts Institute of
Technology   213, 276
materials   7, 13, 37, 144, 218,
357, 421, 474
Material Addition Manufacturing
18
Material Deposition
Manufacturing   241
Material Incress Manufacturing
18

Materialise, N.V.   29, 63, 67,
    108
MCP-HEK Tooling GmbH   21,
    257, 258
mechanization   7
medical devices   102, 106, 163,
    237, 262, 423
medical imaging   60, 117, 338
MegaHouse Corporation   62
Meiko Co. Ltd   128, 135
Mercedes-Benz   385
MERGE   42
metal   96, 103, 105, 163, 182,
    366, 412, 472
metal arc spray system   368
metal spraying process   367
micro integrated fluid system   133
micro lithography   98
micro-electrostatic actuator   98
micro-venous valve   98
Microfabrica® Inc.   19, 102
microfabrication   19, 98, 104
MIFS. *See* micro integrated fluid
    system
mint industry   388
missing facets   159, 303, 307,
    309, 310, 354, 355
MIT. *See* Massachusetts Institute
    of Technology
Mitsubishi Corporation   80
Mitsui Zosen Corporation. *See*
    COLAMM
MJM. *See* 3D Systems Inc.
MJS. *See* Fraunhofer-
    Gessellschaft
Mobius' rule   309

mock-up   3, 68, 140, 147, 163,
    177, 361, 363, 381
model   3, 26, 41, 143, 199, 307,
    360, 380, 411, 454
model making   8, 112, 113, 158,
    394
model validation   308
Modern Engineering   385
monomers   43–45
MRI   49, 75, 85, 278, 421
Multi-Jet Modeling System.   *See*
    3D Systems Inc.
multidirectional inclined
    structures   98
multiphase jet solidification. *See*
    Fraunhofer-Gessellschaft

N
Nanyang Technological
    University   387, 388, 417,
    418, 424
NC controller   127
Nippon Synthetic Rubber Co., Ltd
nonmanifold topology   303
nozzle   59, 90, 91, 93, 143, 150,
    239, 288, 289, 291
NTT Data Communication
    Systems   80
nylon   13, 52, 204, 205, 273,
    377, 382, 428, 474

O
Objet Geometries Ltd   19, 52,
    53, 62, 63, 407
optical scanning system   41–43,
    46, 82, 234

Optomec Inc.   237
oxidation   192, 203, 208, 273

P
painting   33, 158, 203, 360
paper   13, 20, 160, 165, 176,
    178, 443, 446
parallel processing   115
particle bonding   207
particles   101, 183, 202,
    207–209, 214, 219, 242, 277,
    288, 367, 378
Part Manager™   42
Parts Now™. *See also* Soligen
    Inc.
PC   28, 70, 85, 138–140, 145,
    177–179, 224, 244, 281, 285,
    288, 290, 340
PCB   1, 9, 457, 458
pellets   13, 18, 20, 290, 371
personal computer. *See* PC
Personal Modeler™. *See* BPM
    Technology
phase-change printing. *See* 3D
    printing
Phenix Systems   21, 262–266,
    299, 472
photo-curable liquid resin. *See*
    resins
photo-optical methods   363
photochemical process   43
photocopying. *See* Xerography
photoinitiator   44
photopolymerization   42–45, 69,
    72, 108, 110, 114, 208

photopolymers   43
physical properties   10, 146, 216,
    246, 360
piezoelectric   98, 101
pilot-production   363, 364
plaster of Paris   392
plastic bottles and containers
    441
polishing   58, 158, 205, 230,
    267, 272, 282, 359, 378
polyamide. *See* nylon
polycarbonate   52, 138, 208,
    209, 382, 399
polyethylene   52, 160, 272
polygonal approximation models
    303
polyline   339–341, 345, 346,
    348, 349
polymerization   43, 45, 58, 100,
    110
polymers   43, 45, 88, 101, 162,
    242, 277
polypropylene   52, 58, 205, 386
polystyrene   223, 225, 230, 234,
    373, 376
Porsche   222
post cancer reconstruction   404
post-curing   47, 66, 68, 75, 79,
    83, 120, 127, 162, 208
postprocessing   26, 31, 32, 34,
    118
pottery   441
powder   13, 137, 199, 371, 472
powder-binder mixture   289,
    291, 379

power supply   77, 105, 108, 119, 125, 224, 237, 243, 244, 259, 269, 281

presentation   47, 59, 60, 62, 75, 79, 83, 116, 145, 360, 361

price   16, 17, 79, 394, 411, 438, 451, 470–472, 474

print-head   215

Pro Engineer   66, 386, 448, 449, 452

product design   47, 68, 118, 441

product development   1, 2, 4–7, 14, 47, 62, 223, 241, 357

profit   16, 438, 452, 453

prosthesis   60, 117, 364, 403, 404, 413, 419–421

protein–protein interactions   431

prototype   1, 18, 45, 137, 199, 357, 405, 439

prototyping   1, 18, 35, 137, 199, 301, 357, 403, 437

PSL.   *See* Solidimensions Ltd. Plastic Sheet Lamination

Q

QuickCast™. *See* 3D Systems Inc.

QuickSlice. *See* Stratasys Inc.

R

Radio Data System   384

radiographic quality   386

range of materials   47, 58, 138, 208, 234, 271, 357, 474

rapid prototyping   1, 13, 25, 35, 137, 199, 301, 357, 403, 437

Rapid Prototyping Interface. *See* RPI

Rapid Prototyping Reports   440

Rapid Prototyping, Tooling and Manufacturing   18

rapid tooling   52, 164, 189, 205, 235, 248, 254, 256, 258, 261, 363, 365–367, 369, 377, 386

RapidCasting™   209

RapidTool™   209, 374

RDS. *See* Radio Data System

recoating system   38, 40, 41, 46, 82, 83, 119

reconstructive surgery   406, 425, 435

refractive index   111

Rensselaer Design Research Center   346

Rensselaer Polytechnic Institute   346

repair software   302

Report on Census of Industrial Production   440

REPTO. *See* Real Engine Power Take-Off

resins   13, 40, 43, 45, 52, 67, 69, 73, 82, 118, 127, 192, 216, 222, 359, 365, 371

road width   143

round-off errors   305

RP. *See* rapid prototyping

RPI (Rapid Prototype Interface)
    346–348, 356
runners   365, 369, 376

S
sand blasting. *See* abrasive jet
    deburring
Sanders Design International
    (SDI)   148
Sanders Prototype Inc.   148
scaphocephary   406
Schneider Prototyping GmbH   117
SCS. *See* Sony Corporation
selection of RP systems   449
Selective Laser Sintering. *See* 3D
    Systems Inc.
Selective Mask Sintering.   *See*
    Sintermask Technologies
    AB SMS
self-supporting   115
semi-liquid state   142
service bureau   61, 351, 380,
    381, 437–446, 448–450
service bureau brochure   446
SGC. *See* Cubital Ltd.
shell punctures   304
shrinkage   75, 116, 145, 161, 163,
    180, 185, 188, 227, 231, 234,
    246, 286, 359, 371, 375, 455
silicone carbide   221
Silicone Graphics   229
sinter bonding   207
sintering. *See* sinter bonding
sintering process   182, 184, 207,
    208, 227, 229, 231, 262, 274,
    275, 417

Sintermask Technologies AB
    21, 267, 268
Skin and Core   231, 232
SLA. *See* 3D Systems Inc.
SLC (Stereolithography Contour)
    54, 56, 149, 338, 339, 341,
    342, 356
Slice™   42
slicing algorithms   305
Slowinski Inc.   152, 153
SLS. *See* DTM Corporation
Smart Crossblade   210, 211
Solid Creation System. *See*
    D-MEC Ltd
Solid Freeform Fabrication   18
Solid Freeform Manufacturing
    22
Solid Ground Curing. *See* Cubital
    Ltd
solid model   11, 41, 54, 66, 70,
    81, 118, 178, 219, 306, 307,
    378, 413
Solidimension Ltd   166, 172
Solidscape, Inc.   148, 149,
    152, 153, 197, 412,
    471, 476
Soligen Inc.   286, 288
SOMOS photopolymers   118
SOUP. *See* CMET
Sparx AB   267
spectacles frame   128
sprues   288
Stereolithography. *See* 3D
    Systems Inc.
Stereolithography Contour. *See*
    SLC

stitching algorithms   304
STL files   11, 28–30, 41, 42, 60,
   167, 178, 302, 338, 339, 380,
   407, 474
Stratasys Inc.   137
stress analysis. *See also*
   applications
stress distribution   358, 363
stress, residual   162, 190
Sundstrand Aerospace   380
Sundstrand Power Systems
   381
support   5, 30, 42, 136, 199, 341,
   366, 466
supports   5, 28, 41, 120, 181,
   215, 345, 409, 460, 476
surface finish   13, 33, 36, 47, 53,
   124, 189, 204, 219, 234, 272,
   476
surface modeling   336
surface normals   304
surgical operation planning   358,
   364
surgical procedures   364, 418,
   435
Swainson, Wyn Kelly   110
Synchrotron radiation X-ray
   lithography   98

T
tank size   64, 67
Teijin Seiki Co. Ltd   80, 117
temperature   43, 52, 89–95, 142,
   160, 180, 193, 202, 207, 230,
   232, 234, 246, 251, 266, 288,
   290, 379, 425

tensile strength   180, 253
tessellation   302–306, 333
tessellation algorithms   302
TESTA Architecture/Design   209
thermal energy   143, 193
thermoplastic   52, 88, 148–152,
   159, 172, 188, 191, 200, 205,
   225, 236, 284, 366
thermopolymer   170, 171
tissue engineering   84, 87, 89,
   242, 276, 421–423, 435,
   475
tissue engineering scaffolds   276,
   421, 423, 475
titanium   206, 233, 237, 240,
   244, 246, 251, 258, 261, 292
topological data   302
Toyota   146, 182
toys   441, 474
TPM. *See* tripropylene glycol
   monomethyl ether
tracheobronchial stents   419
transparent substrate. *See* mask
   plate
triangulation   162, 335, 336
trimmed surfaces   303
tripropylene glycol monomethyl
   ether   32

U
ultra violet light. *See* UV light
University of Texas at Austin
   199
UNIX   41, 66, 76, 341
UV lamp. *See also* colliminated
   UV lamp

UV light   30, 54, 57, 58, 74, 113,
    114
UV range   35, 43
Uziel, Yehoram   286

V
vacuum molding   120, 122
valid tessellated model
    306–310, 355
validity checks   303
varnishing   392
vat   35–41, 81, 110, 111, 119
vector   159, 162, 301, 330, 355
View™   42
viscosity   90, 94, 119, 143
Vista™   42
visual inspection   42, 453, 460
visualization   3, 75, 79, 83, 89,
    95, 112, 163, 176, 182, 189,
    216, 259, 357, 360–363, 388,
    405, 407, 429
Volkswagen   235, 236, 383
Voxeljet Technology GmbH   21,
    283, 285
VW. *See* Volkswagen

W
Wall M.B.   455
Walter Reed Army Medical
    Centre   49
warpage   162, 163
watches   441
wax   13, 47, 94, 96, 112,
    114–116, 120, 122, 128, 149,
    164, 192, 209, 216, 219, 235,
    373, 382, 387, 403
wireframe   9
Wohlers, Terry   442
work volume   77, 110, 112, 125,
    135, 154, 170, 197, 265, 290,
    293

X
Xerography   115, 178
Xerox   50, 51, 146, 147

Z
Z Corporation   20, 213, 215,
    216, 222, 276, 298, 468, 470
Zephyr™ recoating system   41
zirconia   264